SOURCEBOOK OF ADVANCED ORGANIC LABORATORY PREPARATIONS

SOURCEBOOK
OF ADVANCED
ORGANIC
LABORATORY
PREPARATIONS

Stanley R. Sandler
Elf Atochem North America, Inc.
King of Prussia, Pennsylvania

Wolf Karo
Polysciences, Inc.
Warrington, Pennsylvania

ACADEMIC PRESS, INC.
Harcourt Brace Jovanovich, Publishers
San Diego   New York   Boston   London   Sydney   Tokyo   Toronto

This book is a guide to provide information concerning its subject matter. Synthesis of chemicals is a rapidly changing field. The reader should consult current product specifications for state-of-the-art instructions and applicable government safety regulations. The Publisher and the authors do not accept responsibility for any misuse of this book, including its use as a source of product specifications or other specific information.

This book is printed on acid-free paper. ∞

Academic Press, Inc.
1250 Sixth Avenue, San Diego, California 92101-4311

United Kingdom Edition published by
Academic Press Limited
24–28 Oval Road, London NW1 7DX

Library of Congress Cataloging-in-Publication Data

Sandler, Stanley R., date
    Sourcebook of advanced organic laboratory preparations / Stanley
R. Sandler, Wolf Karo.
        p.   cm.
    Includes indexes.
    ISBN 0-12-618506-9
    1. Organic compounds–Synthesis.   I. Karo, Wolf, date.
II. Title.
QD262.S24   1992
547.2'078–dc20   92-13631
                                                                          CIP

PRINTED IN THE UNITED STATES OF AMERICA
92   93   94   95   96   97   EB   9   8   7   6   5   4   3   2   1

CONTENTS

PREFACE

The purpose of this laboratory text is to provide a ready source of reliable procedures for those involved in organic synthesis in either academic or industrial laboratories. It will be useful to instructors and students seeking to find additional reliable organic or polymer preparations. The introductory material in each chapter gives a brief synopsis of the synthesis of a given functional group or class of compounds. Industrial chemists will also appreciate the convenient source of procedures as well as the tables found in the text. The numerous polymer preparations will also be helpful to those needing a reliable procedure for their synthesis.

This text gives selections of some of the procedures found in two series of books by the authors. These books are listed in the Introduction to this text and should be referred to for additional details and references that cover both patents and journals.

Unique features of this book are the Name Reaction Index with procedures for most reactions and the Appendix titled *Documentation of Product and Process Research and Development*, which explains, with examples, methods for proper documentation in the laboratory notebook.

The procedures described in this sourcebook are based primarily on the literature. Many were developed long before our current concerns of exposure to chemicals, "right-to-know" laws, occupational health and safety, environmental effects, pollution, material disposal, and so on. As a result, all of the syntheses given herein must be re-evaluated in the light of ever-changing regulations (see, for example, "U.S. Courts Overturn Regulatory Shortcuts," *Chemical Engineering News*, July 13, 1992, page 7.)

In this book, many examples have been given that use *benzene* as a reagent or solvent. However, since benzene is a cancer-suspect agent, its use must be severely restricted and efforts must be made to eliminate this solvent from the workplace completely. In some cases, possible substitutes for benzene may be *toluene*, *heptane*, or *cyclohexane*. In other situations, totally different systems may need to be developed. Similar problems exist for other materials. The reader must, therefore, be constantly alert to the shifting regulations on handling materials and the real and/or anticipated health hazards. While the so-called "Material Safety Data Sheets" required for all chemicals are of some value, the reader should keep in mind that the information given there is not always based on solid experimental data.

We would like to express our appreciation to our families for their encouragement in this work. We also would like to thank the staff of Academic Press for all their efforts in guiding this work through the publication process.

INTRODUCTION

*S*ourcebook of *Advanced Organic Laboratory Preparations* has been designed as a convenient source for synthetic procedures of wide utility for both students and industrial chemists.

In the case of students, this laboratory preparations manual can be used to find additional experiments to illustrate concepts in synthesis and to augment existing laboratory texts. A Name Reaction Index is also included to direct the reader to the location where specific reactions appear in this manual.

The industrial chemist is frequently required to prepare a variety of compounds, and this manual can serve as a convenient guide to choose a synthetic route.

Each chapter describes the synthesis of a given class of compound or functional group type (hydrocarbons, olefins, acetylenes, carboxylic acids, polymers of various types, mercaptans, etc.) and gives a brief summary of the available routes that can be used. A few representative preparations are illustrated because of their wide applicability. For more details the reader is referred to review the following six books from which these preparations came:

Organic Functional Group Preparations, Volumes I, II, and III, second editions, by Stanley R. Sandler and Wolf Karo, Academic Press, Inc. (1983, 1986, 1989).

Polymer Syntheses, Volume I, 2d ed. and Volumes II and III by Stanley R. Sandler and Wolf Karo, Academic Press, Inc. (1992, 1977, 1980).

One unique feature of this laboratory manual is that emphasis is placed not only on simple organic compounds but also on polymers of high molecular weight. This topic usually is not covered in as much detail in other laboratory texts of this type. In the present

1

manual the topic of polymers is introduced early on as in the case of Chapter 1, Hydrocarbons, as well as throughout the other chapters. This laboratory manual assumes that the student has already mastered the introductory laboratory techniques in an introductory course on organic chemistry involving the typical glassware normally used in these preparations. Careful record keeping is a must and is covered in the Appendix. Experience in the various analytical techniques described is also assumed. Experience in distillations, both at atmospheric pressure and under reduced pressure, is also assumed.

1. SCALE OF OPERATIONS AND MONITORING OF REACTIONS

Most preparations cited can be scaled *down* provided microware is available. The advantage is that less waste disposal is required.

Preparations should *not* be scaled up unless this is done *gradually* to determine if the exothermic reaction can be safely tolerated by the equipment being used. All glassware should be free of cracks.

The reactions in most cases can be easily monitored by gas chromatography, infrared spectroscopy, ultraviolet spectroscopy, and thin layer chromatography. Where available a nuclear magnetic resonance (NMR) instrument can also be very effectively used to follow the course of the reaction and to determine the structures of the products.

A. SAFETY

All experiments should be carried out in a good fume hood with due personal protection involving safety eye glasses, a laboratory coat or apron, and the use of gloves.

All experiments should first be studied in detail and outlined. A senior chemist or laboratory supervisor should oversee that all details have been understood before starting the laboratory preparation. The nature of each chemical used should be thoroughly understood by examining in detail the appropriate Material Safety Data Sheet (MSDS) and then signed off before using.

Several **CAUTIONS** are a must to review and are mentioned here again for emphasis:

1. Read all the Material Safety Data Sheets for the raw materials that you plan to use. Contact the supplier if you have any questions.

2. Read all the toxicity information.
3. Use good personal protective equipment such as eye protection, laboratory coats, and gloves (respirators when necessary). The type of gloves and eye protection must be suitable for the operation.
4. Use a well-ventilated hood.
5. Use traps to control toxic vapors or other volatile by-products.
6. Always work with someone else nearby in the same laboratory. *Never work alone!*
7. Read and reread all preparations. Write out equations and understand the chemistry involved.
8. Never scale-up experiments unless you are sure this will not lead to a highly exothermic uncontrollable reaction.
9. When uncertain, ask questions of your instructor or other professional who is qualified.

B. WASTE DISPOSAL

For the disposal of waste chemicals consult your supervisor or the Safety Director.

Broken glassware and paper towels used to mop up chemicals should be segregated also by placing them in special containers or plastic bags. Check with your laboratory instructor for the proper guidelines that you are to follow.

1

HYDROCARBONS: PARAFFINIC AND AROMATIC

1. INTRODUCTION

Hydrocarbons are conveniently prepared in the laboratory by reduction, condensation, elimination, or hydrolysis reactions. Isomerization, oxidation, and photochemical reactions are less common on a preparative scale.

From S. R. Sandler and W. Karo, *Organic Functional Group Preparations*, Vol. I, 2d ed. (New York, 1983), 1–37, by permission of Academic Press, Inc.

The reduction methods depend on converting a given functional group to a methylene group. For example, olefins, aromatic rings, alcohols, aldehydes, ketones, and halides give hydrocarbons on reduction. These methods allow the preparation of hydrocarbons of known structure. The Clemmensen (zinc amalgam and hydrochloric acid) and Wolff–Kishner (hydrazine and base) methods can be used to reduce aldehydes and ketones. Catalytic hydrogenation methods can be used to reduce olefins and aromatic compounds. The catalytic hydrogenation method can also be used for ketones, provided that a high-pressure apparatus is available. Nickel and platinum are the most commonly used catalysts.

The use of sodium borohydride with palladium chloride has been described for the reduction of olefins in excellent yields. The method is quite reliable and it has been applied as an analytical technique for the quantitative estimation of the degree of unsaturation of a compound or of a mixture.

$$RCH{=}CH_2 + NaBH_4 \xrightarrow[\text{H}^+]{\text{PdCl}_2} RCH_2{-}CH_3 \tag{1}$$

Condensation reactions are used to synthesize a hydrocarbon from two or more compounds which may or may not be the same as described in Eq. (2)

$$R{-}Y + RX \longrightarrow R{-}R + XY \tag{2}$$

where X or Y may be a hydrogen, halogen, diazo, or organometallic group. The Friedel–Crafts, the Wurtz, the Wurtz–Fittig, organometallic coupling, Ullmann, and Pschorr syntheses are some representative condensation reactions. The Friedel–Crafts reaction and the coupling of organometallics with halides are the most useful laboratory syntheses of hydrocarbons, especially branched-chain hydrocarbons such as neohexane.

The hydrolysis of the Grignard reagent is a useful method of preparing hydrocarbons from halides [Eq. (3)].

$$ROH \xrightarrow{\text{HX}} RX \xrightarrow{\text{Mg}} RMgX \xrightarrow{\text{H}_2\text{O}} RH \tag{3}$$

The yields by this method are usually excellent and pure hydrocarbons are obtained. The chloromethylation reaction can thus be used as a step in the addition of a methyl group to an aromatic nucleus.

Carboxylic acids eliminate carbon dioxide when heated with soda lime or electrolyzed to produce paraffins (Kolbe reaction). The former is a more useful method in the laboratory.

Aldehyde, diazo, and sulfuric acid groups are a few of the other groups that can be eliminated and replaced by hydrogen to give hydrocarbons.

Cyclodehydration of aromatic alcohols and ketones gives tetralins, anthracenes, phenanthrenes, and other ring systems.

The Jacobsen reaction involves the isomerization by sulfuric acid of an aromatic system containing several alkyl halo groups to give vicinal derivatives.

Olefins can also be hydroborated to organoboranes which are then converted to the hydrocarbon by refluxing with propionic acid [1]. This procedure is a convenient noncatalytic laboratory method for the hydrogenation of olefins.

$$3 \, RCH{=}CH_2 + NaBH_4 + BF_3 \longrightarrow (RCH_2CH_2)_3B \xrightarrow{C_2H_5COOH} 3 \, RCH_2CH_3 \quad (4)$$

Terminal olefins are readily hydroborated but internal olefins require additional reaction time and heating prior to refluxing with propionic acid. Substituents such as active sulfur, chlorine, or nitrogen are not affected by this hydrogenation procedure.

2. REDUCTION REACTIONS

1-1. Conversion of 1-Hexene to n-Hexane by Hydroboration Method [1]

$$3 \, CH_3(CH_2)_3{-}CH{=}CH_2 + NaBH_4 + BF_3 \longrightarrow [CH_3(CH_2)_3CH_2CH_2]_3B \longrightarrow$$
$$3 \, CH_3(CH_2)_4CH_3 \quad (5)$$

To a three-necked flask equipped with a mechanical stirrer, dropping funnel, and reflux condenser with attached drying tube are added 16.8 gm (0.20 mole) of 1-hexene and 2.0 gm of sodium borohydride (0.055 mole) in 55 ml of diglyme. While stirring under nitrogen, 10.0 gm (0.075 mole) of boron trifluoride etherate in 25 ml of diglyme is added during a period of 1.5 hr. Then 22.2 gm (0.3 mole) of propionic acid is added and the mixture is refluxed for 2 hr while ether and the product distill over. The product is washed with sodium bicarbonate solution, then water, dried, and fractionally distilled to yield 15.6 gm (91%) of *n*-hexane, bp 68°–69°C (738 mm), n_D^{20} 1.3747.

A. THE WOLFF–KISHNER METHOD

The Wolff–Kishner procedure depends on reacting an aldehyde or ketone with hydrazine in the presence of base to yield the corresponding hydrocarbon.

$$RR'C{=}O + H_2N{-}NH_2 + KOH \longrightarrow RCH_2R' \qquad (6)$$

Semicarbazones or azines undergo the same reaction upon being heated with base.

Huang–Minlon modified and improved the original procedure by using diethylene glycol as a reaction medium at 180°–200°C and at atmospheric pressure for 2–4 hr to obtain 60–90% yields of hydrocarbon products [2]. It should be mentioned that aromatic nitro compounds yield amines in this procedure.

Steric effects influence the efficiency of reaction in many cases. For example 2,3,5,6-tetramethylacetophenone is not reduced, whereas the 2,3,4,5-isomer gives 75% reduction to the corresponding ethylbenzene. A special procedure for sterically hindered ketones has recently been published which is especially useful for 11-oxo steroids.

Cram and co-workers added the pure aldehyde or ketone hydrazones slowly to a potassium hydroxide solution in dimethyl sulfoxide over a period of 8 hr at 25°C and found that yields of the methylene compound varied from 65% to 90% with some azine as a by-product. The procedure has the advantage that a reaction temperature of only 25°C is required. However, the disadvantage is that the pure hydrazones have first to be isolated.

1-2. Preparation of 2-(*n*-Octyl)naphthalene by the Huang–Minlon Modification [2].

$$\text{(7)}$$

To a reaction flask containing a solution of 25 gm of sodium in 700 ml of diethylene glycol are added 100 gm of 2-naphthyl heptyl ketone and 50 ml of 90% hydrazine hydrate. The mixture is heated for 3 hr under reflux to form the hydrazone and then for an additional 12 hr at 220°C. During this period, the upper layer changes color from orange-red to almost colorless. The mixture is cooled, acidified with dilute hydrochloric acid, and extracted with benzene. The benzene layer is concentrated and the residue is crystallized from acetone at −60°C to yield 75% of 2-(n-octyl)naphthalene, mp 13°C. In order to obtain purer products, it is recommended that the crude product be distilled under reduced pressure first and then recrystallized.

B. THE CLEMMENSEN METHOD

The Clemmensen reaction of carbonyl compounds, which requires refluxing for 1–2 days with amalgamated zinc (from 100 gm of mossy zinc, 5 gm mercuric chloride, 5 ml of concentrated hydrochloric acid, and 100–150 ml water) and hydrochloric acid, yields hydrocarbons.

$$RR'C{=}O + Zn(Hg) + HCl \longrightarrow RR'CH_2 \qquad (8)$$

The yields of paraffins and alicyclic hydrocarbons are poor, and the products are frequently contaminated with olefins. Aromatic ketones are reduced in much better yields.

1-3. Preparation of 2-(n-Octyl)naphthalene [2]

$$(9)$$

Naphthyl heptyl ketone (20 gm) is added to a mixture of 100 gm of granulated amalgamated zinc, 100 ml of concentrated hydrochloric acid, and 75 ml of water. The mixture is refluxed for 18 hr. During this time an additional 90 ml of acid is added every 6 hr. The reaction mixture is cooled, decanted from the zinc, and the residue is washed with benzene. The reaction mixture is extracted with benzene, and the combined extracts are washed with water, dried, and concentrated. The residue is then recrystallized from acetone at

$-60°C$ to yield 58% of 2-(n-octyl)naphthalene, mp 13°C. Purer products are obtained by a distillation of the product prior to recrystallization.

1-4. Conversion of Ketones to Alcohols and Subsequent Reduction to Alkanes—Reduction of Benzophenone to 1,1-Diphenylethane [3]

$$(C_6H_5)_2C{=}O \xrightarrow{CH_3Li} (C_6H_5)_2C\!\!\begin{array}{c} \diagup CH_3 \\ \diagdown OLi \end{array} \xrightarrow{Li/NH_3} (C_6H_5)_2CH{-}CH_3 \qquad (10)$$

A dry (all glassware is oven dried) 500-ml, 3-necked flask is equipped with a dry ice condenser, glass-coated magnetic stirring bar, pressure-equalizing dropping funnel, and rubber septum on one neck. Then the flask is flushed with argon and the argon is kept throughout the reaction (controlled via a gas bubbler). Through the septum neck are injected 30 ml of anhydrous ether and 19.5 ml of 1.89 M (0.037 mole) methyllithium in ether (Foote Mineral Co.). From the dropping funnel is added over a 20-min period a solution of 4.54 gm (0.025 mole) of benzophenone in 35 ml of anhydrous ether. The mixture is stirred for $\frac{1}{2}$ hr and dry ice and isopropanol are added to the condenser. In place of the septum is attached a side arm attached with Tygon tubing to a tank of anhydrous ammonia. Then 75 ml of ammonia is condensed into the cooled (dry-ice bath) flask, and this is followed by the addition of 0.53 gm of lithium wire segments (0.5 cm). After 15 min the dark blue color is discharged by adding portionwise 5 gm of ammonium chloride over a 15-min period. Then the ammonia is allowed to evaporate in the hood and to the residue is added 100 ml of sat. aq. NaCl and then 100 ml of ether. The ether extract is separated and the sodium chloride solution is further extracted twice with 50 ml of ether. The combined ether extracts are dried and concentrated to yield 4.4–4.5 gm of crude product. This is then dissolved in 150 ml of petroleum ether and filtered through 60 gm of Woelm alumina (grade III) to yield after evaporation 4.2–4.3 gm (92–95%) of 1,1-diphenylethane, bp 100°C (0.25 mm), n_D^{28} 1.5691.

C. REDUCTION OF HALIDES

Halides are converted to the corresponding hydrocarbons by one of several methods. The Grignard reaction of a halide and subsequent reactions, with water produce a hydrocarbon. Magnesium, lithium, zinc, and acetic acid, lithium aluminum hydride, lithium hydride, sodium hydride, sodium borohydride, sodium in alcohol, magnesium in methanol or isopropanol, and nickel aluminum alloy in aqueous alkali can be used for the reduction of halides. The last three methods have been used specifically for aryl halogen atoms. More recently Pd/C has been used to reduce aryl halogen atoms in the presence of triethylammonium formate.

Sodium borohydride efficiently reduces organic halides, especially if they can form stable carbonium ions. For example, *tert*-cumyl chloride (2-phenyl-2-chloropropane) and benzhydryl chloride are converted into cumene (82%) and diphenylmethane (72%), respectively. The following conditions are used for the reduction of these compounds: 50°C reaction temperature, 1–2 hr reaction time, 65 vol% diglyme, 0.5 mole of the organic halide, 4.0 mole of sodium borohydride, and 1.0 mole of sodium hydroxide (added in order to minimize the hydrolysis of sodium borohydride). The use of DMSO as a solvent has also been reported.

Sodium borohydride is more convenient to handle than lithium aluminum hydride because it is not sensitive to water. In fact, sodium borohydride reduction can be carried out in aqueous solution, whereas lithium aluminum hydride requires strictly anhydrous conditions. Saturated aqueous solutions of sodium borohydride at 30°–40°C are stable in the presence of 0.2% sodium hydroxide.

More recently sodium cyanoborohydride ($NaBH_3CN$) has been reported to be effective in reducing primary, secondary, and tertiary halides and tosylates.

Tri-*n*-butyltin hydride has also been reported selectively to reduce polyhalides such as benzotrichloride or dihalocyclopropanes.

$$R_3SnH + R'X \longrightarrow R_3SnX + R'H \qquad (11)$$

■ **CAUTION:** Lithium aluminum hydride may spontaneously ignite when rubbed or ground vigorously in air. A nitrogen or argon atmosphere is recommended for safe grinding operations in a hood. A safer product is reported to be available from Metallgesellschaft A. G., Frankfurt, Germany (See *Chem. & Eng. News*, May 11, 1981, p. 3).

Catalytic trialkyltin hydride reductions have been reported which use sodium borohydride.

1-5. Lithium Aluminum Hydride Reduction of 1-Bromooctane [4a, 4b]

$$LiAlH_4 + LiH + 5 C_8H_{17}Br \longrightarrow 5 C_8H_{18} + 2 LiBr + AlBr_3 \tag{12}$$

To a flask equipped with a stirrer, dropping funnel, reflux condenser, drying tube, and thermometer are added 5 gm of lithium aluminum hydride (0.13 mole) and 12 gm of lithium hydride (1.5 moles), and the flask is cooled. Tetrahydrofuran (THF) (300 ml) is added with stirring and then the contents are heated to reflux. 1-Bromooctane (193 gm, 1 mole) is added dropwise at such a rate that a moderate reflux is maintained without external heating. The mixture is then refluxed for an additional hour, cooled to 10°C, and cautiously hydrolyzed with 100 ml of a 60/40 mixture of THF/water so that the temperature is kept below 20°C. The mixture is then transferred to a 2-liter beaker containing 80 ml of sulfuric acid in crushed ice water. The organic layer is separated, dried over potassium carbonate, and distilled to yield 109.4 gm (96%) of *n*-octane, bp 125°C, n_D^{10} 1.3975.

1-6. Sodium Borohydride Reduction of Gaseous Methyl Chloride [5]

$$NaBH_4 + 4 CH_3Cl \xrightarrow{\text{Diglyme}} 4 CH_4 + NaBCl_4 \tag{13}$$

To a flask are added 25 ml of diglyme, 0.0874 gm (23 mmole) of sodium borohydride, and 85 ml (3.8 mmole) of methyl chloride. The contents are allowed to stand for 1 hr. Recovery of the volatile products using a liquid nitrogen trap yields 78 ml of methane (3.5 mmole, 92%) and 7 ml (0.3 mmole) of recovered methyl chloride. (At the start, flask is connected to liquid N_2 trap.)

3. CONDENSATION REACTIONS

A. FRIEDEL–CRAFTS ALKYLATION

Mild catalysts such as boron trifluoride (with an alcohol), hydrogen fluoride (with an olefin), or ferric chloride (with an alkyl halide) may produce almost pure para dialkylation products or 1,2,4-trialkylation compounds. An excess of aluminum chloride at elevated temperatures favors meta-dialkyl or symmetrical trialkyl derivatives.

The catalyst quantity varies with the alkylating agent. Trace amounts of aluminum chloride are required only for the alkylations involving alkyl halides or olefins. However, with alcohols or their derivatives large amounts of catalyst are required to offset the deactivating effect on the catalyst by hydroxyl groups from alcohol or water.

1-7. 1,3,5-Triethylbenzene by the Friedel–Crafts Ethylation of Benzene [6]

$$\text{benzene} + 3\,C_2H_5Br + 2\,AlCl_3 \longrightarrow \text{1,3,5-triethylbenzene} + 3\,HBr \qquad (15)$$

To an ice-cooled three-necked flask equipped with a stirrer, dropping funnel, condenser, and ice-water trap is added 267 gm (2 moles) of anhydrous aluminum chloride. Then $\frac{1}{2}$ to $\frac{2}{3}$ of the total ethyl bromide (335 gm, 3 moles) is added to moisten the aluminum chloride. The benzene (78 gm, 1 mole) is added dropwise over a period of $\frac{1}{2}$ hr at 0° to -5°C. The remaining ethyl bromide is added cautiously over a period of $\frac{1}{2}$ hr. The reaction mixture is stirred and allowed to come to room temperature slowly. After 24 hr the yellow intermediate is decomposed by pouring it, with stirring, into 500 ml of crushed ice and 50 ml of concentrated hydrochloric acid contained in a 4-liter beaker. More ice is added and the organic layer is separated, washed with sodium hydroxide solution, dried, and distilled through a Vigreaux column to yield 138–146 gm (85–90%) of 1,3,5-triethylbenzene, bp 215°–216°C.

The ethyl bromide that is carried over with the hydrogen bromide is caught in an ice-water trap and an amount equal to it is added later in the synthesis.

4. POLYMERIZATION (ADDITION) REACTIONS

As a part of a discussion of the methods of synthesis of hydrocarbons, it is pertinent to mention hydrocarbons produced by polymerization reactions (addition reactions).

$$CH{=}CH_2 \xrightarrow[\substack{\text{or ionic} \\ \text{catalysis}}]{\text{Free radical}} \left[-CH-CH_2- \right]_n \qquad (16)$$
$$\underset{R}{} \qquad\qquad \underset{R}{}$$

Polyethylene, polystyrene, polypropylene, polyisobutylene, and polybutadiene are a few of the hydrocarbon polymers that are of great commercial importance today. Copolymers of mixtures of two or more monomers have also been prepared.

The polymerization reactions are effected either thermally, free radically, cationically, or anionically depending on the monomers involved. Ziegler–Natta type catalysts are used to give stereospecific polymers of either the isotactic or syndiotactic conformation.

The references will lead the reader to more theoretical discussions and sources for more synthetic methods. (See Introduction.)

The polymerization of styrene by two different techniques is described below.

1-8. Synthesis of Polystyrene by Thermal Activation [7]

$$C_6H_5CH{=}CH_2 \longrightarrow \left[-CH-CH_2- \right]_n \qquad (17)$$
$$\underset{C_6H_5}{}$$

To a test tube is added 25 gm of styrene monomer; the tube is flushed with nitrogen and stoppered. The test tube is immersed in an oil bath at 125°–130°C for 24 hr, cooled, and broken open to recover a clear glasslike mold of polystyrene. If all the oxygen has been excluded, the polystyrene will be free of yellow stains, especially on the surface. The polystyrene can be purified and freed of residual monomer by dissolving it in benzene and reprecipitating it in a stirred solution of methanol. The solids are filtered and dried in a vacuum oven at 50°–60°C to give a 90% yield (22.5 gm). The molecular weight is about 150,000 to 300,000 as determined by viscometry in benzene at 25°C.

1-9. Emulsion Polymerization of Styrene [8]

To a standard resin kettle equipped with a stirrer, condenser, and dropping funnel are added 71.2 gm of distilled styrene and 128.2 gm of distilled conductivity water. To the water the following are added separately: 31.4 ml of 0.68% potassium persulfate solution (recrystallized $K_2S_2O_8$ used) and 100 ml of a 3.56% soap solution (commercial soap was used from Proctor and Gamble Co. which was composed of the sodium salts of stearic, palmitic, and oleic acids). Nitrogen is used to purge the system of dissolved air, and then the temperature is raised to 50°C and kept at this temperature for 24 hr. A 90% conversion to polystyrene emulsion will have occurred at this point.

5. COUPLING REACTIONS

1-10. Grignard Coupling—Synthesis of Neohexane [9]

$$CH_3MgCl + Cl-\underset{\underset{CH_3}{|}}{\overset{\overset{CH_3}{|}}{C}}-CH_2CH_3 \longrightarrow CH_3-\underset{\underset{CH_3}{|}}{\overset{\overset{CH_3}{|}}{C}}-CH_2-CH_3 + MgCl_2 \quad (18)$$

To methylmagnesium chloride [from the action of gaseous methyl chloride on 121.5 gm (5 moles) of magnesium in 1600 ml of dry di-*n*-butyl ether] is added 610 ml (5 moles) of *tert*-amyl chloride in 1 liter of di-*n*-butyl ether at 5°C over a period of 8 hr. Decomposition with ice and distillation of the ether yielded the crude product at bp 37°–50°C. Washing with sulfuric acid and redistillation yielded pure neohexane, bp 49°–50°C (740 mm), n_D^{20} 1.3688, in 36–39% yield.

Methylmagnesium iodide can be prepared more conveniently from liquid methyl iodide and used in place of methylmagnesium chloride in the above coupling reaction.

Organolithium reagents can also react with hydrocarbons to give 15–50% yields of alkylated products. For example, *tert*-butyllithium is reported to react with naphthalene in decalin solution at 165°C for 41 hr to give 30% yield of mono-*tert*-butylnaphthalene and a 50% yield of di-*tert*-butylnaphthalene. The position of substitution was not reported. The generality of this reaction has not been explored beyond benzene, phenanthrene, and perylene.

$$+ \textit{tert-}\text{Butyl Li} \longrightarrow \quad (19)$$

Grignard reagents also react with inorganic halides to give coupled products. Examples of such halides are cupric chloride, lead chloride, silver bromide, silver cyanide, nickel chloride, palladium chloride, chromic chloride, iron halide, ruthenium halide, and rhodium halide.

$$2 \, RMgX + MX_2 \longrightarrow R\!-\!R + 2 \, MgX_2 + M \tag{20}$$

1-11. Ullmann Synthesis of 2,2′-Diethylbiphenyl [10]

$$(21)$$

To a Pyrex flask are added 150 gm (0.647 mole) of o-ethyliodobenzene and 150 gm of copper bronze. The mixture is heated for 3 hr at approximately 240°C. The product 2,2′-diethylbiphenyl is isolated by extraction with boiling chlorobenzene and then by concentration under reduced pressure. The residue is distilled from sodium. The product is obtained in 60% yield (42.5 gm), bp 142°–143°C (14–15 mm), n_D^{25} 1.5620.

Grignard reagents also couple in good yields with dimethyl and diethyl sulfate to give alkyl aromatics. The alkyl esters of arylsulfonic acids also couple, as do the sulfates.

REFERENCES

1. H. C. Brown and K. Murray, *J. Am. Chem. Soc.* **81,** 4108 (1959).
2. B. Bannister and B. B. Elsner, *J. Chem. Soc.* p. 1055 (1951).
3. S. D. Lipsky and S. S. Hall, *Org. Synth.* **55,** 7 (1976).
4a. J. E. Johnson, R. H. Blizzard, and H. W. Carhart, *J. Am. Chem. Soc.* **70,** 3664 (1948).
4b. B. Loubinoux, R. Vanderesse, and P. Canbere, *Tetrahedron Lett.* p. 3951 (1977).
5. H. C. Brown and P. A. Tierney, *J. Am. Chem. Soc.* **80,** 1552 (1958).
6. J. F. Norris and D. Rubinstein, *J. Am. Chem. Soc.* **61,** 1163 (1939).
7. Authors' Laboratory.
8. I. M. Kolthoff and W. J. Dale, *J. Am. Chem. Soc.* **69,** 441 (1947).
9. F. C. Whitmore, H. I. Bernstein, and L. W. Nixon, *J. Chem. Soc.* **60,** 2539 (1938).
10. P. M. Everitt, D. M. Hall, and E. E. Turner, *J. Chem. Soc.* p. 2286 (1956).

2

OLEFINS

1. INTRODUCTION

Olefins are commonly prepared by elimination reactions (loss of water, hydrogen halides, acids, etc.) and condensation reactions. Methods utilizing oxidation, reduction, isomerization or rearrangement, free radical, photolytic, and enzyme reactions are less commonly used in the laboratory to prepare a center of unsaturation.

The elimination reactions are summarized by Eq. (1)

$$R-\underset{\underset{Z}{|}}{C}H-CH_3 \xrightarrow[\text{heat}]{\text{Acids, bases, or}} RCH{=}CH_2 + ZH \qquad (1)$$

where Z may be a hydroxyl, halogen, ester, ether, methyl xanthate (Chugaev reaction), carbamate, carbonate, sulfite, amine, quaternary

From S. R. Sandler and W. Karo, *Organic Functional Group Preparations*, Vol. I, 2d ed. (New York, 1983), 38–81, by permission of Academic Press, Inc.

ammonium hydroxide (Hofmann degradation), amine oxide, or one of many other labile groups.

Another elimination reaction involves disubstituted derivatives as in Eq. (2)

$$R\!-\!\underset{\underset{Z}{|}}{CH}\!-\!\underset{\underset{Z}{|}}{CH_2} \longrightarrow RCH\!=\!CH_2 \tag{2}$$

where Z may be a hydroxyl or halogen.

Condensation reactions, such as the Boord synthesis, are good methods for converting aliphatic aldehydes [Eq. (3)] to substituted olefins via the preparation of dibromoethyl ethers, Grignard coupling, and elimination of bromoethoxy zinc. (See the referenced text for more detailed equations.)

$$RCH_2CH\!=\!O \longrightarrow RCH\!=\!CR'R'' \tag{3}$$

Aldehydes can also be converted to olefins by reaction with active methylene compounds [Eq. (4)] by the Knoevenagel, Perkin, Claisen, and aldol condensation reactions

$$RCH_2CH\!=\!O + CH_2(R')_2 \longrightarrow RCH_2CH\!=\!C(R')_2 \tag{4}$$

where R's are carboxylic acid or ester groups, nitro, nitrile, carboxylic anhydride groups, aldehydes, and ketones or any other strongly electron-withdrawing substituents.

The Wittig reaction [Eq. (5)] is a convenient laboratory method useful for the conversion of aldehydes and ketones to olefins via the reaction of triphenylphosphinemethylenes

$$RCH\!=\!O + (C_6H_5)_3P\!=\!CX_2 \longrightarrow RCH\!=\!CX_2 \tag{5}$$

where X is hydrogen, alkyl, alkyl carboxylate, and halogen.

A convenient modification of the Wittig method uses phosphonate carbanions, $(RO)_2P^-\ OCX_2$, in place of $(C_6H_5)_3P\!=\!CX_2$.

The condensation of acetylenes with carbon monoxide, hydrogen halides, alcohols, acids, amines, mercaptans, halogens, etc., gives extremely useful unsaturated compounds as generalized by Eq. (6).

$$HC\!\equiv\!CH + HZ \longrightarrow CH_2\!=\!CHZ \tag{6}$$

However, as a result of the hazards involved in handling acetylene, a great many laboratory workers avoid its use. Nevertheless, Reppe's pioneering experimental work has shown that under the proper conditions the hazards associated with acetylene may be minimized and that acetylene is quite useful for vinylation reactions. Some industrial processes today utilize acetylene reactions to prepare vinyl ethers, acrylic acid, vinyl fluoride, 2-butene-1,4-diol, and some other olefins used for the preparation of plastics.

2. ELIMINATION REACTIONS

A. DEHYDRATION OF ALCOHOLS

The acid-catalyzed or thermal elimination of water from alcohols is a favorite laboratory method for the preparation of olefins. Isomeric mixtures usually arise with the acid-catalyzed method. The order of reactivity in dehydration usually follows the order of stability of the intermediate (transient) carbonium ion, i.e., tertiary > secondary > primary. The acid-catalyzed procedure is illustrated below, where a 79–87% yield of cyclohexene is obtained [1–3].

Tertiary arylcarbinols have been reported to be converted within 30 sec to the corresponding alkenes (70% yield) with warm 20% sulfuric–acetic acid (by volume) [4]. The yields are much lower with aliphatic tertiary or secondary arylcarbinols. Some tertiary alcohols, such as those obtained from tetralone and the Grignard reagent, dehydrate on simple distillation and in the presence of anhydrous cupric sulfate as a catalyst [5].

Secondary and tertiary alcohols can be dehydrated in dimethyl sulfoxide when heated to 160°–185°C for 14–16 hr to give olefins in yields of 70–85% [6]. The solution is diluted with water, extracted with petroleum ether (30°–60°C), dried, and then distilled. Other acid catalysts that have been reported for dehydration of alcohols are: anhydrous or aqueous oxalic [7, 8], or phosphoric acid [9], and potassium acid sulfate [10, 11]. In addition, acidic oxides such as phosphorus pentoxide [10–12] and acidic chlorides such as phosphorus oxychloride or thionyl chloride [13] have been reported to be effective as catalysts for the dehydration reaction.

2-1. Preparation of Cyclohexene [3]

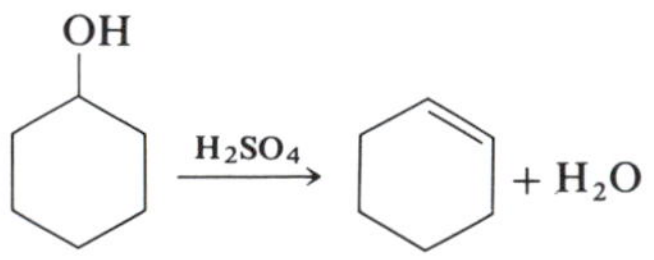

$$\text{(7)}$$

To 400 gm (4 moles) of cyclohexanol, in a flask set up for distillation of the contents into an ice-cooled receiver, is added 12 ml of concentrated sulfuric acid. The flask is heated to 130°–150°C by means of an oil bath for 5–6 hr. Water and crude cyclohexene are distilled out of the reaction mixture. The cyclohexene is salted out of the distillate, dried, and fractionated to give 260–285 gm (78–87%), bp 80°–82°C.

B. DEHYDROHALOGENATION REACTIONS

The dehydrohalogenation reaction is complex because the nucleophile B, can remove the β-proton to produce elimination, it can attack the α-carbon to give the SN_2 product or give α-elimination. Normally α-elimination leads to the same product as obtained by β-elimination.

With simple unbranched alkyl halides, the use of alcoholic bases gives the Saytzeff olefin. Steric effects on the β-position usually increases Hofmann elimination.

$$2 \ CH_3CH_2CH_2CH_2CH_2Br + C_2H_5OK \xrightarrow{C_2H_5OH} CH_3CH_2CH=CH-CH_3$$

$$\text{B} \qquad\qquad\qquad \text{69\% Saytzeff olefin}$$
$$(18\% \ cis, 51\% \ trans)$$

$$+ \ CH_3CH_2CH_2CH=CH_2$$
$$\text{31\% Hofmann olefin} \tag{8}$$

$$(CH_3)_3C-CH_2-\underset{\underset{Br}{|}}{\overset{\overset{CH_3}{|}}{C}}-CH_3 + C_2H_5-OK \xrightarrow{C_2H_5OH} (CH_3)_3C-CH_2-\overset{\overset{CH_3}{|}}{C}=CH_2 \tag{9}$$
$$\text{81\%}$$

**2-2. Preparation of 3-Chloro-2-methyl- and
3-Chloro-4-methyl-α-methylstyrene [14]**

$$\tag{10}$$

Four hundred and eight grams (2.0 moles) of chloropropylated o-chlorotoluene (from propylene chlorhydrin, o-chlorotoluene, boron trifluoride, and phosphorus pentoxide) is refluxed with an 85% solution of potassium hydroxide in methanol [392 gm (7 moles) of

KOH in 1850 ml of methanol]. The methanol is removed by distillation and the remaining liquid is washed with water, dried with calcium
chloride, and distilled through an efficient column under reduced
pressure to give 90 gm (26%) of 3-chloro-2-methyl-α-methylstyrene,
bp 64°–65°C (4 mm), n_D^{25} 1.5340 and 152 gm (48%) of 3-chloro-
4-methyl-α-methylstyrene, bp 73°–74°C (4 mm), n_D^{25} 1.5520. The purity of these materials should be checked by gas chromatography.

C. THE WITTIG SYNTHESIS OF OLEFINS [15–20]

In 1953 Wittig and Geissler discovered that methylenetriphenyl-
phosphorane reacted with benzophenone to give 1,1-diphenylethyl-
ene and triphenylphosphine oxide in almost quantitative yield. The
phosphorane was prepared from triphenylmethylphosphonium bromide and phenyllithium.

$$[(C_6H_5)_3P-CH_3]^+Br^- \xrightarrow{\ C_6H_5-Li\ } (C_6H_5)_3P=CH_2 \xrightarrow{\ (C_6H_5)_2C=O\ }$$

$$(C_6H_5)_2C=CH_2 + (C_6H_5)_3PO \quad (11)$$

The advantage of the Wittig method is that a carbonyl group is
replaced specifically with a carbon–carbon double bond. Furthermore, the reaction is carried out under mild alkaline conditions at
low temperatures, which allow sensitive olefins to be prepared easily.

Other bases such as butyllithium, sodium amide, and alkali alk-
oxides can be substituted for phenyllithium. The solvent may be ether,
tetrahydrofuran, or dimethylformamide. Polar solvents give better
yields than do solvents such as benzene.

The preparation of 1,2-distyrylbenzene [21] in 84% yield is described as a representative example of the unmodified Wittig reaction.

2-3. Preparation of 1,2-Distyrylbenzene [21]

$$+ 2 LiBr + 2 C_2H_5OH + 2(C_6H_5)_3PO \quad (12)$$

A solution of 66.1 gm (0.25 mole) of *o*-xylylene dibromide and 142.5 gm (0.55 mole) of triphenylphosphine in 500 ml of dimethylformamide (DMF) is heated under reflux. After the first 10–15 min, a colorless crystalline solid begins to separate and the refluxing is continued for 3 hr. The mixture is cooled, filtered, and the solid is washed with DMF and ether. After air-drying, 175.9 gm (89.4%) of pure *o*-xylylene bis(triphenylphosphonium) dibromide, mp > 340°C is obtained.

To a solution of 42.5 gm (0.054 mole) of *o*-xylylene bis(triphenylphosphonium)dibromide and 12.6 gm (0.119 mole) of benzaldehyde in 150 ml of absolute alcohol is added 500 ml of 1.4 M lithium ethoxide in ethanol. After standing at room temperature for 30 min. the solution is refluxed for 2 hr to yield a red-orange solution. Concentrating the mixture to 100 ml and adding 300 ml of water causes the precipitation of a yellow oil which is extracted with ether. Upon concentrating the ether solution, a mobile oil is isolated which is purified by column chromatography using alumina (Fisher A540, 2.5 × 55 cm). Elution with 250–300 ml of low-boiling petroleum ether gives an oil which solidifies on further evaporation of the residual solvent. The combined solids are recrystallized from ethanol to yield 12.7 gm (84%) of colorless crystals melting at 117°–119°C.

D. CONDENSATION OF ALDEHYDES AND KETONES WITH THEMSELVES OR WITH OTHER ACTIVE METHYLENE COMPOUNDS

$$2\,RCH_2CHO \xrightarrow[OH^-]{Aldol} \left[RCH_2-\overset{\overset{\displaystyle OH}{|}}{CH}-\underset{\underset{\displaystyle R}{|}}{CH}CHO \right] \longrightarrow RCH_2-CH=\underset{\underset{\displaystyle R}{|}}{C}-CHO \tag{13}$$

$$ArCHO + (RCH_2CO)_2 + RCOONa \xrightarrow{Perkin} ArCH=\underset{\underset{\displaystyle R}{|}}{C}-COONa \tag{14}$$

$$ArCHO + CH_3COOC_2H_5 \xrightarrow[base]{Claisen} ArCH=CH-COOC_2H_5 \tag{15}$$

$$RCH_2CHO + CH_2(COOH)_2 \xrightarrow{Knoevenagel} RCH_2-CH=CH-COOH \tag{16}$$

$$RCHO + R'CH_2NO_2 \xrightarrow{OH^-} RCH=\underset{\underset{\displaystyle R'}{|}}{C}-NO_2 \tag{17}$$

The aldol, Perkin, and Knoevenagel condensations of active methylene compounds with aldehydes give olefins which are probably derived from the intermediate alcohols, as is true in the aldol condensation shown above. The Perkin, Knoevenagel, and Claisen reactions are described in further detail in Chapter 9, Carboxylic Acids.

The aldol condensation [22] yields alcohols which in some cases dehydrate easily at room temperature upon acidification by acetic acid. For example, the condensation of benzaldehyde with butyraldehyde gives α-ethylcinnamaldehyde in 58% yield [23].

2-4. Aldol Condensation—Preparation of α-Ethylcinnamaldehyde [23]

$$C_6H_5CHO + C_2H_5CH_2CHO \longrightarrow C_6H_5CH{=}\underset{\underset{C_2H_5}{|}}{C}{-}CHO \qquad (18)$$

To 15 gm of a 50% solution by weight of potassium hydroxide and 175 gm of ethanol is added 110 gm (1.1 mole) of benzaldehyde. The reaction mixture is stirred and cooled to 5°C while 50 gm (0.69 mole) of butyraldehyde is added over a period of 3 hr. Then the reaction mixture is allowed to remain overnight at room temperature. Acidification, filtration, and then distillation yield 65 gm (58%) of α-ethylcinnamaldehyde, bp 111°–112°C (7 mm), n_D^{25} 1.5822.

REFERENCES

1. E. J. Corey and A. K. Long, *J. Org. Chem.* **41**, 2208 (1978).
2. J. Rencroft and P. G. Sammes, *Q. Rev., Chem. Soc.* **25** (1), 135 (1971).
3. G. H. Coleman and H. F. Johnstone, *Org. Synth. Collect. Vol.* **1**, 183 (1941).
4. E. W. Garbisch, Jr., *J. Org. Chem.* **26**, 4165 (1961).
5. W. Karo, R. L. McLaughlin, and H. F. Hipsher, *J. Am. Chem. Soc.* **75**, 3233 (1953).
6. V. J. Traynelis, W. L. Hergenrother, J. R. Livingston, and J. A. Valicenti, *J. Org. Chem.* **27**, 2377 (1962).
7. R. B. Carlin and D. A. Constantine, *J. Am. Chem. Soc.* **69**, 50 (1947).
8. R. E. Miller and F. F. Nord, *J. Org. Chem.* **15**, 89 (1950).
10. G. B. Backman and L. L. Lewis, *J. Am. Chem. Soc.* **69**, 2022 (1947).
11. J. R. Dice, T. E. Watkins, and H. L. Schuman, *J. Am. Chem. Soc.* **72**, 1738 (1950).
12. N. Campbell and D. Kidd, *J. Chem. Soc.* p. 2154 (1954).
13. W. S. Allen and S. Bernstein, *J. Am. Chem. Soc.* **77**, 1028 (1955).
14. G. B. Bachman and H. M. Hellman, *J. Am. Chem. Soc.* **70**, 1772 (1948).

15. G. Wittig and G. Geisler, *Justus Liebigs Ann. Chem.* **580,** 44 (1953).
16. G. Wittig and U. Schoellkopf, *Chem. Ber.* **87,** 1318 (1954).
17. G. Wittig and W. Haag, *Chem. Ber.* **88,** 1654 (1955).
18. U. Schoellkopf. *Angew. Chem.* **71,** No. 8. 260 (1959).
19. A. Maercker, *Org. React.* **14,** 270 (1965).
20. M. Schlosser, G. Muller, and K. F. Christman, *Angew. Chem. Int. Ed. Engl.* **5,** 667 (1966).
21. C. E. Griffin, K. R. Martin, and B. E. Douglas, *J. Org. Chem.* **27,** 1627 (1962).
22. M. Baches, *Bull. Soc. Chim. Pr.* **1,** No. 5, 1101 (1934).
23. W. M. Kraft, *J. Am. Chem. Soc.* **70,** 3570 (1948).

3

ACETYLENES

1. INTRODUCTION

The two most important synthetic methods for introducing an acetylenic group into the molecule involve the elimination of hydrogen halides [Eq. (1)] or condensation with acetylenic derivatives.

$$\left.\begin{array}{l} RCX{=}CH_2 \\ RCX_2{-}CH_3 \end{array}\right\} \xrightarrow{\text{Base}} RC{\equiv}CH \tag{1}$$

Condensation reactions of alkyl halides and carbonyl compounds with organometallic derivatives of acetylene or with acetylene itself are quite useful in the laboratory and in industry.

$$R_2C{=}O + HC{\equiv}CH \longrightarrow R_2\underset{\underset{\displaystyle OH}{|}}{C}{-}C{\equiv}CH \tag{2}$$

From S. R. Sandler and W. Karo, *Organic Functional Group Preparations*, Vol. I, 2d ed. (New York, 1983), 82–97, by permission of Academic Press, Inc.

$$RC{\equiv}C{-}M + R'X \longrightarrow RC{=}CR' + MX \tag{3}$$

where M = metal.

Hydrogen halides can be eliminated either from 1,1-, or 1,2-dihalogenated hydrocarbons or from 1-halo olefins to yield acetylenes in good yields.

$$\left.\begin{array}{l} \underset{\underset{X}{|}}{RC}{=}CH, \quad RCH{=}CH{-}X \\[2em] \underset{\underset{X}{|}\;\underset{X}{|}}{RCH}{-}CH_2, \quad RCH_2CHX_2 \end{array}\right\} \xrightarrow{\text{Base}} RC{\equiv}CH \tag{5}$$

$$\underset{\underset{X}{|}}{RCH}{=}C{-}R \quad RCX_2CH_2R, \quad \underset{\underset{X}{|}\;\underset{X}{|}}{RCH}{-}CHR \xrightarrow{\text{Base}} RC{\equiv}CR \tag{6}$$

The most frequently used bases in the above dehydrohalogenations are finely divided potassium hydroxide and sodium amide. Alcoholic potassium hydroxide tends to cause the isomerization of 1-acetylenes to internal acetylenes (Favorskii rearrangement). Aromatic acetylenes are not effected. Sodium amide may cause the reverse rearrangement from internal acetylenes to 1-acetylenes. Impure sodium amide may be an ineffective reagent and should not be used since dangerous (explosive) peroxides may be present.

2. ELIMINATION REACTIONS

3-1. Preparation of 1-Butyne from 1,2-Dibromobutane

$$\underset{\underset{Br}{|}\;\underset{Br}{|}}{CH_3CH_2{-}CH{-}CH_2} + KOH \xrightarrow{C_2H_5OH} CH_3CH_2{-}C{\equiv}CH \tag{7}$$

To a flask equipped with an addition funnel, mechanical stirrer, and a condenser (whose exit is connected to a dry ice trap) and containing 145 gm (2.6 moles) of potassium hydroxide and 145 ml of 95% alcohol is added 100 gm (0.46 mole) of 1,2-dibromobutane, dropwise. The mixture is heated in an oil bath and the evolved ethylacetylene is passed through the reflux condenser into a cold trap ($-18°C$) to yield 17 gm (31%) of product, bp 18°C.

3-2. Preparation of Propyne from 1-Bromo-1-propene [2]

$$CH_3—CH{=}CH—Br + KOH \xrightarrow{C_4H_9OH} CH_3C{\equiv}CH + KBr + H_2O \qquad (8)$$

In a flask equipped as in Procedure 3-1, a stirred refluxing solution of 454 gm (8.1 moles) of potassium hydroxide in 1 liter of n-butyl alcohol is treated dropwise over a period of 4 hr with 242 gm (2 moles) of 1-bromo-1-propene. Propyne, 68 gm (85%), bp 27°–31°C, is obtained by trapping the vapors from the condenser in the dry ice trap.

An important reaction for the preparation of substituted acetylenes involves the condensation of metallo acetylenes with alkyl halides or with carbonyl compounds.

Sodium acetylide (prepared from sodium amide) is useful for the condensations with primary alkyl halides. However, secondary, tertiary, and primary halides branched at the second carbon atom are dehydrohalogenated to olefins by the reagent. Iodides react at a faster rate than bromides and the latter faster than chlorides. Chlorides are rarely used. The bromides are more common for preparative reactions. Sodium acetylide can also react with carbonyl compounds to yield acetylenic carbinols.

The synthesis of lithium acetylide–ethylenediamine complex has been reported; it is a white, free-flowing powder that is safe and stable up to about 45°C. This complex reacts with ketones to give excellent yields of ethynyl carbinols. The complex can either be prepared or obtained from a commercial source.

3. CONDENSATION REACTIONS

3-3. Preparation of 3-Nonyne [3]

$$C_5H_{11}—C{\equiv}CH + NaNH_2 \xrightarrow{NH_3} C_5H_{11}C{\equiv}C—Na \xrightarrow{C_2H_5Cl} C_5H_{11}C{\equiv}C—C_2H_5 \qquad (9)$$

To a flask containing a cold ($-50°C$) well-stirred mixture of 1 liter of ammonia and 40 gm (1.0 mole) of sodamide is added 50 gm (0.52 mole) of 1-heptyne. The mixture is stirred for 1 hr and ammonia is added in order to maintain the volume. To the mixture is added 75 gm (1.2 mole) of ethyl chloride and stirring is continued for 3 hr. The reaction is worked up by very carefully adding water to separate the oil, washing the latter with water, drying, and distilling to yield 15 gm (23%) of 3-nonyne, bp 151°–154°C.

NOTE: The sodamide should be freshly prepared or obtained from a good commercial source. Long exposure to air and oxygen reduces the effectiveness of the material and may also cause the formation of explosive peroxides, oxides, etc.

3-4. Preparation of 1-Ethynyl-1-cyclohexanol [4]

$$HC\equiv C-Na \ + \ \underset{}{\bigcirc}\!\!=\!O \xrightarrow[\text{2. H}_3\text{O}^+]{\text{1. NH}_3} \overset{OH \ C\equiv CH}{\bigcirc} \tag{10}$$

To a flask containing a mixture of 5.1 moles of sodium acetylide, prepared during a period of 3 hr from 117 gm of sodium and acetylene in 3 liters of liquid ammonia at $-50°C$, is added dropwise 5.1 moles (500 gm) of cyclohexanone. The mixture is stirred overnight while a slow stream of dry acetylene is passed through the solution. The ammonia is then evaporated, the residue acidified with 200 gm of tartaric acid in 500 ml of water, and the mixture extracted with ether. The ether layer is dried and evaporated. The residue is fractionated to yield 518 gm (82%) of 1-ethynyl-1-cyclohexanol, bp $74°-77°C$ (15 mm), n_D^{20} 1.4823 and mp $31°-32°C$.

REFERENCES

1. J. R. Johnson and W. L. McEwen, *J. Am. Chem. Soc.* **48,** 469 (1926).
2. G. B. Heisig and H. M. Davis, *J. Am. Chem. Soc.* **57,** 339 (1935).
3. T. H. Vaughn, R. R. Vogt, and J. A. Nieuwland, *J. Am. Chem. Soc.* **56,** 2120 (1934).
4. N. A. Milas, N. S. MacDonald, and D. M. Black, *J. Am. Chem. Soc.* **70,** 1829 (1948).

4

ALCOHOLS
AND PHENOLS

1. INTRODUCTION

The more common and synthetically useful methods for the preparation of alcohols depend on hydrolysis, condensation (Grignard reagents condensing with carbonyl compounds or alkylene oxides), and reduction reactions of carbonyl or alkylene oxide compounds. Oxidation reactions are also of importance and will be described below. The more specialized reactions will be treated briefly.

Hydrolysis reactions involving alkyl or activated aryl halides, and sulfonates, or diazonium compounds afford alcohols in good yields. The products in some cases may be contaminated by olefin by-products. The Bucherer reaction of an aromatic amine can be consid-

From S. R. Sandler and W. Karo, *Organic Functional Group Preparations*, Vol. I, 2d ed. (New York, 1983), 98–128, by permission of Academic Press, Inc.

ered a hydrolysis reaction. The generalized hydrolysis-type reaction may be represented by Eq. (1).

$$R-Z + H_2O \longrightarrow ROH + ZOH \tag{1}$$

The hydration of olefins can be considered a hydrolysis reaction since the olefin on reaction with sulfuric acid yields an alkyl sulfuric acid which on subsequent hydrolysis yields the alcohol. Sulfuric acid adds to olefins in accordance with the Markovnikoff rule as illustrated in Eq. (2) for isobutylene.

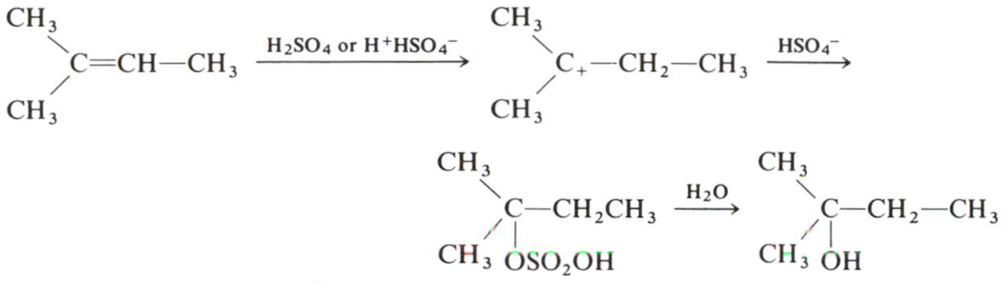

$$
\begin{array}{c}
CH_3 \\
\diagdown \\
C=CH-CH_3 \xrightarrow{H_2SO_4 \text{ or } H^+HSO_4^-} \\
\diagup \\
CH_3
\end{array}
\begin{array}{c}
CH_3 \\
\diagdown \\
C_+-CH_2-CH_3 \xrightarrow{HSO_4^-} \\
\diagup \\
CH_3
\end{array}
$$

$$
\begin{array}{c}
CH_3 \\
\diagdown \\
C-CH_2CH_3 \xrightarrow{H_2O} \\
\diagup \quad | \\
CH_3 \; OSO_2OH
\end{array}
\begin{array}{c}
CH_3 \\
\diagdown \\
C-CH_2-CH_3 \\
\diagup \quad | \\
CH_3 \; OH
\end{array}
\tag{2}
$$

Alkyl halides, aldehydes, ketones, and esters can be converted to alcohols of a higher carbon content by condensation reactions involving the Grignard reagent or a related organometallic (lithium compounds). The general scope of the reaction is illustrated below in Eq. (3).

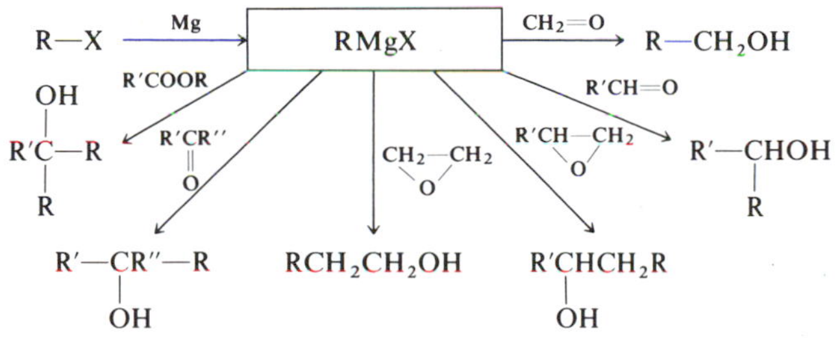

$$\tag{3}$$

The Friedel–Crafts condensation of ethylene oxide with aromatics gives aryl ethanols in good yields.

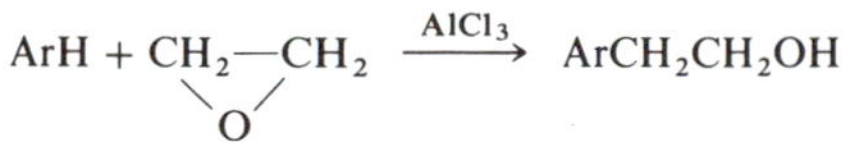

$$ArH + CH_2-CH_2 \xrightarrow{AlCl_3} ArCH_2CH_2OH \tag{4}$$
$$\diagdown \diagup$$
$$O$$

Some of the more specialized condensation reactions that are useful are the aldol condensation [Eqs. (5, 6)], Reformatskii [Eq. (7)], acyloin [Eq. (8)], and benzoin condensations [Eq. (9)], alkylation of phenols by alcohols [Eq. (10)], Prins reaction [Eqs. (11, 12)], Tollens hydroxymethylation reaction [Eq. (13)], the condensation of ketones [Eq. (14)] and aldehydes [Eq. (15)] with acetylenes, and the Pinacol reaction [Eq. (16)].

$$2\ RCH_2C{=}O \xrightarrow{\ OH^-\ } RCH_2-\underset{HO}{\overset{R}{C}}-\underset{R}{CH}-\overset{R}{C}{=}O \quad R{=}H,\ \text{alkyl or aryl} \tag{5}$$

$$RCH{=}O + R'CH_2NO_2 \xrightarrow{\ OH^-\ } RCH-\underset{OH}{\overset{R'}{CH}}-NO_2 \tag{6}$$

$$R_2C{=}O + \underset{Br}{RCH}-COOC_2H_5 + Zn\ or\ Mg \longrightarrow R_2C-\underset{OH}{\overset{R}{CH}}-COOC_2H_5 \tag{7}$$

$$2\ RCOOC_2H_5 + Na \longrightarrow \begin{matrix} R-C-ONa \\ \| \\ R-C-ONa \end{matrix} \longrightarrow \begin{matrix} R-CH-OH \\ | \\ R-C{=}O \end{matrix} \tag{8}$$

$$2\ ArCH{=}O + NaCN \longrightarrow \underset{OH}{ArCH}-\overset{O}{\underset{}{CAr}} \tag{9}$$

$$C_6H_5OH + ROH \xrightarrow{\ AlCl_3\ } R-C_6H_4OH \tag{10}$$

$$R_1-\underset{CH_3}{C}{=}\overset{R_2}{\underset{H}{C}} + CH_2{=}O + H_3O^+ \longrightarrow R_1-\underset{HO}{C}\underset{CH_3}{}-\underset{R_2}{CH}-CH_2OH \tag{11}$$

$$\text{(benzene)} + CH_2{=}O + H_3O^+ \longrightarrow \text{(benzene)}\overset{OH}{\underset{CH_2OH}{}} \tag{12}$$

$$CH_3CH{=}O + 4\ CH_2{=}O + H_2O \xrightarrow[Heat]{Cu(OH)_2} C(CH_2OH)_4 + HCOOH \tag{13}$$

$$R_2C{=}O + HC{\equiv}CR' \xrightarrow{\ KOH\ } R_2C-\underset{OH}{C}{\equiv}CR' \tag{14}$$

$$RCH{=}O + HC{\equiv}CR' \xrightarrow{\text{KOH}} \underset{\underset{OH}{|}}{RCH}{-}C{\equiv}CR' \tag{15}$$

$$2\,R_2C{=}O \longrightarrow \underset{\underset{HO}{|}\ \underset{OH}{|}}{R_2C}{-}CR_2 \tag{16}$$

The reduction of carbonyl compounds or alkylene oxides by organometallic hydrides yields alcohols of known structure in good yields.

$$\underset{\underset{O}{\|}}{R{-}C}{-}R' + NaBH_4 \longrightarrow \underset{\underset{OH}{|}}{R{-}CH}{-}R' \tag{17}$$

$$RCH{=}O + NaBH_4 \longrightarrow RCH_2{-}OH \tag{18}$$

$$\underset{\diagdown\!O\!\diagup}{R{-}CH{-}CH_2} + LiAlH_4 \longrightarrow \underset{\underset{OH}{|}}{R{-}CH}{-}CH_3 \tag{19}$$

2. CONDENSATION REACTIONS

4-1. Synthesis of 1-Octanol Using a Grignard Reaction [1]

$$C_5H_{11}CH_2MgBr + \underset{\diagup\,O\,\diagdown}{CH_2{-}CH_2} \longrightarrow C_5H_{11}CH_2CH_2CH_2OH \tag{20}$$

To a solution of *n*-hexylmagnesium bromide in 700 ml of ether [prepared from 2 moles (330 gm) of *n*-hexylbromide, 2 moles (48 gm) of magnesium turnings, and 700 ml of anhydrous ether] is added 95 gm (2.2 moles) of liquid ethylene oxide over a period of 1 hr while the ether refluxes vigorously as a result of the heat of reaction. After the addition has been completed, the ether is distilled off until 275 ml have been collected. Then 330 ml of dry benzene is added and the distillation is continued until the temperature reaches 65°C. The mixture is then refluxed for 1 hr and then hydrolyzed with ice water containing 10% sulfuric acid. The benzene layer is separated, washed twice with 10% sodium hydroxide solution, and then distilled to remove the benzene solvent. The residue is distilled under reduced pressure to yield 185 gm (71%) of 1-octanol, bp 105°C (15 mm).

4-2. Synthesis of 2-(2,4-Dimethylphenyl)ethanol by the Friedel–Crafts Reaction [2]

$$
\text{(21)}
$$

To a 3-liter, three-necked round-bottomed flask with a stirrer, dropping funnel, and gas outlet tube are added 5 moles of dry *m*-xylene (618 ml, 99%, Aldrich Chemical Co.) and 120 gm (0.9 mole) of aluminum chloride. The flask is cooled in an ice–salt water bath while 2 moles (88 gm) of ethylene oxide dissolved in 5 moles (613 ml) of *m*-xylene is added over a period of 2 hr. The reaction mixture is allowed to stand at room temperature overnight and then 100 ml of concentrated hydrochloric acid in 400 ml of ice water is added. The xylene layer is separated, washed with aqueous sodium hydroxide solution, and then distilled to yield 125 gm (41% based on ethylene oxide) of 2-(2,4-dimethylphenyl)ethanol, bp 100°–104°C (1 mm Hg), n_D^{25} 1.5310.

3. REDUCTION REACTIONS

4-3. Synthesis of Dimethylphenylethanols Using Sodium Borohydride [3]

$$
\text{(22)}
$$

To a 1-liter flask containing 300 ml of isopropanol and 10 gm of sodium hydroxide is added 20 gm (0.54 mole) of sodium borohydride with cooling. To the solution is added dropwise a solution of 280 gm (1.9 moles) of 3,4-dimethylacetophenone in 150 ml of isopropanol at such a rate that the solution gently refluxes. After the addition, the solution is allowed to stand overnight. The solution is then poured into a 5-liter flask and with stirring 1 liter of 1 *N* sodium hydroxide is added. The oily layer is separated, dried, and vacuum distilled. If a clear separation does not occur, then the solution is saturated with salt. The alcohol product is distilled through a 2-ft column to yield 177 gm (62%) of product, bp 101°–103°C (5 mm Hg), n_D^{20} 1.5284.

4. REARRANGEMENT REACTIONS

The Claisen rearrangement [4, 5] of allyl phenyl ethers to o-allyl-phenol in the presence of base is a reaction giving good yields. Ortho substituents on the allyl phenyl ethers do not effect the predominant formation of the ortho isomer in the rearranged product.

4-4. Claisen Rearrangement—Preparation of Allyl Phenyl Ether and Its Rearrangement to 2-Allylphenol [4, 5]

To a flask containing 188 gm (2.0 moles) of phenol are added 242 gm (2.0 moles) of allyl bromide, 280 gm (2.0 moles) of anhydrous potassium carbonate, and 300 gm of acetone. The mixture is refluxed on the steam bath for 8 hr, cooled, diluted with an equal volume of water, and extracted with ether. The ether extract is washed twice with 10% aqueous sodium hydroxide, dried, the solvent stripped off, and the residue distilled under reduced pressure to yield 230 gm (86%) of allyl phenyl ether, bp 85°C (19 mm).

$$\text{(23)}$$

The allyl phenyl ether is rearranged by boiling at 195°–200°C at atmospheric pressure under nitrogen until the refractive index of the liquid remains constant (5 to 6 hr to get n_D^{24} 1.55). The crude material is dissolved in 20% sodium hydroxide solution and extracted twice with 30°–60°C petroleum ether. The alkaline solution is acidified, extracted with ether, dried, the solvent stripped off, and the remaining liquid distilled under reduced pressure to yield 73% of 2-allylphenol, bp 103°–105.5°C (19 mm), n_D^{24} 1.5445.

REFERENCES

1. T. H. Vaughn, R. J. Spahr, and J. A. Nieuwland, *J. Am. Chem. Soc.* **55,** 4208 (1933).
2. S. R. Sandler, Borden Chem. Co. Central Research Lab., Philadelphia, Pennsylvania, unpublished results (1960).
3. D. S. Tarbell, *Org. React.* **2,** 8 (1944).
4. L. Claisen, *Justus Liebigs Ann. Chem.* **418,** 97 (1919).

5

ETHERS
AND OXIDES

1. INTRODUCTION

E thers and oxides differ in the type of chemical bonding of carbon to oxygen. Oxides are three-membered rings and are attached to adjacent carbon atoms in a given system whereas ethers, if they are cyclic, are not attached to adjacent carbon atoms as shown below.

From S. R. Sandler and W. Karo, *Organic Functional Group Preparations*, Vol. I, 2d ed. (New York, 1983), 129–146, by permission of Academic Press, Inc.

Ethers   Oxides (Epoxides)

The common methods used to produce ethers in the laboratory are the Williamson synthesis and the dehydration of alcohols using acids or inorganic oxides such as alumina.

$$RONa + R'X \longrightarrow ROR' + NaX \tag{1}$$

$$2\,ROH \xrightarrow[\text{Heat}]{H^+ \text{ or Alumina}} ROR + H_2O \tag{2}$$

Other methods involve the reaction of dialkyl sulfates with sodium alcoholates (or sodium phenolates) and the addition of olefins to alcohols or phenols using acid catalysts.

$$(RO)_2SO_2 + R'ONa \longrightarrow R'OR + ROSO_3Na \tag{3}$$

$$ROH + R'CH{=}CH_2 \longrightarrow \underset{\underset{OR}{|}}{R'CH}{-}CH_3 \tag{4}$$

The laboratory reactions with dialkyl sulfates is not preferred since it has been reported that these reagents may cause carcinogenic reactions.

Phenolic ethers can also be formed via the Claisen rearrangement as described earlier in Chapters 2 and 4 of this book.

Ethers are also formed by the base- or acid-catalyzed condensation of oxides with themselves or with alcohols or phenols to give monomeric or polymeric systems.

$$\underset{\underset{\text{Ethylene oxide}}{}}{\overset{\text{CH}_2\text{---CH}_2}{\diagdown\text{O}\diagup}} \longrightarrow \begin{cases} \xrightarrow{\text{H}^+} \overset{\text{CH}_2\text{---CH}_2}{\underset{\text{CH}_2\text{---CH}_2}{O \qquad\qquad O}} \\ \qquad\qquad\qquad \text{Dioxane} \\ \\ \xrightarrow{\text{CH}_3\text{OH}} \text{CH}_3\text{OCH}_2\text{CH}_2\text{OH} \\ \qquad\qquad\qquad \text{Methyl cellosolve} \end{cases}$$

$$\xrightarrow{\text{Base}} \text{HO(CH}_2\text{CH}_2\text{O)}_n\text{---H} \tag{5}$$

Carbowax

Oxides are usually formed in the laboratory by the peroxidation of olefins with H_2O_2, peracids, or by the dehydrohalogenation of halohydrins.

$$\overset{\diagdown}{\underset{\diagup}{C}}{=}\overset{\diagup}{\underset{\diagdown}{C}} + [O] \longrightarrow \overset{\diagdown}{\underset{\diagup}{C}}\overset{}{\underset{O}{---}}\overset{\diagup}{\underset{\diagdown}{C}}$$

$$\overset{\diagdown}{\underset{\underset{\text{OH}}{|}}{C}}{---}\overset{\diagup}{\underset{\underset{\text{X}}{|}}{C}} + \text{B}^- \tag{6}$$

■ **CAUTION:** Ethers tend to absorb and react with oxygen from the air to form peroxides. These peroxides must be removed prior to distillation or concentration to prevent an explosive detonation. In some cases, heat, shock, or friction can also cause a violent decomposition.

Some epoxides have been reported to be carcinogenic, especially those containing electron-withdrawing groups.

Due to their volatility, the lower members of the ethers and oxides are extremely flammable and care should be taken in handling these materials. Do not distill ethers to dryness.

2. CONDENSATION REACTIONS

A. THE WILLIAMSON SYNTHESIS

$$\text{RX} + \text{R}'\text{OH} \longrightarrow \text{ROR} + \text{HX} \tag{7}$$

The Williamson synthesis usually involves the use of the sodium salt of the alcohol and an alkyl halide. Primary halides give the best yields since secondary and tertiary halides readily dehydrohalogenate to give olefins. However, triarylmethyl chlorides react with alcohols directly to give 97% yields of ethers [1]. Alkyl phenyl ethers are prepared from aqueous or alcoholic solutions of alkali phenolates and alkyl halides. Benzyl halides are easily replaced by alkoxy groups in high yields. Polar solvents such as dimethylformamide favor the reaction. Phase transfer catalysis has been reported to aid this reaction.

5-1. Preparation of Triphenylmethyl Ethyl Ether [1]

$$(C_6H_5)_3C—Cl + C_2H_5OH \longrightarrow (C_6H_5)_3C—OC_2H_5 + HCl \qquad (8)$$

To an Erlenmeyer flask containing 100 ml of absolute ethanol is added 27.9 gm (0.10 mole) of triphenylmethyl chloride. The flask is heated to get rid of hydrogen chloride. Upon cooling, 28.0 gm (97%) of the trityl ethyl ether separates, m.p. 83°C.

5-2. Preparation of Allyl Phenyl Ether [2]

$$C_6H_5OH + CH_2{=}CH—CH_2Br \xrightarrow[\text{Acetone}]{K_2CO_3} C_6H_5OCH_2—CH{=}CH_2 \qquad (9)$$

In a flask equipped with a reflux condenser are placed 18.8 gm (0.20 mole) of phenol, 24.2 gm (0.23 mole) of allyl bromide, 28.0 gm (0.20 mole) of potassium carbonate, and 200 ml of acetone. The contents are refluxed for 10 hr, cooled, treated with 200 ml of water, and extracted three times with 25-ml portions of ether. The combined ether extracts are washed with 10% aqueous sodium hydroxide and three times with 25-ml portions of a saturated NaCl solution, dried, and distilled to yield 22 gm (82%) of the product, bp 119.5%– 120.5°C (30.2 mm), n_D^{25} 1.5210, ν_{max} 882 cm^{-1}.

B. ETHERS AND OXIDES

5-3. Preparation of Isobutyl Ethyl Ether [3]

$$
\begin{array}{c}
CH_3 \\
| \\
CH_3—CH—CH_2ONa + (C_2H_5)_2SO_4 \longrightarrow
\end{array}
$$

$$
\begin{array}{c}
CH_3 \\
| \\
CH_3—CH—CH_2OC_2H_5 + Na(C_2H_5)SO_4 \qquad (10)
\end{array}
$$

To a flask is added 93 gm (1.25 moles) of dry isobutyl alcohol followed by 12.5 gm (0.54 gm atom) of sodium (small pieces). The exothermic reaction causes the mixture to reflux. After the reaction ceases it is heated by means of an oil bath at $120°-130°C$ for $2\frac{3}{4}$ hr. After this time, some of the sodium still remains unreacted. The mixture is cooled to $105°-115°C$ and 77.1 gm (0.5 mole) of diethyl sulfate is added dropwise over a 2-hr period. The reaction is exothermic and steady refluxing occurs while the addition proceeds. The mixture is refluxed for 2 hr after all the diethyl sulfate has been added. The reaction mixture is cooled to room temperature. Then to it are added an equal weight of crushed ice and a slight excess of dilute sulfuric acid. The ether is steam-distilled off, separated, washed three times with 30% sulfuric acid, washed twice with water, and then dried over potassium carbonate. The dried product is refluxed over sodium ribbon and then fractionally distilled to give 35.7 gm (70%) of iso-butyl ethyl ether, bp $78°-80°C$, n_D^{25} 1.3739.

Alcohols also react with epoxides to give hydroxy ethers by a trans opening of the ring.

$$ROH + CH_2\!\!-\!\!CH\!\!-\!\!CH_3 \xrightarrow[\text{or } H^+]{\text{NaOCH}_3} ROCH_2\!\!-\!\!CH\!\!-\!\!CH_3 \qquad (11)$$
$$\underset{O}{\diagdown\diagup} \qquad\qquad \underset{OH}{|}$$

Cyclohexene oxide reacts with refluxing methanol in the presence of a catalytic amount of sulfuric acid to give *trans*-2-methoxycyclo-hexanol in 82% yield. Unsymmetrical epoxides such as propylene oxide give a primary or secondary alcohol, depending on the reaction conditions. Base catalysis favors secondary alcohol formation where-as acid or noncatalytic conditions favor a mixture of the isomeric ethers. Epichlorohydrin can be used in a similar manner to give chlorohydroxy ethers.

5-4. Preparation of 1-Ethoxy-2-propanol [4]

$$CH_3\!\!-\!\!CH\!\!-\!\!CH_2 + C_2H_5OH \xrightarrow{\text{NaOH}} CH_3\!\!-\!\!CH\!\!-\!\!CH_2OC_2H_5 \qquad (12)$$
$$\underset{O}{\diagdown\diagup} \qquad\qquad\qquad \underset{OH}{|}$$

To a mixture of 2560 gm (55.5 moles) of absolute ethanol and 10 gm of sodium hydroxide at $76°-77°C$ is added 638 gm (11 moles) of propylene oxide over a period of 4 hr. The mixture is boiled for

2 additional hr until the temperature becomes steady at 80°C. Distillation of the neutralized liquid yields 770 gm (81.4%) of 1-ethoxy-2-propanol, bp 130°–130.5°C.

3. OXIDATION REACTIONS

A. PEROXIDATION OF OLEFINS TO GIVE OXIRANES (EPOXIDES) [5–8]

■ **CAUTION:** All organic peracid reactions should be conducted behind a safety shield because some reactions proceed with uncontrollable violence. Reactions should first be run on a small scale, e.g., 0.1 mole or less, before scaling the preparation up. Efficient stirring and cooling should be provided.

Peracids and other peroxides can be destroyed by the addition of ferrous sulfate or sodium bisulfite [7].

Peracid-containing mixtures should not be distilled until the peracids have been eliminated.

The preparation of peracetic, performic, perbenzoic, and mono-perphthalate acids have been described. *m*-Chloroperbenzoic acid is available; it has the advantage of being more stable than perbenzoic acid. Higher aliphatic peracids have been prepared in sulfuric acid as a solvent with 50% hydrogen peroxide.

5-5. Preparation of 1-Hexene Oxide [9]

$$CH_3-(CH_2)_3-CH=CH_2 + \underset{Cl}{\overset{\overset{\displaystyle O}{\parallel}}{\underset{}{C}OOH}} \longrightarrow CH_3-(CH_2)_3-\overset{O}{\overset{\displaystyle \diagup \diagdown}{CH-CH_2}} + \underset{Cl}{\overset{\overset{\displaystyle O}{\parallel}}{C-OH}} \tag{13}$$

To a round-bottomed flask are added 24.4 gm (0.119 mole) of *m*-chloroperbenzoic acid (85% pure) and 10.0 gm (0.119 mole) of 1-hexene in 300 ml of anhydrous diglyme. The flask is then placed in the refrigerator for 24 hr and afterward the reaction mixture is subjected to distillation through a 2-ft helices-packed column to give 7.05 gm (60%) of 1-hexene oxide, bp 116°–119°C, n_D^{20} 1.4051.

5-6. Preparation of Isophorone Oxide [10]

$$
\underset{(CH_3)_2}{\overset{O}{\bigwedge}}\underset{CH_3}{}\quad\xrightarrow[\text{NaOH}]{\text{H}_2\text{O}_2}\quad\underset{(CH_3)_2}{\overset{O}{\bigwedge}}\underset{CH_3}{\bigwedge O}\tag{14}
$$

To a flask containing a stirred mixture of 55.2 gm (0.4 mole) of isophorone and 115 ml (1.2 moles) of 30% aqueous hydrogen peroxide in 400 ml of methanol at 15°C is added 33 ml (0.2 mole) of 6 N aqueous sodium hydroxide over a period of 1 hr at 15°–20°C. The resulting mixture is stirred for 3 hr at 20°–25°C and then poured into 500 ml of water. The product is extracted with ether, dried over anhydrous magnesium sulfate, and distilled to yield 43.36 gm (70.4%), bp 70°–73°C (5 mm), n_D^{25} 1.4500–1.4510.

REFERENCES

1. A. C. Nixon and G. E. K. Branch, *J. Am. Chem. Soc.* **58,** 492 (1936).
2. W. N. White and B. E. Norcross, *J. Am. Chem. Soc.* **83,** 3268 (1961).
3. E. M. Marks, D. Lipkin, and B. Bettman, *J. Am. Chem. Soc.* **59,** 946 (1937).
4. H. C. Chitwood and B. T. Freure, *J. Am. Chem. Soc.* **68,** 680 (1946).
5. D. Swern, *J. Am. Chem. Soc.* **69,** 1692 (1947).
6. D. Swern, *Chem. Rev.* **45,** 1 (1949).
7. D. Swern, *Org. React.* **7,** 378 (1953).
8. E. Searles, "Preparation, Properties, Reactions and Uses of Organic Peracids and Their Salts." F.M.C. Corporation, Inorg. Chem. Div., New York, 1964; B. Phillips, "Peracetic Acid and Derivatives." 2nd ed. Union Carbide Chemicals Co., New York, 1957.
9. D. J. Pasto and C. C. Cumbo, *J. Org. Chem.* **30,** 1271 (1965).
10. H. O. House and R. L. Wasson, *J. Am. Chem. Soc.* **79,** 1488 (1957).

6

HALIDES

1. INTRODUCTION

$\mathbf{M}$ost halogenation reactions involve either condensation or elimination reactions. Practically any organic compound can be halogenated by these reactions.

$$X\text{—}R\text{—}Z \xleftarrow{-HX} RZ + X_2 \longrightarrow RX + ZX$$
$$\longrightarrow R\text{—}Z\text{—}X \tag{1}$$

Olefins either add halogen to give 1,2-dihaloalkanes or react at the allylic hydrogen to give allyl halides.

From S. R. Sandler and W. Karo, *Organic Functional Group Preparations*, Vol. I, 2d ed. (New York, 1983), 147–179, by permission of Academic Press, Inc.

Selective chlorination reactions involving alcohols, ethers, and carboxylic acids have been reported to give ω-1 monochlorination products in good yields.

The reaction of halomethanes with olefins under basic conditions yields either mono or *gem*-dihalocyclopropanes in Eqs. (2) and (3).

$$CHX_3 + K\text{—}O\text{-}\textit{tert}\text{-butyl} + \begin{array}{c}\backslash\\C\end{array}\!\!=\!\!\begin{array}{c}/\\C\end{array}\!\!\backslash \longrightarrow$$

$$\begin{array}{c}\backslash\\C\end{array}\!\!-\!\!\begin{array}{c}/\\C\end{array}\!\!\backslash + KX + \textit{tert}\text{-Butyl alcohol} \qquad (2)$$

$$CH_2X_2 + n\text{-Butyllithium} + \begin{array}{c}\backslash\\C\end{array}\!\!=\!\!\begin{array}{c}/\\C\end{array}\!\!\backslash \xrightarrow{\text{Ether}}$$

$$\begin{array}{c}\backslash\\C\end{array}\!\!-\!\!\begin{array}{c}/\\C\end{array}\!\!\backslash + LiX + n\text{-Butyl halide} \qquad (3)$$

Haloalkanes also add free radically to olefins to give linear addition products and in some cases telomers.

$$CCl_4 + R\text{—}CH\!\!=\!\!CH_2 \longrightarrow R\text{—}\underset{\underset{Cl}{|}}{CH}\text{—}CH_2CCl_3 \qquad (4)$$

Aromatic hydrocarbons undergo electrophilic substitution reactions where electron-donating substituents favor the reaction and influence the orientation. The halogen source may be Cl_2, Br_2, I_2, mixed halogens, PCl_5, PCl_3, P + halogen, $SOCl_2$, N-bromosuccinamide, and others.

The chloromethylation reaction is important in adding $Cl\text{—}CH_2\text{—}$ to aromatics and heterocycles.

The interest in fluorocarbons, with their high temperature stability and chemical resistance, has generated extensive research in the preparation of these compounds.

2. CONDENSATION REACTIONS

The condensation reactions for the introduction of a halogen atom involve the reaction of a halogen source HX, PX_3, PX_5, $SOCl_2$,

RCOX, SF$_4$, X$_2$, HOX, or RX with alcohols, ethers, diazonium compounds, Grignard reagents, silver salts of acids, acids, amides, aromatic compounds, aldehydes, ketones, olefins, and amines. Many other organic compounds also undergo these reactions.

A. CONVERSION OF ALCOHOLS TO ALKYL HALIDES

$$ROH + HX \longrightarrow RX + H_2O \tag{5}$$

$$3\,ROH + PX_3 \longrightarrow 3\,RX + H_3PO_3 \tag{6}$$

$$ROH + KI + H_3PO_4 \longrightarrow RI + KH_2PO_4 + H_2O \tag{7}$$

$$ROH + P_2I_4 \longrightarrow RI \tag{8}$$

$$ROH + CH_3I + (C_6H_5O)_3P \longrightarrow RI + C_6H_5{-}OH + (C_6H_5O)_2POCH_3 \tag{9}$$

$$ROH + SOCl_2 \longrightarrow RCl + SO_2 + HCl \tag{10}$$

$$ROH + RSO_2Cl \longrightarrow RCl \tag{11}$$

$$3\,ROH + \text{cyanuric chloride} \longrightarrow 3\,RCl + \text{cyanuric acid} \tag{12}$$

$$ROH + CHCl_3 \xrightarrow[\text{H}_2\text{O}]{\text{NaOH}} RCl \tag{13}$$

6-1. Preparation of n-Butyl Bromide [1]

$$2\,CH_3CH_2CH_2CH_2OH + 2\,NaBr + H_2SO_4 \longrightarrow$$
$$2\,CH_3CH_2CH_2CH_2Br + Na_2SO_4 + 2\,H_2O \tag{14}$$

To a 2-liter round-bottomed flask equipped with a stirrer, dropping funnel, and reflux condenser and containing 270 ml of water is added with stirring 309 gm (5 moles) of sodium bromide powder. (The reverse addition causes caking.) To this solution is first added 178 gm (2.4 moles) of n-butyl alcohol, and then 218 ml of concentrated sulfuric acid is slowly added dropwise. The mixture is stirred vigorously or shaken to prevent the sulfuric acid from forming a layer. The mixture is refluxed for 2 hr and then distilled to yield the product. The water-insoluble layer is separated, washed with water, then washed with a sodium carbonate solution (5 gm/100 ml water), separated, dried over 5–10 gm of calcium chloride, and distilled to yield 298 gm (90%), bp 101°–104°C, n_D^{20} 1.4398.

B. HALOGENATION REACTIONS

Halogenation of organic compounds is a common method of introducing a halogen functional group. Bromine and chlorine react in the liquid or gas phase. Fluorine, however, is too reactive and oxidation reactions occur. Sulfur tetrafluoride is an important reagent for the conversion of carboxyl group to CF_2 or CF_3 groups.

Hydrofluoric acid and lead dioxide or nickel dioxide react with toluenes bearing electronegative substituents (nitro, cyano, carboethoxy, etc.) to give the corresponding benzyl and/or benzal fluoride.

Aromatic compounds are conveniently chlorinated or brominated in the laboratory. For example, durene is chlorinated at 0°C in chloroform to 57% monochlorodurene (mp 47°–48°C). In the absence of catalysts and in sunlight, alkylbenzenes are brominated or chlorinated in the side chain.

Trichlorocyanuric acid has been reported to be an effective laboratory chlorinating reagent for nuclear or side-chain halogenation of aromatic systems.

Naphthalene is brominated at room temperature in the absence of a catalyst to α-bromonaphthalene in 75% yield, whereas in the presence of an iron catalyst at 150°–165°C, β-bromonaphthalene is formed in 57% yield. Bromine, bromine monochloride [2], iodine monobromide, and N-bromosuccinimide have been employed as brominating agents.

Bromine monochloride reacts with p-nitrophenol to give 2,6-dibromo-4-nitrophenol, indicating that the more electrophilic bromine group reacts preferentially.

para-Bromination of aromatic amines can be achieved by 2,4,4,-6-tetrabromo-2,5-cyclohexadien-1-one.

6-2. Bromination of p-Nitrophenol in Aqueous Solution with Bromine Chloride To Give 2,6-Dibromo-4-nitrophenol [2]

$$\text{(15)}$$

To a flask containing a vigorously stirred mixture of 13.9 gm (0.1 mole) of p-nitrophenol, 21.6 gm (0.21 mole) of sodium bromide, and 450 ml of water at 45°C (the heating mantle is removed) is slowly added a stream of chlorine gas. A white precipitate forms and the temperature rises to 50°–55°C. The chlorine gas is added just so long as no bromine color appears above the solution. An oil deposits on the side of the flask but it changes to a white solid as the reaction progresses. After 15–20 min the bromine color appears above the solution and the original white suspension turns orange. The chlorine gas is stopped and stirring is continued for 10–15 min at 50°C. If the bromine color fades, more chlorine gas is added until the color reappears. Stirring is then continued for an additional 15 min. When no further fading of the bromine color results, the suspension is cooled to 20°C and filtered. The light yellow precipitate is washed with cold water and dried for 16–20 hr in a vacuum oven at 60°C to give 28.6 gm (96%), mp 141°–142°C.

C. REACTION OF OLEFINS WITH HALOGENS AND HALOGEN DERIVATIVES

Bromine readily adds to a double bond at −20° to 20°C to give dibromides in high yield, e.g., allyl bromide gives 98% 1,2,3-tribromopropane. Chloroform, carbon tetrachloride, acetic acid, or ether are recommended solvents for the addition of halogen to olefins. Heat or sunlight favors dehydrohalogenation reactions.

Bromine adsorbed on a molecular sieve has been reported to be useful for the selective bromination of olefins. For example, when a cyclohexene-styrene mixture is treated with a 5 Å sieve that was previously saturated with bromine, only α,β-dibromostyrene (95% yield) is obtained and no trace of dibromocyclohexane is formed. This selectivity may be due to the fact that 5 Å molecular sieves favor the free radical chain reaction which without molecular sieves give this same selectivity.

Chlorine adds trans to a double bond at low temperatures. Elevated temperatures favor substitution reactions. Sulfuryl chloride and phosphorus pentachloride have been used as chlorination agents.

The chlorination of olefins using molybdenum chlorides and cupric chloride has also been reported. In addition, other transition metal chlorides and bromides show a similar reactivity to halogenate olefins.

Conjugated diolefins undergo 1,4-addition of halogen at low temperatures. Bromine gives mainly 1,4-addition, whereas chlorine in the liquid or vapor phase gives equal amounts of 1,2- and 1,4-addition products.

Mixtures of bromine and chlorine add to olefins such as cyclohexene, styrene, ethylene, stilbene, and cinnamic acid to give the bromochlorides. The products isolated are those expected for the electrophilic addition of a bromine atom and a nucleophilic addition of a chlorine atom.

Recently the feasibility of the addition of elemental fluorine to rather sensitive olefins has been demonstrated. The vicinal difluorides produced were predominantly cis.

6-3. Bromination of Cyclohexene to 1,2-Dibromocyclohexane [3]

$$\text{(16)}$$

To a 2-liter flask equipped with a stirrer, separatory funnel, thermometer, and containing a solution of 123 gm (1.5 moles) of cyclohexene in 300 ml of carbon tetrachloride and 15 ml of absolute alcohol is added dropwise 210 gm (67 ml, 1.3 moles) of bromine in 145 ml of CCl_4 at such a rate (3 hr) that the temperature does not exceed $-1°C$. Higher temperatures cause substitution to occur. The carbon tetrachloride is distilled off using a water bath and the residue washed with 20 ml of 20% alcoholic potassium hydroxide, washed with water, dried, and immediately distilled under reduced pressure to yield 303 gm (85%) of product, bp $99°-103°C$ (16 mm), n_D^{25} 1.5495.

Dihalocarbenes react with olefins to give 1,1-dihalocyclopropanes [4, 5].

$$:CHX \text{ or } :CX_2 + \begin{matrix} \diagdown \\ \diagup \end{matrix} C = C \begin{matrix} \diagup \\ \diagdown \end{matrix} \longrightarrow \begin{matrix} \diagdown \\ \diagup \end{matrix} C - C \begin{matrix} \diagup \\ \diagdown \end{matrix} \text{ or } \begin{matrix} \diagdown \\ \diagup \end{matrix} C - C \begin{matrix} \diagup \\ \diagdown \end{matrix} \qquad \text{(17)}$$

Halocarbenes can be generated from haloforms and base, methylene halide and lithium alkyls, and by the decarboxylation of sodium

METHOD OF PREPARATION AND PHYSICAL PROPERTIES OF SEVERAL 1,1-DIHALOCYCLOPROPANES[a]

1,1-Dibromocyclopropane (except where noted)	Olefin (moles)	Solvent (ml) potassium-tert-butoxide, moles	Haloform (moles) C = $CHCl_3$ B = $CHBr_3$	Yield (%)	Product Bp[b] (mm) (°C)	n_D (temp., °C)
6,6-Dichlorobicyclo[3.1.0]-hexane	Cyclopentene (1.0)	None/0.75	0.75C	20	87°–90° (61)	1.4907–1.4941[c] (27.2°)
6,6-Dibromobicyclo[3.1.0]-hexane	Cyclopentene (1.0)	None/0.75	0.82B	42	63°–69° (2)	1.5560–1.5594 (18°)
7,7-Dichlorobicyclo[4.1.0]-heptane	Cyclohexene (1.0)	n-Pentane(400)/ 1.0	1.0C	18	67°–68° (6.0)	1.5038 (20°)
7,7-Dibromobicyclo[4.1.0]-heptane	Cyclohexene (1.0)	Cyclohexane (100)/0.88	1.18B	35	98°–100° (6–6.5)	1.5579[d] (23°)
2,2-Dimethyl	Isobutylene (1.0)	None/0.4	0.3B	28	47°–48° (11)	1.5136 (23°)
cis-2,3-Dimethyl	cis-2-Butene (1.0)	None/0.4	0.3B	90	55°–56° (11–12)	1.5188–1.5206 (23°)
trans-2,3-Dimethyl	trans-2-Butene (1.0)	None/0.4	0.3B	90	55°–56° (11)	1.5110 (25°)
2,2,3-Trimethyl	2-Methyl-2-butene (1.0)	None/0.6	0.5B	50	48°–50° (3.8)	1.5167 (23°)
2,2,3,3-Tetramethyl	2,3-Dimethyl-2-butene (1.0)	n-Pentane(100)/ 1.0	0.9B	60	mp 75°–76°	—
2-Phenyl-	Styrene (1.0)	None/0.25	0.25B	55	118°–120° (5.7)	1.5996 (23°)
2,2-Diphenyl-	1,1-Diphenylethylene (0.14)	n-Pentane(100)/ 0.25	0.26B	63	mp 146°–147°[e]	—
2-sec-Butyl	4-Methyl-1-pentene (1.0)	None/0.25	0.25B	52	50° (1.0)	1.4992 (23°)[f]

[a] Reprinted in part from S. R. Sandler, *J. Org. Chem.*, **32**, 3876 (1967). Copyright 1967 by the American Chemical Society. Reprinted by permission of the copyright owner.

[b] Boiling points and melting points are uncorrected.

[c] Analysis of $C_6H_8C_2$—calculated: C, 47.60; H, 5.30; found: C, 47.98; H, 5.18.

[d] Analysis of $C_7H_{10}Br_2$—calculated: C, 33.10; H, 4.00; found: C, 33.06; H, 4.15.

[e] Recrystallization from isopropanol.

[f] Analysis of $C_7H_{12}Br_2$—calculated: C, 32.80; H, 4.68; found: C, 33.22; H, 4.76.

trichloroacetate. Difluorocarbene can also be generated by the photolysis or pyrolysis of difluorodiazirene. Mixed halocarbenes have also been reported.

6-4. General Procedure for the Preparation of 1,1-Dihalocyclopropanes [5]

The *gem*-dihalocyclopropanes are generally prepared by adding 1.0 mole of haloform to 1.0 mole of dry potassium *tert*-butoxide (M.S.A. Research Corp.) and 1.0 mole or more of the olefin in 200–300 ml of *n*-pentane, at 0°–10°C. After the addition has been completed, the temperature is raised to room temperature and the mixture stirred for several hours. Water is added and the organic layer separated, washed with water, dried, and concentrated at atmospheric pressure. The crude material is weighed, analyzed by gas–liquid chromatography, and distilled under reduced pressure to yield the pure product. Several examples are summarized in Table I. The detailed preparation of 1,1-dibromo-2,2-diphenylcyclopropane is given as an example (6–5).

6-5. Preparation of 1,1-Dibromo-2,2-diphenylcyclopropane [5, 6]

$$(C_6H_5)_2C\!\!=\!\!CH_2 + CHBr_3 + KO\text{-}tert\text{-}Bu \longrightarrow$$

$$(C_6H_5)_2\underset{\displaystyle \underset{Br\quad Br}{\diagdown C \diagup}}{C\text{---}CH_2} + KBr + tert\text{-}Butyl\ alcohol \qquad (18)$$

To a flask containing 100 ml of dry pentane, 25 gm (0.14 mole) of 1,1-diphenylethylene, and 28 gm (0.25 mole) of potassium *tert*-butoxide at 0°C is added dropwise 66 gm (0.26 mole) of bromoform over a 1-hr period at 0°–10°C. During the reaction a yellow solid is formed which upon filtration at the end is obtained in 63% yield (30.8 gm), mp 140°–147°C (from isopropanol).

REFERENCES

1. H. L. Goering, S. J. Cristol, and K. Dittmer, *J. Am. Chem. Soc.* **70**, 3314 (1948).
2. C. O. Obenland, *J. Chem. Educ.* **41**, 566 (1964).
3. H. R. Snyder and L. A. Brooks, *Org. Synth. Collect. Vol.* **1**, 171 (1943).
4. P. S. Skell and A. Y. Garner, *J. Am. Chem. Soc.* **78**, 5430 (1956).
5. S. R. Sandler, *J. Org. Chem.* **32**, 3876 (1967).
6. S. R. Sandler, *Org. Synth.* **56**, 32 (1977).

7

ALDEHYDES

1. INTRODUCTION

The laboratory preparations of aldehydes involve oxidation, reduction, condensation, and elimination reactions.

The oxidation of primary alcohols using chromic acid, chromic oxide–pyridine complex, manganese dioxide, etc., gives aldehydes in good yields. Allylic alcohols are readily oxidized at room temperature with active manganese dioxide.

Olefins, alkyl groups (Étard reaction), and alkyl halides (Sommelet reaction) may also be oxidized to aldehydes. Acetylenes can be converted to trivinylboranes which in turn are easily oxidized to aldehydes.

From S. R. Sandler and W. Karo, *Organic Functional Group Preparations*, Vol. I, 2d ed. (New York, 1983), 180–205, by permission of Academic Press, Inc.

Several reduction methods are available to convert nitriles (Stephen reaction) or acyl chlorides (Rosenmund reduction) or amides to aldehydes. The use of lithium aluminum hydride at low temperatures or lithium triethoxyaluminum hydride appears to be a more effective means of reduction than the stannous chloride–HCl used in the Stephen method or palladium and hydrogen used in the Rosenmund reduction.

The Gattermann condensation reaction and its modifications, the Vilsmeier reaction as well as the use of modified Friedel–Crafts and Grignard reactions, offer useful methods for the laboratory synthesis of aromatic and heterocyclic aldehydes.

The Darzens reaction and the McFadyen–Stevens reaction are also useful aldehyde syntheses involving elimination reactions.

Since aldehydes are usually highly reactive compounds, these preparations should be carried out under conditions that will reduce product losses by further reactions (such as oxidation to carboxylic acids).

Recently, the use of phase transfer catalysts has facilitated the oxidation of alcohols and primary alkyl halides to aldehydes.

Benzal halides are hydrolyzed to aldehydes via water and an iron catalyst. Aromatic aldehydes are also obtained by the partial oxidation of an aromatic side chain using chromic acid and acetic anhydride followed by hydrolysis of the diacetate.

Industrially, olefins are converted to aldehydes via the OXO process (olefin, carbon monoxide, hydrogen, and a catalyst). In addition, olefins can be oxidized to give the aldehyde (propylene→acrolein).

2. OXIDATION REACTIONS

7-1. Preparation of Propionaldehyde [1]

$$CH_3CH_2CH_2OH + K_2CrO_7 + H_2SO_4 \longrightarrow CH_3CH_2CH{=}O \qquad (1)$$

To a well-stirred mixture of 100 gm (1.66 mole) of boiling isopropanol is added dropwise a solution of 164 gm (0.55 mole) of potassium dichromate and 120 ml of concentrated sulfuric acid in 1 liter of water. The addition takes 30 min and the aldehyde distills off as it is formed. The propionaldehyde is redistilled to yield 44–47 gm (45–49%), bp 48°–55°C, n_D^{20} 1.3636.

Ratcliffe has reported that 1-decanol is oxidized to 1-decanal in 63–66% yield, using dipyridine–chromium(VI) oxide prepared directly in methylene chloride at 20°C. This procedure has the advantage over the earlier reported procedure by Sarett and co-workers and Collins in that pure dipyridine–chromium(VI) oxide does not have to be separately prepared. The latter is troublesome since it is hygroscopic and has a propensity to enflame during preparations.

7-2. Preparation of Acrolein by Oxidation of Allyl Alcohol with Manganese Dioxide [2]

$$CH_2{=}CH{-}CH_2OH + MnO_2 \longrightarrow CH_2{=}CH{-}CHO \qquad (2)$$

To a flask containing 2 gm (0.034 mole) of allyl alcohol is added a suspension of 20 gm (0.219 mole) of commercial manganese dioxide in 100 ml of petroleum ether (bp 40°–60°C). The mixture is stirred for $\frac{1}{2}$ hr. The mixture is quickly filtered with suction and the precipitate is washed with fresh solvent. The combined filtrates are concentrated and distilled to afford a 60% yield of acrolein, isolated as the 2,4-dinitrophenylhydrazone, 4.7 gm, mp 165°C. Similarly cinnamyl alcohol is converted to cinnamaldehyde in 75% yield.

Active manganese dioxide gives better results. This reagent may be prepared from manganese sulfate. Manganese sulfate, 111.0 gm (0.482 mole, tetrahydrate) in 1.5 liters of water, and 1170 ml of a 40% sodium hydroxide solution are added simultaneously over a period of 1 hr to a hot stirred solution of 960 gm (5.91 mole) of potassium permanganate in 6 liters of water. After 1 hr the precipitate of brown manganese dioxide is filtered, washed with water until the wash water is colorless, dried at 100°–120°C, and ground into a fine powder to give 920 gm.

3. CONDENSATION REACTIONS

7-3. Preparation of 3-Formylindole by the Vilsmeier Method [3]

$$(3)$$

To a flask containing 16 gm (0.22 mole) of dimethylformamide cooled to 10°–20°C and protected from moisture is added 5.0 ml (0.055 mole) of phosphorus oxychloride. Indole (5.85 gm, 0.059 mole) in 4 gm of dimethylformamide is then added slowly with stirring at such a rate as to keep the temperature at 20°–30°C. The mixture is kept at 35°C for 45 min and then it is poured into crushed ice. The clear solution is heated at 20°–30°C and 9.5 gm (0.24 mole) of sodium hydroxide in 50 ml water is added at such a rate that the solution remains acidic until approximately three-fourths of the alkali solution has been added. The last quarter is quickly added and the solution is boiled for 1 min. The resulting white crystals are filtered off, washed with five 25-ml portions of water, and dried to constant weight at 100°C at 10 mm to yield 6.93 gm (95.5%), mp 197°–199°C.

REFERENCES

1. C. D. Hurd, R. N. Meinert, and L. V. Spence, *J. Am. Chem. Soc.* **52,** 138 (1930).
2. J. Attenburrow, A. F. B. Cameron, J. H. Chapman, R. M. Evans, B. A. Hems, A. B. A. Jansen, and T. Walker, *J. Chem. Soc.* p. 1094 (1952).
3. G. F. Smith, *J. Chem. Soc.* p. 3842 (1954).

8

KETONES

1. INTRODUCTION

The oxidation of secondary alcohols by sulfuric–chromic acid at 20°–40°C gives good yields of ketones and is a widely used synthetic method. Several other oxidizing agents have been found to be quite useful. Mild oxidizing agents such as dimethyl sulfoxide and air have also been used.

$$R_2CHOH \longrightarrow R_2C{=}O \tag{1}$$

Triethylene compounds activated by halogen, carbonyl, double bond, aromatic rings, heterocyclic rings, etc., can be oxidized to ketones or quinones using several oxidants.

The hydroboration of olefins and their subsequent oxidation offers a novel method of oxidizing olefins to ketones.

From S. R. Sandler and W. Karo, *Organic Functional Group Preparations*, Vol. I, 2d ed. (New York, 1983), 206–234, by permission of Academic Press. Inc.

Ozonolysis of olefins and the use of other oxidants such as permanganate, permanganate–periodate, dichromate–sulfuric acid, and
hydrogen peroxide–lead tetraacetate are also useful in preparing
ketones by cleaving the olefin into two fragments.

The Oppenauer oxidation is a preferred method for oxidizing sensitive alcohols such as the sterols. Modifications have been reported
which permit the reaction to be carried out at room temperature.

The condensation reactions involving the Grignard reagent and
the Friedel–Crafts method are perhaps the most popular laboratory
methods for introducing a ketone group into a molecule.

$$
\text{RMgX} + \begin{cases} \text{R'COCl} \\ \text{R'CN} \\ \text{(R'CO)}_2\text{O} \\ \text{R'CONH}_2 \end{cases} \longrightarrow \overset{\overset{\text{O}}{\|}}{\text{RCR'}} \tag{2}
$$

$$
\text{R}_2\text{Cd} + 2\,\text{RCOCl} \longrightarrow 2\,\text{RCOR} + \text{CdCl}_2 \tag{3}
$$

$$
\text{RCOCl} + \text{ArH} \xrightarrow{\text{AlCl}_3} \text{RCOAr} \tag{4}
$$

Active methylene groups of esters, ketones, and other compounds
can be alkylated, acylated, or self-condensed prior to hydrolysis or
cleaved to give substituted mono- or diketones.

The thermal decarboxylation of acids is not a preferred laboratory method since high temperatures are involved and low yields of
unsymmetrical ketone are obtained. However, new modifications
have been described to extend this reaction to the preparation of
unsymmetrical ketones.

The pinacol rearrangement is an effective method for rearrang-
ing completely alkylated 1,2-glycols. The reaction has been extended
to 1,1'-dihydroxyalkanes to yield spiro ketones.

2. OXIDATION REACTIONS

The oxidation of secondary alcohols to ketones in good yields is effected by sulfuric–chromic acid mixtures. For water-soluble alcohols
the reaction is carried out in aqueous solution at $20°–40°C$. Insoluble aromatic alcohols are oxidized in an acetic acid solvent. Some
other oxidation reagents that have been used are nitric acid, copper
sulfate in pyridine, cupric acetate in 70% acetic acid, ferric chloride

in water, chromic anhydride in glacial acetic acid, chromic oxide in pyridine, chromic acid in aqueous ether solutions, dimethyl sulfoxide and air, and dinitrogen tetroxide in chloroform.

More recently it was found that methyl sulfide–N-chlorosuccinimide in toluene at $0°C$ gives good yields of ketones by the oxidation of primary and secondary alcohols. However, allylic and dibenzylic alcohols give halides. The latter alcohols can be oxidized to ketones with a dimethyl sulfoxide–chlorine reagent.

8-1. Preparation of 4-Methylcyclohexanone [1]

$$\text{(5)}$$

To a flask containing a solution of 367 gm (1.4 moles) of sodium dichromate in $1\frac{3}{4}$ liters of water at $80°C$ is added 798 gm (7.0 moles) of 4-methylcyclohexanol. To the stirred mixture is added dropwise a solution of 367 gm (1.4 moles) of sodium dichromate and 1078 gm (11.0 moles) of sulfuric acid in $1\frac{3}{4}$ liters of water at such a rate to maintain the temperature at $80°C$ (approximately 12 hr). The ketone and water are distilled from the reaction mixture, the ketone is separated from the aqueous phase, and then distilled to give 549 gm (70%) of 4-methylcyclohexanone, bp $171.4°–171.5°C$ (760 mm), n_D^{20} 1.4463; d_{20}^{20} 0.917.

3. CONDENSATION REACTIONS

The Grignard reaction and Friedel–Crafts reaction are the two condensation reactions used most frequently to prepare ketones in the laboratory.

The Grignard reagent reacts with nitriles to form ketimine salts which on hydrolysis yield ketones. Low molecular weight aliphatic nitriles give ketones contaminated with hydrocarbons derived from the acidic α-hydrogen of the nitrile. The difficulty can be overcome to some extent by discarding the ethereal solution containing the hydrocarbon by-products before the hydrolysis of the ketimine salts. Therefore, the Grignard procedure is successful with aromatic nitriles or high molecular weight aliphatic nitriles. Low molecular weight aliphatic nitriles respond favorably with aromatic Grignard reagents.

Tetra-organotin compounds have been used to react with acid chlorides to give ketones. Benzylchlorobis(triphenylphosphine)palladium(II) is used to catalyze the reaction.

8-2. Preparation of Acetophenone by the Grignard Method [2]

$$C_6H_5MgBr + (CH_3CO)_2O \xrightarrow[\text{Ether}]{80°C} C_6H_5\overset{\displaystyle O}{\overset{\|}{C}}CH_3 \qquad (6)$$

To a flask containing 40 gm (0.39 mole) of acetic anhydride in 100 ml of anhydrous ether cooled to $-80°C$ using dry ice–acetone is added 0.2 mole of an ether solution of phenylmagnesium bromide. The reaction mixture is stirred for 2–3 hr and then hydrolyzed with ammonium chloride solution. The ether layer is washed with water and alkali in order to remove the excess acetic anhydride and acetic acid. The remaining ether is dried and fractionally distilled to yield 16.8 gm (70%) of acetophenone, bp 202°C, n_D^{20} 1.5339.

The Friedel–Crafts acylation method is an excellent method for introducing a ketone group into an aromatic hydrocarbon molecule. Some of the acylating agents used vary in reactivity as follows: $RCO^+BF_4^- > RCO_2ClO_3 > RCOOSO_3H > RCO$ halogen $> RCOOCOR' > RCOOR' > RCONR_2 > RCHO > RCOR'$. The acylation of polyalkylbenzenes with acetyl chloride, and aluminum chloride in carbon tetrachloride has been reported to give good yields. Aluminum chloride is the most effective catalyst.

The acylation of heterocycles has been reported. Diketones are prepared by the Friedel–Crafts method using adipyl chloride, benzene, and aluminum chloride.

8-3. Preparation of Dimethylacetophenones by the Friedel–Crafts Acylation Method [3]

$$\underset{}{\text{(ring)}}\text{—CH}_3 + \text{AlCl}_3 + \text{CH}_3\text{COCl} \xrightarrow{\text{CCl}_4} \underset{\text{COCH}_3}{\text{(ring)}}\text{—CH}_3 + \text{AlCl}_3 + \text{HCl} \qquad (7)$$

The method of Mowry [4] was used to react each of the xylene isomers with acetyl chloride to yield the corresponding aceto-phenones.

To a 5-liter three necked flask equipped with a stirrer, dropping funnel, condenser, and drying tube is added $1\frac{1}{2}$ liter of dry, freshly distilled carbon tetrachloride. To this is added 454 gm (3.3 moles) of reagent grade granular aluminum chloride and the mixture is cooled to about 5°C by means of an ice-water bath. Acetyl chloride (275 gm, 249 ml, 3.5 moles) are added dropwise to the cooled mixture over a 5–10 min interval. This addition is then followed with the drop-wise addition of the appropriate xylene isomer [o-xylene, 300 ml (2.5 moles), m-xylene 300 ml (2.5 moles), or p-xylene 275 ml (2.3 moles)] at 10°–15°C, which takes about 1–2 hr. The mixture is stirred at room temperature for 2 hr and then allowed to stand overnight at room temperature.* The mixture is poured into a mixture of 5 kg of ice and 700 ml of concentrated hydrochloric acid. The lower carbon tetra-chloride layer is separated, washed twice with 250 ml portions of water, once with 500 ml of 2% sodium hydroxide solution, and then several times with water until the washings are neutral. The carbon tetrachloride is distilled off at atmospheric pressure and the residue is fractionally distilled to give 280 gm (75%) of 3,4-dimethylaceto-phenone, bp 103°C (6 mm), n_D^{20} 1.6381; 140 gm (43%) of 2,5-dimethyl-acetophenone. bp 85°–86°C (3 mm), n_D^{20} 1.5291; or 225 gm (61%) of 2,4-dimethylacetophenone, bp 90°C (6 mm), n_D^{20} 1.5340 depending on which xylene isomer was used.

REFERENCES

1. F. K. Signaigo and P. L. Cramer, J. Am. Chem. Soc. 55, 3326 (1933).
2. M. S. Newman and W. T. Booth, J. Am. Chem. Soc. 67, 154 (1945).
3. S. R. Sandler, unpublished results (1960).
4. D. T. Mowry, M. Renoll, and W. F. Huber, J. Am. Chem. Soc. 68, 1105 (1946)

---

*The reaction may be worked up immediately but this is a convenient place to stop the reaction.

9

CARBOXYLIC
ACIDS

From S. R. Sandler and W. Karo, *Organic Functional Group Preparations*, Vol. I, 2d ed. (New York, 1983), 235–287, by permission of Academic Press, Inc.

1. INTRODUCTION

The common laboratory methods for the synthesis of carboxylic acids are oxidation, oxidation–reduction reactions, carbonation of organometallics, condensation reactions, and hydrolysis reactions.

Oxidation reactions are useful for the conversion of aliphatic side chains of aromatic compounds, primary alcohols, aldehydes, ketones, olefins, and a combination of one or more of the latter groups to a carboxylic acid.

$$\left.\begin{array}{l} ArCH_3 \\ RCH_2OH \\ RCH{=}O \\ R_2C{=}O \\ RCH{=}CH_2 \end{array}\right\} \xrightarrow{(O)} RCOOH \qquad (1)$$

The haloform reaction is a good method for the oxidation of aliphatic methyl ketones or of methyl carbinols to carboxylic acids. The reaction is also applicable to aromatic methyl ketones and aryl methyl carbinols obtained by the Friedel–Crafts reaction and by the Grignard reaction, respectively.

$$\text{(Ph)} + CH_3COCl + AlCl_3 \longrightarrow \text{(Ph–COCH}_3) \xrightarrow{NaOCl} \text{(Ph–COOH)} \qquad (2)$$

$$\text{(Ph–CH{=}O)} + CH_3MgX \longrightarrow \text{(Ph–CHCH}_3,\ OH) \xrightarrow{NaOCl} \text{(Ph–COOH)} \qquad (3)$$

Oxidation–reduction reactions such as the Cannizzaro reactions are useful for the conversion of aldehydes lacking an α-hydrogen atom to a mixture of the acid and alcohol. The reaction is useful in those cases where an acid oxidation medium would degrade or re-arrange the molecule.

Carbonation of the Grignard reagent and hydrolysis of the magnesium halide derivative is one of the most generally applicable laboratory methods for the preparation of carboxylic acids.

$$RX \xrightarrow[\text{ether}]{Mg} RMgX \xrightarrow{CO_2} RCOOMgX \xrightarrow{H_2O} RCOOH + MgXOH \qquad (4)$$

where R = aliphatic or aromatic.

Carbonation of aryl or alkyl lithium compounds also affords carboxylic acids in good yields.

$$C_6H_5Li + CO_2 \longrightarrow C_6H_5COOLi \xrightarrow{H_2O} C_6H_5COOH + LiOH \qquad (5)$$

Condensation reactions such as the Reformatskii reaction, Perkin reaction, malonic ester synthesis, and the Diels–Alder reaction afford acids or their esters in good yields.

The hydrolysis of nitriles in the aliphatic or aromatic series yields carboxylic acid. This method is useful for the conversion of primary aliphatic nitriles and aromatic nitriles to the carboxyl derivative. Since the nitriles are generally prepared from primary halides, the hydrolysis of nitriles represents another method of converting readily accessible organic raw materials to carboxylic acids.

$$RX + KCN \longrightarrow RCN \xrightarrow[H^+ \text{ or } OH^+]{H_2O} RCOOH \qquad (6)$$

$$X-CH_2-R-CH_2-X + 2\,KCN \longrightarrow NC-CH_2-R-CH_2-CN$$

$$\xrightarrow[H_2O]{H^+ \text{ or } OH^+} \underset{\displaystyle}{HO\overset{O}{\overset{\|}{C}}-CH_2-R-CH_2\overset{O}{\overset{\|}{C}}OH} \qquad (7)$$

Other hydrolysis reactions are also used to prepare acids from esters, amides, acid chlorides, anhydrides, and haloketones.

More specialized reactions having limited synthetic value in the laboratory are listed briefly with references so that the reader may obtain more detailed information on the particular reaction. Some of these reactions are of industrial importance in the preparation of particular acids.

2. OXIDATION REACTIONS

A. ALKYL SIDE CHAINS

The liquid phase oxidation of aromatic alkyl side chains may be affected by aqueous sodium dichromate or with dilute nitric acid by refluxing the mixture for a prolonged period as in the case for the conversion of *o*-xylene to *o*-toluic acid. The procedure is illustrated in Eq. (8), a reaction which gives a 55% yield of *o*-toluic acid. With aqueous sodium dichromate *o*-xylene, yields phthalic acid.

9-1. Preparation of *o*-Toluic Acid [1]

$$(8)$$

In a 5-liter, round-bottomed flask are placed 1.6 liters of water, 800 ml of concentrated nitric acid, and 400 ml (3 moles) of xylene (90%). A reflux condenser with an outlet to a gas absorption trap is attached to the flask. The mixture is refluxed by heating in an oil bath at 145°–155°C for 55 hr. At the end of this time, the organic layer has settled to the bottom of the flask. The hot reaction mixture is poured onto 1 kg of ice, and the precipitate is filtered off, washed with cold water, and filtered again. The wet product is dissolved by warming in 1 liter of 15% solution of sodium hydroxide. Residual *o*-xylene is separated by an ether extraction. The aqueous layer is then decolorized with Norite and acidified with concentrated hydrochloric acid. The crude material is recrystallized from aqueous ethanol to yield a tan product melting at 99°–101°C, in the amount of 218–225 gm (53–55%).

B. ALDEHYDES

Sulfuric acid–potassium dichromate can also be used to oxidize aldehydes to acids, as is illustrated by the oxidation of furfural to furoic acid [2].

9-2. Preparation of Furoic Acid [2]

$$\text{(furan)}-CH=O \longrightarrow \text{(furan)}-COOH \tag{9}$$

In a round-bottomed flask equipped with a mechanical stirrer, dropping funnel, and reflux condenser are placed 100 gm of furfural, 100 gm of potassium dichromate, and 10 gm of water. On a steam bath the flask is heated to 100°C while a mixture of 200 gm of sulfuric acid and 100 gm of water is added during 30–45 min. The heat of reaction is such that the steam bath is removed after a short time. When reaction is complete, the reaction mixture is cooled and nearly neutralized with sodium hydroxide. Then it is completely neutralized with sodium carbonate. The chromium hydroxide that is filtered off weighs 56 gm after drying. The filtrate is made acid with sulfuric acid and the dark brown precipitate of furoic acid is filtered. More furoic acid is obtained by concentrating the filtrate. One hundred and five grams of crude material is collected. The product is recrystallized from water to yield 87 gm of white crystals of furoic acid (75%), mp 131.5°C.

An alternative procedure for the preparation of furoic acid by oxidation of furfural with hydrogen peroxide in pyridine has been reported.

3. OXIDATION OF OLEFINS

The preparation of 5-methylhexanoic acid by the ozonization of 6-methyl-1-heptene and decomposition of the ozonide by acidic hydrogen peroxide gives yields which average 67% [3].

9-3. Preparation of 5-Methylhexanoic Acid [3]

$$CH_3-\underset{\underset{CH_3}{|}}{CH}-CH_2-CH_2-CH_2CH=CH_2 \xrightarrow{O_3}$$

$$CH_3-\underset{\underset{CH_3}{|}}{CH}-CH_2-CH_2-CH_2-\underset{\underset{O_3}{\diagdown\diagup}}{CH-CH_2} \xrightarrow[H_2SO_4]{H_2O_2} CH_3\underset{\underset{CH_3}{|}}{CH}-CH_2CH_2-CH_2-COOH \tag{10}$$

A solution of 0.5 mol of 6-methyl-1-heptene in 200 ml of methylene chloride is cooled to $-78°C$ and subjected to a stream of 6% ozonized oxygen at 20 litres/hr for 12 hr.

■ **CAUTION:** Ozonides can explode. Use a hood and work behind a safety barrier with suitable personal protective equipment.

The ozonide is dissolved in 100 ml acetic acid, the methylene chloride removed carefully by suction, and the solution is slowly dropped into a mixture of 114 gm of 30% hydrogen peroxide, 5 ml of concentrated sulfuric acid, and 200 ml of water. Behind a shield, cautious heating is applied progressively. This phase needs careful attention because the reaction becomes vigorous and requires intermittent cooling. Refluxing is continued for 2 hr. After cooling to ice temperature, extraction with ether is performed. The ether layer is then extracted with a solution of sodium hydroxide. The latter is acidified, extracted with ether, dried, and distilled. The acid is collected at $200°-210°C$. Redistillation yields a fraction bp $204°-207°C$ (752 mm), n_D^{20} 1.4220, d_D^{20} 0.9162; amide mp $99.5°-100°C$; p-bromophenacyl ester mp $72.5°-73°C$. The yields average 67%.

4. OXIDATION OF KETONES AND QUINONES

9-4. Preparation of Diphenic Acid [4]

$$\text{phenanthraquinone} \xrightarrow[\text{or } H_2O_2 + HOAc]{K_2CrO_7 + H_2SO_4} \text{diphenic acid} \tag{11}$$

To a cold solution of 200 gm of potassium dichromate, 500 gm of water and 300 gm of concentrated sulfuric acid is added 50 gm of phenanthraquinone. The mixture, in a round-bottomed flask fitted with an air condenser, is cautiously heated for 1 hr to avoid oxidation to carbon dioxide and water. The reaction mixture is then maintained between $105°$ and $110°C$ for approximately 20 hr with occasional agitation. The reaction mixture is then cooled, poured into cold water, and the precipitated product filtered off. By repeated washes with cold 5% sulfuric acid, the product may be substantially freed of excess oxidizing agent and reduced side products. The product is then washed and recrystallized from water or glacial acetic acid. Yield, 85%, mp 228°C (uncorrected).

NOTE: Roberts and Johnson recommend that commercial phenanthraquinone be free from chromium compounds and that it be allowed to stand for several hours in concentrated sulfuric acid prior to oxidation. They also claim that sodium bichromate affords a better yield of product than does the potassium salt [22].

A. THE HALOFORM REACTIONS

Acetyl groups and methylcarbinols are converted to carboxyl groups by substitution of the three hydrogens of the methyl groups by halogen which is subsequently hydrolyzed.

$$RCOCH_3 + 3\,NaOH + 3\,X_2 \longrightarrow [RCOCX_3] + 3\,H_2O + 3\,NaX$$

$$\Big\downarrow{\scriptstyle NaOH} \qquad\qquad (12)$$
$$RCOONa + CHX_3$$

In order to ensure good yields, it is desirable that no similarly replaceable hydrogens be present in the R group. However, methylene groups are not easily affected by the reagents. For example, β-phenylisovaleric acid is obtained in 84% yield from 4-methyl-4-phenyl-2-pentanone by this method.

The reagents that are used for the haloform reaction are chlorine in sodium hydroxide at 55°–80°C, aqueous sodium or potassium hypochlorite, commercial bleaching agents, iodine in sodium hydroxide, or bromine in sodium hydroxide at 0°C.

An example of the haloform reaction is the oxidation of propiophenone to benzoic acid in 96% yield [5].

9-5. Preparation of Benzoic Acid [5]

$$\underset{\displaystyle C_6H_5-\overset{\textstyle O}{\overset{\|}{C}}-CH_2-CH_3}{} + NaOBr \longrightarrow C_6H_5COOH + CH_3COOH + NaBr \qquad (13)$$

To 300 ml of a rapidly stirred sodium hypobromite solution (0.512 mole) at 22°C is added 20.1 gm (0.15 mole) of propiophenone over a 5-min period. Stirring is rapid so that the solution exists as an emulsion throughout the reaction. Best yields are obtained under these conditions. Stirring is continued for $2\frac{1}{2}$ hr and the mixture kept at 24°–25°C by immersing the flask in an ice bath when necessary. The unreacted hypobromite is destroyed with sodium bisulfite and the basic solution is extracted with ether to remove the unreacted ketone. Acidification of the aqueous phase with concentrated hydrochloric acid yields 17.6 gm (96%) of benzoic acid, mp 121.5°–122°C alone and when mixed with an authentic sample.

B. THE WILLGERODT REACTION

The Willgerodt reaction is useful in the preparation of arylacetic acids and amides from substituted methyl aryl ketones or vinyl aromatic compounds. The aliphatics and acetylenes give lower yields. The conversion is effected by heating aromatic compounds at 160°–200°C in an aqueous solution under pressure using ammonium polysulfide. In the Kindler modification, the ketone or styrene is refluxed with a mixture of sulfur and an amine, usually morpholine, to give a thioamide, $ArCH_2CSNR_2$.

$$\left.\begin{array}{l} RCOCH_3 \\ RCH{=}CH_2 \\ RC{\equiv}CH \end{array}\right] \longrightarrow RCH_2CSNH_2 \longrightarrow RCH_2COOH \qquad (14)$$

Schwenk and Block [23] also suggested the use of morpholine as an amine, which permits operations in ordinary laboratory equipment. The reaction appears to be quite general for aromatic monomethyl ketones. Substitutions such as nitro, amino, hydroxy, or second acetoxy groups interfere with the standard reaction, probably because these functional groups are capable of reacting with sulfur, polysulfides, or other components of the reaction mixture.

A more recent modification of the Willgerodt reaction describes the preparation of acids from aromatic hydrocarbons using aqueous base and sulfur [6]. An example is the preparation of isophthalic acid from *m*-xylene using aqueous ammonia, sulfur, and water.

9-6. Preparation of Isophthalic Acid [6]

$$\text{(CH}_3\text{-C}_6\text{H}_4\text{-CH}_3) + \text{NH}_4\text{OH} + \text{H}_2\text{O} + \text{S} \longrightarrow \text{(HOOC-C}_6\text{H}_4\text{-COOH)} \tag{15}$$

(after acidification)

A 2.5-liter autoclave is charged with 42.5 gm (0.4 mole) of *m*-xylene (98%), 243 gm of 28% aqueous ammonia (4 moles of NH_3), 250 gm of water (23.6 moles H_2O), and 96 gm (3.0 gm atoms) of sulfur, and heated to 316°C before shaking is begun. After 30 min, the maximum pressure of 176 atm is obtained, indicating completion of the reaction. Shaking is continued an additional 5 min. Steam distillation of the reaction mixture removes hydrogen sulfide and ammonia and reduces the pH to 7. No xylene is recovered. Sulfur from the polysulfide decomposition is filtered off and 2 moles of sodium hydroxide is added to saponify amides. The solution is steam-distilled until the vapors test neutral to moist pH paper and discarded. Then the remaining solution of nonvolatiles is adjusted to pH 7 with dilute hydrochloric acid. After thorough washing and drying further acidification gives 57.0 gm (86%) of isophthalic acid, neutral equivalent 82.9.

5. BIMOLECULAR OXIDATION–REDUCTION REACTIONS

Aldehydes that lack an α-hydrogen and therefore cannot undergo an aldol condensation undergo the Cannizzaro reaction in the presence of a strong base, giving the alcohol and the corresponding carboxylic acid. Furfural in the presence of sodium hydroxide yields 72–76% furfuryl alcohol and 73–76% furoic acid upon acidifying [7].

9-7. Cannizzaro Reaction: Preparation of Furoic Acid [7]

$$2\,(\text{furyl-CH=O}) + \text{NaOH} \longrightarrow (\text{furyl-CH}_2\text{OH}) + (\text{furyl-COONa}) \tag{16}$$

One kilogram (10.4 moles) of redistilled furfural is placed in a flask which is surrounded by an ice bath and cooled to 5°–8°C. While stirring, 625 gm of 33.3% of sodium hydroxide (5.2 moles) is added dropwise at such a rate to keep the reaction temperature below 20°C. After the addition, the mixture is stirred for an additional hour. The mixture is cooled to 0°C and filtered. The precipitate is pressed as dry as possible on a suction filter. The sodium 2-furancarboxylate is transferred to a beaker, triturated therein with a 200–250-ml portion of cold water, cooled to −5°C, and again filtered. The trituration of the solid with water is repeated.

The combined filtrates are distilled at 25 mm nearly to dryness using a heating bath, the temperature of which is kept below 145°C. The water is distilled away under vacuum and the residue of furfuryl alcohol is shaken with sodium bisulfite solution to remove any remaining furfural. Fractionation yields 367–390 gm (72–76%), bp 83°C (24 mm), n_D^{21} 1.4869.

All the solid residues (sodium 2-furancarboxylate) are dissolved in warm water and filtered from a small amount of dark insoluble material. The filtrate is acidified with concentrated hydrochloric acid, cooled to 0°C, and filtered. The solid is washed twice with a little ice water and dried. A yield of 420–440 gm (73–76%) of white furoic acid is obtained, mp 132°C (recrystallized from carbon tetrachloride).

The reaction of α-diketones with strong bases yields the rearranged α-hydroxy carboxylic acids. This reaction is known as the benzilic acid rearrangement and is illustrated below using benzoin [8].

9-8. Benzilic Acid Rearrangement [8]

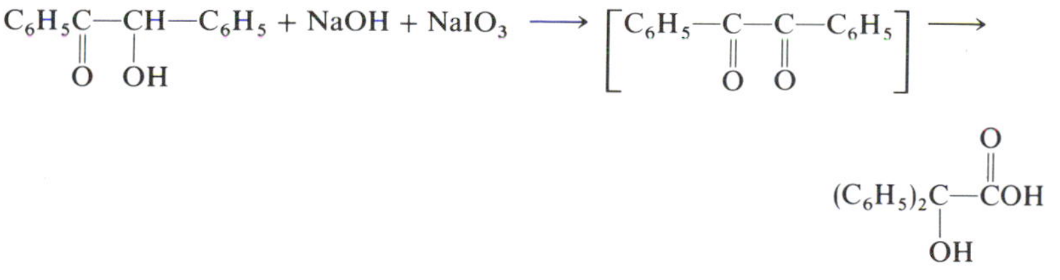

$$C_6H_5C{-}CH{-}C_6H_5 + NaOH + NaIO_3 \longrightarrow \left[C_6H_5{-}C{-}C{-}C_6H_5 \right] \longrightarrow$$
$$\underset{O\ \ OH}{} \qquad \underset{O\ \ O}{}$$

$$(C_6H_5)_2C{-}\overset{O}{\overset{\|}{C}}OH \quad (17)$$
$$\underset{OH}{}$$

To a solution of 20 gm of sodium hydroxide and 7 gm of sodium iodate in hot water is added 20 gm of benzoin. The mixture is stirred and a purple color becomes evident. The mixture is heated while being stirred until the purple color fades and then sufficient water is added to dissolve the solids and form a clear solution. Concentrated hydrochloric acid is added, and the iodine and traces of benzoic acid are expelled by boiling. On cooling, the solution is filtered, and 20 gm of dry crude benzilic acid is obtained. Recrystallization from benzene yields 18 gm (90%) of benzilic acid, mp 150°C.

In the above example, benzoin is oxidized directly to benzilic acid via the benzil intermediate. A similar rearrangement occurs when α-epoxy ketones are refluxed with 30% aqueous sodium hydroxide.

6. CARBONATION OF ORGANOMETALLIC REAGENTS

Carbonation of the Grignard reagent and organometallic compounds is a useful laboratory method for the conversion of most halides to acids containing one additional carbon atom. One technique involves pouring the ether solution of the organometallic into excess crushed dry ice. Carbon dioxide is sometimes required for tertiary Grignard reagents. A low temperature and rapid stirring produce high yields of acids. The main by-products are symmetrical ketones and tertiary alcohols formed by the reaction of the organometallic compound with the carboxylic acid salt. Jetwise addition of the organometallic compound to excess powdered dry ice greatly reduces the amount of these products. Yields range from 50% to 85% of the carboxylic acids. Of the other organometallic reagents the lithium and sodium derivatives have been the most popular and have also given good yields of carboxylic acids under suitable conditions.

The carbonation of the Grignard reagent has advantages over the procedure involving hydrolysis of nitriles since the latter is applicable only to primary halides if reasonable yields are to be obtained. The Grignard method has limitations since tertiary acids above dimethylethylacetic acid cannot be prepared by carbonation of the appropriate Grignard compounds because the latter reagents prepared from higher halides react abnormally, yielding mixtures of alkanes and alkenes. The Wurtz type of Grignard reagent coupling becomes increasingly more important as higher halides are involved.

Recently the use of highly reactive magnesium has been reported to be useful for the preparation of Grignard reagents from a variety of aromatic and aliphatic halides, including fluorides. These have been carbonated to convert them to carboxylic acids in fair to good yields. For example, fluorobenzene is converted to benzoic acid in 69% yield and 1-chloronorboronene is converted to norboronene 1-carboxylic acid in 63% yield [24].

9-9. Grignard Reagents—Preparation of α-Methylbutyric Acid [9]

$$C_2H_5CHClCH_3 + Mg \longrightarrow C_2H_5-CH(MgCl)CH_3 \xrightarrow[H_2O]{CO_2} C_2H_5-CH(COOH)CH_3 \tag{18}$$

sec-Butylmagnesium chloride is prepared in 400 ml of ether from 13.4 gm (0.55 gm atom) of magnesium shavings and 46 gm (0.5 mole) of sec-butyl chloride. A stream of carbon dioxide is passed through the solution at $-5°$ to $-12°C$. After $1\frac{1}{2}$ hr the temperature drops from $-5°$ to $-12°C$ and does not rise on increasing the flow rate of carbon dioxide. The drop in temperature is taken as the end point for the carbonation. The reaction mixture is hydrolyzed with 500 ml of 25% sulfuric acid while cooling with ice. The water layer is extracted with ether, the ether washed with 150 ml of 25% sodium hydroxide solution, and the aqueous layer is acidified to yield an oil which is separated. The acid is distilled to yield the product boiling at $173°-174°C$, 39–44 gm (76–86% yield based on sec-butyl chloride used).

9-10. Lithium Reagents—Preparation of Fluorene-9-carboxylic Acid [10]

A solution of 1 mole of n-butyllithium in 500 ml of ether is treated portionwise with 0.75 mole of fluorene. The solution turns an orange color accompanied by vigorous evolution of butane. The mixture is refluxed for 1 hr and then poured jetwise onto crushed dry ice. As soon as the mixture warms up to room temperature, the unreacted lithium is skimmed off and 2 liters of water is added cautiously. The insoluble residue is filtered off and the organic layer is extracted three times with 300-ml portions of luke-warm 2% sodium hydroxide. Acidification of the combined aqueous solutions precipitates the desired acid. The yield is 118 gm (75%), mp $228°-230°C$, based upon fluorene.

7. CARBOXYLATION OF THE AROMATIC NUCLEUS

Heating the alkali salts of resorcinol and α-naphthol with carbon dioxide gives excellent yields of the carboxylic acids (Kolbe–Schmitt reaction) [11].

$$\text{(structure)} + CO_2 \longrightarrow \text{(structure)} \qquad (20)$$

When salicylic acid is heated to 240°C with potassium carbonate, an 80% yield of p-hydroxybenzoic acid results because of a carboxyl group migration.

$$\text{(structure)} + K_2CO_3 \longrightarrow \text{(structure)} \qquad (21)$$

It has been reported that the reaction of aromatic compounds with sodium palladium(II) malonate in a mixed solvent of acetic acids and acetic anhydride or carbon tetrachloride gives aromatic acids in good yields, together with lower yields of aromatic dimers.

9-11. Kolbe–Schmitt Reaction—Preparation of β-Resorcyclic Acid [11]

$$\text{(structure)} \xrightarrow{CO_2 + KHCO_3} \text{(structure)} \qquad (22)$$

To a 5-liter flask equipped with a reflux condenser is added a solution containing 200 gm (1.8 moles) of resorcinol, 1 kg (9.9 moles) of potassium bicarbonate (or sodium carbonate in equivalent amount), and 2 liters of water. The mixture is heated slowly at 80°–100°C in an oil bath for 4 hr. Then the temperature of the bath is raised so that the contents reflux vigorously for 30 min while a rapid stream of carbon dioxide is passed through.

The hot solution is acidified by adding 900 ml of concentrated hydrochloric acid from the dropping funnel, which is connected to the gas inlet tube reaching the bottom of the flask.

After being cooled to room temperature, the flask is chilled in an ice bath. The resorcylic acid crystallizes in colorless prisms, which, on exposure to air, may turn pink due to free resorcinol. The crude yield is 225 gm. Extraction of the filtrate with ether yields an additional 35 gm of crude resorcylic acid. The crude resorcylic acid (260–270 gm) is dissolved in 1 liter of boiling water, boiled with 25 gm of activated carbon (Norite), filtered, and cooled in an ice–salt water bath. The solution is stirred vigorously and the crystalline product is obtained. Yield 160–170 gm (57–60%), mp 216°–217°C.

A. THE FRIEDEL–CRAFTS REACTION

Direct introduction of the carboxyl group into the aromatic ring has also been accomplished with phosgene and oxalyl chloride. Thus, for example, 9-anthroic acid is prepared in 67% yield from anthracene by heating to 240°C with oxalyl chloride in nitrobenzene [12].

Certain activated carbon atoms may be conveniently acylated with oxalyl chloride to afford an intermediate which is readily decarboxylated to afford a carboxylic acid.

9-12. The Use of Oxalyl Chloride—Preparation of 9-Anthroic Acid [12, 25]

$$(23)$$

A mixture of 50 gm of anthracene, 45 gm aluminum chloride, and 200 ml of dry nitrobenzene are stirred while 30 ml of oxalyl chloride is added dropwise. After the addition the mixture is heated over a period of 5–6 hr while the temperature is raised from 120° to 240°C. After steam distillation to remove the nitrobenzene, 150 ml of 10 N sodium hydroxide solution and enough water to make 700 ml of reaction mixture are added. The mixture is refluxed for $\frac{1}{2}$ hr. Insoluble materials (from which 11 gm of anthracene can be isolated) are removed by filtration. The filtrate is extracted with ligroin (bp 30°–60°C), treated with activated charcoal, and filtered hot. The charcoal is washed with 2 N sodium carbonate solution. Acidification of the combined filtrate and carbonate washings gives 41.6 gm (67%) of 9-anthroic acid, mp 208°–212°C.

8. CONDENSATION REACTIONS

A. THE PERKIN REACTION

The Perkin reaction is the base-catalyzed reaction of an active methylene group of an anhydride and an aldehyde group. Basic catalysts such as the sodium salt of the acid corresponding to the anhydride, potassium carbonate, or tertiary amines may be used satisfactorily. This reaction is very useful for the preparation of substituted cinnamic acids such as those containing halo, methyl, and nitro groups [13].

The preparation of 2-methyl-3-nitrocinnamic acid is representative of the reaction and its conditions.

9-13. Preparation of 2-Methyl-3-nitrocinnamic Acid [13]

A mixture of 75 gm of m-nitrobenzaldehyde, 98 gm of propionic anhydride, and 48 gm of sodium propionate is heated at 170°C in an oil bath for 5 hr. The reaction mixture is poured into water and saturated sodium carbonate solution added to strong alkalinity. The tarry liquid is boiled with decolorizing charcoal for 10 min and then filtered. The alkaline mixture is poured into dilute hydrochloric acid.

$$\text{CH}{=}\text{O} \quad + (\text{CH}_3\text{CH}_2\text{CO})_2\text{O} + \text{CH}_3\text{CH}_2\text{COONa} \xrightarrow[\text{2. H}_3\text{O}^+]{\text{1. 170°C}} \quad \begin{matrix}\text{CH}_3\\ |\\ \text{CH}{=}\text{C}{-}\text{COOH}\end{matrix} \qquad (24)$$

A white curdy precipitate results. This is filtered by suction and allowed to dry overnight. The crude acid is recrystallized from 85% ethanol to give 70 gm of white needles, mp 199.5°–200.5°C (corr.).

B. KNOEVENAGEL CONDENSATION

The condensation of aldehydes with the active methylene group of malonic acid to give α,β- and β,γ-olefinic acids is called the Knoevenagel condensation. Decarboxylation occurs at room temperature or heating to 100°C to give the unsaturated acids. The reaction is catalyzed by pyridine, and an example is the preparation of 2-hexenoic acid in 64% yield.

Triethanolamine is the best catalyst for the preparation of β,γ-olefinic acids such as 3-hexenoic acid.

$$RCH_2CH{=}O + CH_2(COOH)_2 \xrightarrow{\substack{Pyridine \\ \\ Triethanolamine}} \begin{array}{l} RCH_2CH{=}CH{-}COOH \\ \\ RCH{=}CH{-}CH_2COOH \end{array} \quad (25)$$

Substituted benzaldehydes and malonic acids give cinnamic acids in excellent yields.

9-14. Preparation of 2-Hexenoic Acid [14]

$$CH_3CH_2CH_2CH{=}O + CH_2(COOH)_2 + Pyridine \longrightarrow$$
$$CH_3CH_2CH_2CH{=}CH{-}COOH \quad (26)$$

To 200 gm (1.92 moles) of dry malonic acid in 200 ml of anhydrous pyridine is added 158 ml (1.75 moles) of freshly distilled *n*-butyraldehyde. The reaction is allowed to proceed at 25°C for 24 hr, at 40°–45°C for an additional 24 hr, and finally at 60°C for 3 hr. The reaction mixture is chilled, acidified with 6 N sulfuric acid, the nonaqueous phase collected and the aqueous phase extracted with three 100 ml portions of ether. The organic layer and the ether extracts are dried over calcium chloride, filtered, the solvent removed, and the residue is allowed to crystallize at 0°C. The crystals are collected to give 130 gm (64%) of crude 2-hexenoic acid, mp 30°–32°C.

C. OTHER CONDENSATION REACTIONS

The Knoevenagel reaction is related to the general class of reactions in which the condensation of aldehydes with active methylene compounds is catalyzed by base. The Claisen condensation is also an example and, for one typical case, may involve the condensation of ethyl acetate and aromatic aldehydes catalyzed by sodium sand. Benzaldehyde yields ethyl cinnamate in 74% yield [15].

$$\text{ArCH=O} \begin{cases} \xrightarrow[\text{Sodium acetate}]{\text{Acetic anhydride}} \text{(Perkin)} \\ \xrightarrow[\text{Pyridine}]{\text{Malonic acid}} \text{(Knoevenagel--Doebner)} \quad \text{ArCH=CH--COOH} \\ \xrightarrow[\text{Sodium sand}]{\text{Ethyl acetate}} \text{(Claisen)} \end{cases} \quad (27)$$

The Stobbe condensation is used to condense ketones with diethyl succinate by a variety of basic reagents to give isopropylidene succinates.

Acetoacetic ester and pyruvic acid are some other compounds containing active methylene groups that undergo base-catalyzed condensations with aldehydes to give olefinic β-keto esters and α-keto acids, respectively.

A one-carbon homologation of aldehydes and ketones to carboxylic acids has been reported to involve the Horner–Emmons modification of the Wittig reaction using diethyl-*tert*-butoxy(cyano)-methylphosphonate $(\text{EtO})_2 \text{POCH(CN)O}t\text{-Bu}$ to produce ethyl-*tert*-butoxyacrylonitriles. The *tert*-butyl ether group is cleaved by zinc chloride in refluxing acetic anhydride, and the α-acetoxyacrylonitrile is converted to the acid by solvolysis.

9-15. Claisen Condensation: Preparation of Ethyl Cinnamate and Cinnamic Acid [15]

$$\text{C}_6\text{H}_5\text{CHO} + \text{CH}_3\text{COOC}_2\text{H}_5 \xrightarrow{\text{C}_2\text{H}_5\text{ONa}} \text{C}_6\text{H}_5\text{CH=CHCOOC}_2\text{H}_5 + \text{H}_2\text{O} \quad (28)$$

In a two-necked flask equipped with a reflux condenser and mechanical stirrer are placed 400 ml of dry xylene and 29 gm of clean sodium. The sodium is melted under the xylene by means of an oil bath and the stirrer is used to powder the sodium. Care should be taken not to splash any of the sodium onto the walls of the flask above the solvent. The oil bath is removed and the stirring is continued until the sodium powder has been formed completely and no more liquid sodium remains. The xylene is decanted and to the sodium is added 460 ml (4.7 moles) of absolute ethyl acetate containing 3–4 ml of absolute alcohol. The flask is quickly cooled to 0°C and 106 gm (1 mole) of pure benzaldehyde is slowly (1 to $1\frac{1}{2}$ hr) added by means of a dropping funnel. The temperature is kept between 0° to 5°C, being careful not to exceed 5°C for the best yields. The reaction commences as soon as the benzaldehyde is added, as evidenced by the production of a reddish color on the sodium particles. The stirring is continued about 1 hr after the addition to complete the reaction of all the sodium particles. Glacial acetic acid (90–95 ml) is now added slowly and water is cautiously added to the mixture. Care should be exercised in hydrolyzing free sodium that has caked on top of the flask. The ester layer is separated, and the water layer is extracted with about 25–50 ml of ethyl acetate. The combined ester portions are washed with 300 ml of 6 N hydrochloric acid and then dried with sodium sulfate. The ethyl acetate is distilled off and the remaining liquid is distilled under reduced pressure to yield ethyl cinnamate, bp 128°–133°C (6 mm), 120–130 gm (68–74%). During the distillation, a reddish brown semisolid mass sometimes appears in the flask. This mass melts down if the oil bath is heated to 220°–230°C and if the distillation is continued smoothly.

Acid hydrolysis of ethyl cinnamate yields cinnamic acid.

Monoalkylation of malonic ester occurs with primary and some secondary halides in 75–90% yield [16]. An example of the reaction is given for the preparation of ethyl n-butylmalonate [16].

9-16. Malonic Ester Synthesis—Preparation of Ethyl-n-Butylmalonate [16]

$$CH_2(COOC_2H_5)_2 + NaOC_2H_5 \longrightarrow NaCH(COOC_2H_5)_2 \xrightarrow{n\text{-BuBr}}$$

$$n\text{-BuCH}(COOC_2H_5)_2 \quad (29)$$

A sodium ethoxide solution freshly prepared from 2.5 liters of anhydrous ethanol and 115 gm (5 gm atom) of sodium is warmed to 50°C and stirred while 825 gm of diethyl malonate is added. To the clear solution is added slowly 685 gm of n-butyl bromide. The reaction commences almost immediately and considerable heat is generated. The addition rate is adjusted so that the reaction does not become violent. Cooling may be necessary. After the addition, the reaction mixture is refluxed until neutral to moist litmus (about 2 hr). Then a distillation column is attached to the flask and approximately 2 liters of alcohol are distilled off in 6 hr, using a water bath. The residue is treated with about 2 liters of water and shaken thoroughly. The upper layer of *n*-butylmalonic ester is separated and distilled under reduced pressure. First, a low-boiling portion is collected, consisting of alcohol, water, and *n*-butyl bromide; then a small intermediate fraction of unchanged malonic ester comes over; and finally *n*-butylmalonic ester boiling at 130°–135°C (20 mm). The first fraction amounts to less than 100 ml, while the main fractions weigh 860–970 gm (80–90%). All the starting reagents used should be highly purified in order to achieve the maximum yield. The ester can be hydrolyzed to the free acid.

D. ETHYL ACETOACETIC ESTER SYNTHESIS

The alkylation of ethyl acetoacetate with a variety of alkyl halides affords intermediates that may be converted to α-alkyl-substituted acetic acids. The alkylation itself is widely discussed.

Depending on hydrolytic conditions, the alkylation products may be converted to methyl ketones or acids according to the following reaction schemes:

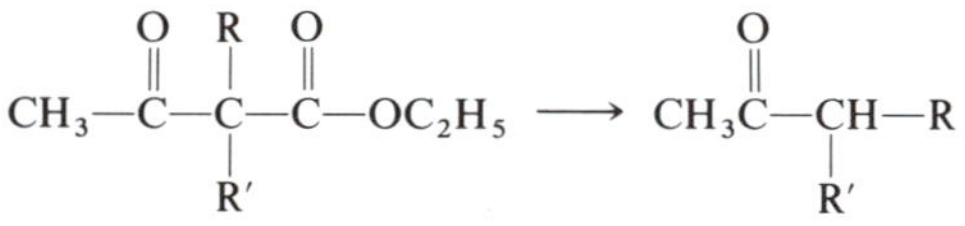

$$
\underset{\substack{| \\ R'}}{CH_3-\overset{O}{\overset{\|}{C}}-\overset{\overset{R}{|}}{C}-\overset{O}{\overset{\|}{C}}-OC_2H_5} \longrightarrow \underset{\substack{| \\ R'}}{CH_3\overset{O}{\overset{\|}{C}}-CH-R} \tag{30}
$$

or

$$
\underset{\substack{| \\ R'}}{CH_3-\overset{O}{\overset{\|}{C}}-\overset{\overset{R}{|}}{C}-\overset{O}{\overset{\|}{C}}-OC_2H_5} \longrightarrow \underset{\substack{| \\ R'}}{R-CH-\overset{O}{\overset{\|}{C}}-OH} \tag{31}
$$

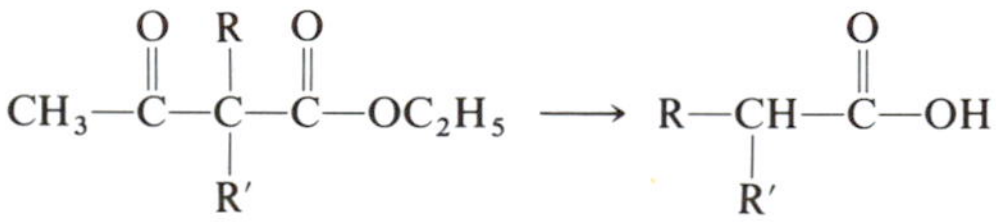

These two reactions are competitive in nature, and reaction conditions must be selected such that the desired product predominates in yield. Generally speaking, the concentration of alkali used in hydrolysis appears to have the most profound effect on the course of the reaction. High concentrations of alkali tend to favor acid formation.

While the temperature of the reaction has an effect on relative yields, this effect appears to be slight [17]. Additional research, using the statistical design of experiments to study simultaneous variation of reaction variable, would be of considerable interest.

9-17. Preparation of Caproic Acid [17]

$$CH_3CH_2CH_2CH_2\underset{\underset{\displaystyle CH_3}{\overset{\displaystyle \|}{\underset{\displaystyle C=O}{|}}}{CH}}{}—CO_2C_2H_5 \xrightarrow[H_2O]{KOH} CH_3CH_2CH_2CH_2CH_2CO_2H$$

and

$$CH_3CH_2CH_2CH_2CH_2\overset{\displaystyle O}{\overset{\displaystyle \|}{C}}—CH_3 \quad (32)$$

To a solution of 59.5 gm of potassium hydroxide in 100 gm of an aqueous solution maintained at 75°C is added over a 1-hr period 12.3 gm of ethyl *n*-butylacetoacetate. Heating is continued for 5 hr with efficient stirring.

The alkaline solution is then diluted with 250 cc of water and the alcohol and ketone by-product are distilled off until no more water-insoluble product distills over.

The alkaline residue is then cooled and acidified by careful addition of 50% sulfuric acid, care being taken that the mixture does not become unduly hot during acidification.

The caproic acid is then extracted with three 100-cc portions of ether, the combined extracts are dried with sodium sulfate, and the ether is distilled off on a steam bath. The sodium sulfate may also be extracted a few times with anhydrous ether. The crude caproic acid remaining is distilled at reduced pressure, the fraction boiling from 105° to 110°C at 16 mm being collected. A yield of 64.6% of the theoretical amount of caproic acid is isolated.

E. ARNDT–EISTERT REARRANGEMENT

Acid chlorides are reacted with diazomethane to yield the diazoketone which upon reaction with silver oxide rearranges to the next higher homolog of the acid. Biphenyl-2-acetic acid is produced in 86% by the method described below [18].

9-18. Preparation of Biphenyl-2-acetic Acid [18]

$$(33)$$

Biphenyl-2-carboxylic acid (1 mole) dissolved in dry benzene is treated with 1.6 mole of oxalyl chloride and kept at 30°C for 1 hr or until no further evolution of gas. The solvent and excess of reagent are removed under reduced pressure at 35°C. Ether is added twice and evaporated under reduced pressure to ensure the removal of the reagent and hydrogen chloride. The acid chloride is added dropwise to an etheral solution of diazomethane (see Chapter 15, Diazo and Diazonium Compounds) of equal molarity and then cooled. After keeping the reaction mixture overnight at room temperature, the ether is removed under reduced pressure. A 61% yield of yellow crystals of ω-diazo-o-phenylacetophenone is obtained, mp 106°C (from alcohol).

To 0.6 mole of silver oxide in 0.84 mole of sodium thiosulfate in 1 liter of warm water is added a 1 M dioxane solution of ω-diazo-o-phenylacetophenone. The mixture is stirred for 3 hr at room temperature while an additional quantity of freshly precipitated silver oxide (equal in amount to that used initially) is added in portions at intervals and the temperature is kept at 50°C for 1 hr. The solution is cooled and filtered and the residue washed with a 1% sodium hydroxide solution. This slowly deposits a flocculent precipitate upon acidifying. The residue is again treated with about one-half the amount of silver oxide used above and worked up as before to yield a further quantity of product. The total yield of biphenyl-2-acetic acid is 86% mp 116°C (from benzene).

Wilds and coworkers have discussed the influence of highly hindered acyl chlorides on the Arndt–Eistert synthesis. The diazomethanes derived from such acyl chlorides fail to rearrange normally with any of the three conventional catalysts, silver oxide–methanol, silver benzoate–triethylamine–methanol, or tertiary amines–high-boiling solvents. Under special reaction conditions, abnormal reaction products were isolated.

F. REFORMATSKII REACTION

The Reformatskii reaction involves the reaction of the product of an α-halo ester and activated zinc in the presence of an anhydrous organic solvent, with a carbonyl compound, followed by hydrolysis. The reaction is very similar in nature to the Grignard reaction except the carbonyl reagent is added at the start. It has been suggested that Grignard reactions might be conducted in a similar manner.

Magnesium has been used in some reactions in place of zinc but poor yields resulted since the more reactive organomagnesium reagents attack the ester group. With zinc, this latter reaction is not appreciable and the organozinc reagent attacks the carbonyl group of aldehydes and ketones to give the β-hydroxy ester.

α-Bromo esters react satisfactorily, but β- and γ-derivatives of saturated esters give poor yields unless activated by an unsaturated group in such a manner as to resemble allylic bromides.

The reaction of α-lithiated acid salts with carbonyl compounds has been reported to offer an improved route to the β-hydroxy acids usually obtained by the hydrolysis of the β-hydroxy ester products of Reformatskii reactions.

9-19. Preparation of Ethyl 4-Ethyl-3-hydroxy-2-octanoate [19]

$$n\text{-}C_4H_9\text{—}\underset{\underset{C_2H_5}{|}}{CH}\text{—}CHO + Br\underset{\underset{CH_3}{|}}{CH}COOC_2H_5 + Zn \longrightarrow$$

$$C_4H_9\underset{\underset{C_2H_5}{|}}{CH}\text{—}\underset{\overset{OZnBr}{|}}{CH}\text{—}\underset{\underset{CH_3}{|}}{CH}\text{—}COOC_2H_5 \quad (34)$$

$$2\ C_4H_9\underset{\underset{C_2H_5}{|}}{CH}\text{—}\underset{\overset{OZnBr}{|}}{CH}\text{—}\underset{\underset{CH_3}{|}}{CH}\text{—}COOC_2H_5 + H_2SO_4 \longrightarrow$$

$$2\ C_4H_9\underset{\underset{C_2H_5}{|}}{CH}\text{—}\underset{\overset{OH}{|}}{CH}\text{—}\underset{\underset{CH_3}{|}}{CH}\text{—}COOC_2H_5 + ZnBr_2 + ZnSO_4 \quad (35)$$

To a flask containing a nitrogen atmosphere are added freshly sandpapered zinc foil strips, and 750 ml of thiophene-free benzene (dried). To further ensure that the flask and contents are dry, 175–200 ml of benzene is distilled off. Distillation is interrupted and the benzene is refluxed while a solution of 64.1 gm (0.5 mole) of 2-ethylhexanal and 271.5 gm (1.5 mole) of ethyl α-bromoporpionate in 500 ml of dried benzene is placed in the dropping funnel. The first 50 ml of the solution is added to the flask at once. In most cases the reaction starts immediately, as evidenced by the darkening of the zinc surface and the formation of a cloudy solution. However, approximately 15 min may elapse before the reaction starts. When the reaction has started, the remainder of the solution is added in about 1 hr, and the solution then refluxed for 2 hr with continuous stirring. When the solution has cooled to room temperature, 750 ml of 12 N sulfuric acid is added and the solution is stirred vigorously for 1 hr. The benzene layer is separated and the aqueous layer extracted several times with benzene. The combined benzene layers are washed with 500 ml of water, saturated sodium bicarbonate solution, and then with water. The benzene layer is dried over anhydrous sodium sulfate and the benzene distilled off under aspirator pressure. The product at this point is ethyl 4-ethyl 3-hydroxy-2-octanoate and is obtained from the distillation in 87% yield (100 gm), bp 122°–124°C (4.9 mm), n_D^{25} 1.4415. The ester can be hydrolized to the free carboxylic acid.

G. DIELS–ALDER REACTION

The Diels–Alder reaction is a 1,4-addition of an olefinic compound to a conjugated diene. The diene system may be part of an aliphatic, aromatic, or heterocyclic nucleus such as furan. The olefinic compound usually contains one or more groups that activate the double bond, although this is not always necessary. For example, ethylene is condensed with butadiene at 200°C to give cyclohexene. Triple bonds may replace double bonds in both the diene and the dieneophile. Cis addition of the dienophile to the diene occurs and several reactions of the above type have been shown to be reversible.

Maleic anhydride condenses with butadiene to give Δ⁴-tetrahydrophthalic anhydride. The latter can be hydrolyzed to the diacid [20].

9-20. Preparation of Tetrahydrophthalic Anhydride [20].

$$(36)$$

A flask containing 500 ml of dry benzene and 196 gm (2 moles) of maleic anhydride is heated with a pan of hot water while butadiene is introduced rapidly (0.6–0.8 liter/min) from a commercial cylinder. The flask is stirred rapidly and the heating is stopped after 3–5 min when the temperature of the solution reaches 50°C. In 15–25 min, the reaction causes the temperature of the solution to reach 70°–75°C. The absorption of the rapid stream of butadiene is nearly complete in 30–40 min. The addition of butadiene is continued at a slower rate for a total of $2\frac{1}{2}$ hr. The solution is poured into a 1-liter beaker which is covered and kept at 0°–5°C overnight. The product is collected on a larger Büchner funnel and washed with 250 ml of 35°–60°C bp petroleum ether. A second crop is obtained by diluting the filtrate with an additional 250 ml of petroleum ether. Both crops are dried to constant weight in an oven at 70°–80°C to yield 281.5–294.5 gm (96–97%, mp 99°–102°C). Recrystallization from ligroin or ether raises the melting point to 103°–104°C.

9. HYDROLYSIS OF ACID DERIVATIVES

A. HYDROLYSIS OF NITRILES

Nitriles can be hydrolyzed to carboxylic acids by refluxing in concentrated solutions of sulfuric acid or sodium hydroxide. For example, a solution of potassium hydroxide and ethylene glycol monoethyl ether is used to prepare 9-phenanthroic acid in 98% yield from 9-cyanophenanthrene [21].

9-21. Preparation of 9-Phenanthroic Acid [21]

$$\text{CN} \xrightarrow[\text{KOH}]{\text{H}_2\text{O}} \text{HOCO} + \text{NH}_3 \qquad (37)$$

To a hot solution of 350 gm of 9-cyanophenanthrene in 1400 ml of ethylene glycol monoethyl ether is added a solution of 350 gm of potassium hydroxide in 160 ml of water. The solution is refluxed while a slow stream of carbon dioxide-free air is bubbled through it until 1 ml of 0.1 N hydrochloric acid is not neutralized within 5 min by the exit air carrying the liberated ammonia. Approximately 6 hr is required to attain this condition. The solution is cooled and poured into a stirred solution of 610 ml of concentrated hydrochloric acid in 5250 ml of water. After standing overnight, the precipitated 9-phenanthroic acid is filtered off and washed thoroughly with water; yield 374 gm (98%), mp 246°–248°C. After one recrystallization from glacial acetic acid, a sample melted at 252°–253°C.

REFERENCES

1. H. E. Zaugg and R. T. Rapala, *Org. Synth. Collect. Vol.* **3,** 820 (1955).
2. C. D. Hurd, J. W. Garrett, and E. N. Osborne, *J. Am. Chem. Soc.* **55,** 1084 (1933).
3. A. L. Henne and P. Hill, *J. Am. Chem. Soc.* **65,** 753 (1943).
4. F. Bischoff and H. Adkins, *J. Am. Chem. Soc.* **45,** 1031 (1923).
5. R. Levine and J. R. Stephens, *J. Am. Chem. Soc.* **72,** 1642 (1950).
6. W. G. Toland, Jr., D. L. Hagmann, J. B. Wilkes, and F. J. Brutschy, *J. Am. Chem. Soc.* **80,** 5423 (1958).
7. C. D. Hurd, J. W. Garrett, and E. N. Osborne, *J. Am. Chem. Soc.* **55,** 1083 (1933).
8. T. W. Evans and W. M. Dehn, *J. Am. Chem. Soc.* **52,** 3645 (1930).
9. H. Gilman and C. H. Kirby, *Org. Synth. Collect Vol.* **1,** 361 (1941).
10. R. R. Burtner and J. W. Cusic, *J. Am. Chem. Soc.* **65,** 264 (1943).
11. M. Nierenstein and D. A. Clibbens, *Org. Synth Collect. Vol.* **2,** 557 (1943).
12. H. G. Latham, Jr., E. L. May, and E. Mosettig, *J. Am. Chem. Soc.* **70,** 1079 (1948).
13. R. W. Maxwell and R. Adams, *J. Am. Chem. Soc.* **52,** 2967 (1930).
14. C. Niemann and C. T. Redemann, *J. Am. Chem. Soc.* **68,** 1933 (1946).
15. C. S. Marvel and W. O. King, *Org. Synth. Collect. Vol.* **1,** 252 (1941).
16. R. Adams, and R. H. Kamm, *Org. Synth. Collect. Vol.* **1,** 250 (1941).
17. N. L. Drake and R. W. Riemenschneider, *J. Am. Chem. Soc.* **52,** 5005 (1930).
18. A. Schönberg and F. L. Warren, *J. Chem. Soc.* p. 1840 (1939).
19. K. L. Rinehart, Jr. and E. G. Perkins, *Org. Synth.* **37,** 37 (1957).
20. A. C. Cope and E. C. Herrick, *J. Am. Chem. Soc.* **72,** 984 (1950).
21. M. A. Goldberg, E. P. Odas, and G. Carsch, *J. Am. Chem. Soc.* **69,** 261 (1947).
22. R. C. Roberts and T. B. Johnson, *J. Am. Chem. Soc.* **47,** 1399 (1925).
23. E. Schwenk and E. Block, *J. Am. Chem. Soc.* **64,** 3051 (1942).
24. R. D. Rieke, S. E. Bales, P. M. Hudnall, and G. S. Poindexter, *Org. Synth.* **59,** 85 (1979).
25. The authors have modified this original preparation in Reference 12 by adding aluminum chloride which was probably mistakenly left out [see original preparations: C. Liebermann and M. Zsuffa, *Ber.* **44,** 208 (1911)].

10

ESTERS

1. INTRODUCTION

The most common laboratory method for the preparation of esters utilizes the condensation between a carboxylic acid and an alcohol catalyzed by acids such as HCl, H_2SO_4, BF_3, *p*-toluenesulfonic acid or methanesulfonic acid. The problem of catalysis is receiving continued attention and several new catalyst systems are briefly mentioned.

The esterification reaction is an equilibrium reaction and it can be displaced toward the product side by removal of water, or by the use of an excess of one of the reactants. The use of acetone dimethylacetal, which reacts with the water formed to produce methanol and acetone, allows the preparation of methyl esters in high yield. Primary and

From S. R. Sandler and W. Karo, *Organic Functional Group Preparations*, Vol. I, 2d ed. (New York, 1983), 288–314, by permission of Academic Press, Inc.

secondary alcohols are esterified in good yield but tertiary alcohols give very low yields. Neopentyl alcohol reacts normally and is esterified with either acids or acid chlorides.

The preferred method for the preparation of tertiary esters is based on the interaction of the acid halides and the tertiary alcohol or an olefin and a carboxylic acid. For example, *tert*-butyl acrylate is made by the condensation of isobutylene and acrylic acid.

Some other common condensation methods may be summarized by Eqs. (1), (2), and (3).

$$RCO(Z) + R'OH \longrightarrow RCOOR' + ZOH$$
where $Z = OH, Cl, OR, (OCOR)$, and lactones   (1)

$$RCHCOOR + RX \longrightarrow RCH-COOR + X^- \qquad (2)$$
$$\underset{R}{|}$$

$$RCOONa \ (or \ Ag) + R'X \longrightarrow RCOOR' \qquad (3)$$

The condensation of carboxylic acids with diazomethane leads to methyl esters.

$$RCOOH + CH_2N_2 \longrightarrow RCOOCH_3 + N_2 \qquad (4)$$

Little use has been made of this technique on a preparative scale since diazomethane is a yellow toxic gas which may explode violently if undiluted or on contact with rough glass surfaces.

Dimethyl sulfate can be used in place of diazomethane to form methyl esters of carboxylic acids through the sodium salt.

$$2\,RCOONa + (CH_3)_2SO_4 \longrightarrow 2\,RCOOCH_3 + Na_2SO_4 \qquad (5)$$

However, dimethyl sulfate is acidic in nature and may not be as satisfactory as diazomethane for methylating sensitive acids. Dimethyl sulfate is also very toxic and is a suspected carcinogen.

Transesterification reactions are valuable methods for the preparation of vinyl esters. In these reactions the alcohol portion of the ester is exchanged [Eq. (6)]. In another type of transesterification, acid moieties are exchanged [Eq. (7)].

$$RCOOR' + R''OH \rightleftharpoons RCOOR'' + R'OH \qquad (6)$$

$$RCOOR' + R''COOH \rightleftharpoons R''COOR' + RCOOH \qquad (7)$$

The other methods of oxidation and reduction are more specialized reactions and are treated briefly.

2. CONDENSATION REACTIONS

A. THE REACTION OF ALCOHOLS WITH CARBOXYLIC ACIDS

$$R'OH + RCOOH \xrightarrow{\text{Acid catalyst}} RCOOR' + H_2O \qquad (8)$$

Primary alcohols give better yields of esters than secondary alcohols and tertiary alcohols, and phenols react only to a very small extent. Acid catalysts are used in small amounts. The mixture is refluxed for several hours and the equilibrium is shifted to the right by the use of a large excess of either the alcohol or acid and the removal of water. Azeotropic distillation of water, the use of a Dean and Stark trap, or a suitable drying agent helps to increase the rate of reaction. No acid catalysts are required for the preparation of esters of formic acid or of benzyl alcohol.

Polyesters are prepared by heating diols and dicarboxylic acids with an acid catalyst.

Some useful acid catalysts are sulfuric acid, hydrogen chloride, p-toluenesulfonic acid, and methanesulfonic acid (Atochem).

In addition, trifluoracetic anhydride has been a useful acid catalyst for the esterification of phenols. The use of heavy metal salts as effective esterification catalysts has been widely reported. For example, $BuSnO_2H$ and Bu_2SnO are equal or superior to p-$MeC_6H_4SO_3H$ as esterification catalysts and do not effect appreciable dehydration of secondary alcohols. It has been reported that catalytic activities generally descend in the order Sn^{+4}, Co^{+3}, Fe^{+3}, Al, Bi, Cr, Sn^{+2}, Cu, Co^{+2}, Pb, Fe^{+2}, Zn, Ni, Mn, Cd, Mg, Ba, K, and ClO_4, SO_4, $PhSO_3$, Cl, Zr, I, NO_3. The use of 1 wt% of $Fe_2(SO_4)_3$ is recommended as a noncorrosive catalyst in steel.

The use of stannous salts of carboxylic acids [$Sn(OOCR)_2$, where $RCOO$ = 2-ethylhexoate, n-octoate, laurate, palmitate, stearate, and oleate] as catalysts for the preparation of polyesters has been reported to give colorless products with low acid numbers. Stannous oxide gives the same results but there is a short induction period before the catalyst becomes effective. The catalysts are present in 5×10^{-4} to 1×10^{-2} mole of catalyst per 100 gm of polyester and the temperature of the reaction is up to 200°C.

Metal oxides of the Group V metals such as tetraalkyltitanate esters, sodium alkoxy titanates, and alkaline earth salts of weak acids are among some other catalysts used to give polyesters without dehydration to the ether or olefin. The Tyzor titanates of DuPont are useful for this application.

Trifluoromethanesulfonic anhydride is an effective esterification catalyst, as is trifluoroacetic anhydride. However, in an organic medium, trifluoromethanesulfonic acid is a much stronger acid than trifluoroacetic acid or perchloric acid. When the anhydride is used, the esterification is exothermic, and heating for only a short period of time may be required.

Dihydric alcohols readily yield cyclic ethers under esterification conditions and the usual catalysts are not effective. The use of boric acid as a catalyst allows one to prepare esters of mono- or dibasic acids. For example, esters of 1,4-butanediol and 2,5-hexanediol can be prepared by this method.

The use of orthophosphoric acid allows one to prepare less-colored esters from "oxo synthesis" alcohols, which may contain sulfur impurities.

Ortho substituents in the aromatic acids retard esterification by the conventional method, but they can be esterified by dissolving in 100% sulfuric acid and pouring the solution into the desired alcohol. This method is applicable to other unreactive systems and was successfully applied to heterocyclic acids, polybasic acids, long-chain aliphatic acids, and several other unreactive substituted aromatic acids.

A simple esterification employing azeotropic removal of water by means of a Dean and Stark trap can be used in an introductory organic laboratory course. This technique is widely used in industry for the preparation of polyesters. The preparation of γ-chloropropyl acetate in 93–95% yield and of n-amyl acetate in 71% yield have been described.

10-1. Preparation of Methyl Acetate [1]

$$CH_3OH + CH_3COOH \underset{}{\overset{H^+}{\rightleftarrows}} CH_3\overset{\overset{\displaystyle O}{\displaystyle \|}}{C}OCH_3 + H_2O \qquad (9)$$

To a flask are added 48 gm (1.5 mole) of absolute methanol, 270 gm (4.5 moles) of glacial acetic acid, and 3.0 gm of concentrated sulfuric acid. The mixture is refluxed for 5 hr and then fractionated to give 112 gm of crude ester, bp $55°-56°$C. The crude ester is washed successively with a saturated salt solution, a sodium bicarbonate solution until the effervescence ceases, and a saturated salt solution, dried, and distilled to yield 92 gm (83%, bp $56°$C (754 mm), n_D^{25} 1.3594.

Although a "simple" procedure for the preparation of methyl esters has been reported using dimethyl sulfate, the 1980 NIOSH *Registry of Toxic Effects of Chemical Substances* lists this chemical on p. 669 as a suspected carcinogen. (It also has a oral LD50 of 250 mg/kg in rats.)

10-2. Preparation of Amyl Acetate [2]

$$CH_3COOH + CH_3CH_2CH_2CH_2CH_2OH \xrightarrow{H^+}$$

$$CH_3CH_2CH_2CH_2CH_2O\overset{\overset{\displaystyle O}{\|}}{C}CH_3 + H_2O \quad (10)$$

To the round-bottomed flask are added 15 gm (0.25 mole) of glacial acetic acid, 17.6 gm (0.20 mole) of *n*-amyl alcohol, 30 ml of benzene, and 0.15 gm of *p*-toluenesulfonic acid catalyst. A Dean and Stark trap is filled with benzene and the contents of the flask are refluxed for 1 hr. The water that is produced remains in the trap as a bottom layer. The reaction mixture is extracted with sodium bicarbonate solution in order to remove the excess acid, washed with water, and then with a saturated sodium chloride solution. The organic layer is fractionated in order to remove the benzene–water azeotrope and the residue is transferred to another flask containing a small loose plug of steel wool in the neck. The product is isolated by simple distillation in 71% yield (20.7 gm), bp $141°-146°$C, n_D^{26} 1.4012.

B. THE REACTION OF ACYL HALIDES WITH HYDROXY COMPOUNDS

$$R'COCl + ROH \longrightarrow R'COOR + HCl \quad (11)$$

Alcohols and phenols react with acid chlorides. The reaction is facilitated by the use of a tertiary amine, pyridine, aluminium alcoholate, or lithium alcoholate to react with the liberated acid. Tertiary alcohols and phenols give good yields of esters. Acid halides of aromatic polycarboxylic acids, olefinic acyl halides, and others give good yields of esters. The alcohol portion of the ester may have an epoxy group (glycidyl) or a cyano group, as in glyconitrile formed from formaldehyde and sodium cyanide.

It has been reported [1] that vinyl esters are obtained by the reaction of aliphatic and aromatic acid halides with acetaldehyde in the presence of pyridine. It was postulated that acetaldehyde enolizes to vinyl alcohol in the presence of pyridine, which subsequently reacts with the acid halide.

$$CH_3CH{=}O \xrightarrow{\text{Pyridine}} [CH_2{=}CHOH] \xrightarrow[\text{pyridine}]{\text{RCOCl}}$$

$$CH_2{=}CH{-}O\overset{\overset{\displaystyle O}{\|}}{C}R + \text{Pyridine hydrochloride} \qquad (12)$$

This reaction has also been referred to in *Organic Syntheses*, "Methods of Preparation" section.

The present authors have independently tried to repeat these experiments with acetyl chloride, benzoyl chloride, and other acid chlorides as described by Sladkov and Petrov but did not obtain any vinyl esters as determined by careful analysis via gas chromatography. Therefore, this procedure is not recommended as a preparative method but as an area for further research.

10-3. Preparation of *tert*-Butyl Acetate [3]

$$(CH_3)C{-}OH + CH_3\overset{\overset{\displaystyle O}{\|}}{C}{-}Cl \xrightarrow{N,N\text{-Dimethylaniline}}$$

$$(CH_3)_3C{-}O\overset{\overset{\displaystyle O}{\|}}{C}CH_3 + N,N\text{-Dimethylaniline hydrochloride} \qquad (13)$$

To a flask equipped with a reflux condenser, ground glass stirrer (or stirrer with Teflon bushings), and a dropping funnel are added 74 gm (1 mole) of *tert*-butyl alcohol, 120 gm (1.0 mole) of dry *N,N*-dimethylaniline, and 200 ml of dry ether. To the stirred mixture is slowly added dropwise 78.5 gm (1 mole) of acetyl chloride at such a rate that the ether vigorously refluxes. (Cooling may be necessary.) After the addition, the mixture is warmed on a water bath for 2 hr and then allowed to stand for several hours. The ether layer is separated from the N,N-dimethylaniline hydrochloride solid precipitate. The ether layer is extracted with portions of 10% sulfuric acid until the extract does not cloud when made alkaline. The ether layer is dried over anhydrous sodium sulfate and then distilled to yield 63–76% (73–88 gm) *tert*-butyl acetate, bp 93°–98.5°C, n_D^{25} 1.3820.

Under special reaction conditions, the reaction of an acyl halide may be used to prepare esters of glycidol. When the procedure outlined above is used, the tertiary amines catalyze the well-known polymerization of the epoxy group. The reaction is quite sudden and highly exothermic. Therefore, both the acyl halide and the tertiary amine are gradually added to glycidol when ester formation is desired. Care must be taken that no excess of the tertiary amine is present in the reaction flask at any time. In this manner, well-crystallized diglycidyl isophthalate was prepared.

C. ESTER INTERCHANGE

$$RCOOR' + R''OH \underset{}{\overset{H^+ \text{ or } OH^-}{\rightleftharpoons}} RCOOR'' + R'OH \qquad (14)$$

The exchange of alcohol fragments is catalyzed by either acid [4] or base. This reaction involves a reversible equilibrium which can be shifted to the right either by employing a large excess of alcohol R''OH or by removing the lower boiling alcohol R'OH. The low-boiling alcohol can also be removed as an azeotrope, as in the preparation of cellosolve acrylate.

10-4. Preparation of Cellosolve Acrylate [4]

$$CH_2{=}CH{-}COOCH_3 + HOCH_2CH_2{-}OC_2H_5 \longrightarrow$$

$$CH_2{=}CH{-}COOCH_2CH_2OC_2H_5 + CH_3OH \quad (15)$$

To a flask equipped with a stirrer, Vigreux column, and distillation head are added 45 gm (0.50 mole) of cellosolve (2-ethoxyethanol), 86 gm (1.0 mole) of methyl acrylate, 2.0 gm of hydroquinone, and 1.0 gm of p-toluenesulfonic acid. The mixture is heated to reflux for about 8 hr. At first the reflux temperature is 81°C. The methanol–methyl acrylate azeotrope is then removed as it is formed at 64°–65°C but not higher. After 9–10 hr, the crude reaction mixture is analyzed by gas chromatography (3 meter Apiezon L 0.3/1.0 on firebrick) which, in this case, indicated the presence of 85% cellosolve acrylate and 15% unreacted cellosolve. Distillation of this material yields 40 gm (56%), bp 174°C (760 mm) and bp 63°C (10 mm), specify gravity 20°/20° 0.9834.

D. THE REACTION OF CARBOXYLIC ACIDS WITH OLEFINS

$$RCOOH + CH_2{=}C(CH_3)_2 \longrightarrow RCOOC(CH_3)_3 \qquad (16)$$

The preparation of esters of tertiary alcohols is accomplished by the reaction of an olefin with a carboxylic acid. For example, isobutylene condenses with malonic acid to give 58–60% of di-*tert*-butyl malonate, with monoethyl maleate to give 53–58% of ethyl *tert*-butyl malonate, and with acrylic acid to give 47% *tert*-butyl acrylate.

Acids also have been esterified with propene and trimethylethylene [4]. These reactions need to be carried out under strictly anhydrous conditions and they are acid-catalyzed (H_2SO_4, or BF_3).

Carboxylic acids also add to acetylenes to give alkenyl esters. The commercial production of vinyl esters involves this synthesis.

$$R'COOH + RC{\equiv}CH \longrightarrow R'COOCR{=}CH_2 \qquad (17)$$

10-5. Preparation of *tert*-Butyl Acetate [5]

$$CH_3COOH + CH_2{=}C(CH_3)_2 \longrightarrow CH_3COOC(CH_3)_3 \qquad (18)$$

To a 500-ml Pyrex pressure bottle containing 26 gm (0.45 mole) of glacial acetic acid and 2 ml of concentrated sulfuric acid is added 50 gm (0.89 mole) of liquid isobutylene (liquified by the passage of the gas through a dry ice trap). The bottle is stoppered and allowed to remain at room temperature overnight. Then it is chilled in an ice–salt water bath, opened, and poured into a cold solution of 40 gm (1.0 mole) of sodium hydroxide in 500 ml of ice water. The organic layer is separated, washed with dilute alkali, dried over potassium carbonate, and distilled through a 6-inch Vigreux column to give 26.5 gm (53%) of *tert*-butyl acetate, bp 94°–97°C (738 mm), n_D^{25} 1.3820.

REFERENCES

1. A. I. Vogel, *J. Chem. Soc.* pp. 624, 644, 654 (1948).
2. W. H. Puterbaugh, C. M. Vanselow, K. Nelson, and E. J. Shrawder, *J. Chem. Educ.* **40,** No. 7, 349 (1963).
3. B. Abramovitch, J. C. Schrivers, B. E. Hudson, and C. R. Hauser, *J. Am. Chem. Soc.* **65,** 986 (1943).
4. S. R. Sandler, Author's laboratory, unpublished results (1966).
5. W. S. Johnson, A. L. McCloskey, and D. A. Dunnigan, *J. Am. Chem. Soc.* **72,** 516 (1950).

11
AMIDES

1. INTRODUCTION

A convenient and, in many cases, most economical method of preparing amides involves the thermal dehydration of ammonium or amine salts of carboxylic acids. Many variations of this reaction have been reported. For example, since the preparation of the amine salt of the gaseous amines is troublesome, mixtures of the carboxylic acids

From S. R. Sandler and W. Karo, *Organic Functional Group Preparations*, Vol. I, 2d ed. (New York, 1983), 315ff, by permission of Academic Press, Inc.

and ureas may be subjected to a combined thermal dehydration and decarboxylation.

By use of entraining agents, amine salts may be dehydrated under modest reaction conditions. In anhydrous THF, many acids may be converted to amides by reaction with appropriate amines in the presence of *o*-nitro-phenylthiocyanate and tri-*n*-butylphosphine.

The use of carbodiimides with amines and carboxylic acids is of particular interest in the preparation of polypeptides as well as in the synthesis of other amides from starting materials bearing other functional groups which may be capable of undergoing competing reactions with one of the reagents (e.g., acyl halides are usually not suitable for the preparation of amides of aminocarbinols, while aminocarbinols with carboxylic acids and a carbodiimide yield only amides).

A very general method for the preparation of amides involves the reaction of ammonia or amines with acyl halides or anhydrides. In this reaction, an excess of the amine may serve to "scavenge" the hydrogen halide (or carboxylic acid in the case of anhydrides) generated. Strong organic bases may also serve to react with the hydrogen halide. The reaction of diacid chlorides with diamines is also of importance in the preparation of polyamides such as Nylon. In the Schotten–Baumann reaction, aqueous inorganic bases such as sodium hydroxide or potassium carbonate are used as hydrogen halide scavengers.

Mixed anhydrides prepared from carboxylic acids and alkyl chlorocarbonates are of paricular value for the preparation of amides of sensitive acids such as *N*-acylated amino acids. The reaction of cyclic dianhydrides with amines yields imides. If diamines are used with dianhydrides, polyimide resins are produced.

The aminolysis of esters is another useful procedure for the preparation of amides, particularly in the presence of glycols as reaction promoters.

Transamidations of amides with amines, as well as of carboxylic acids with amides, have been reported.

Amides may be converted to *N*-alkyl and *N*-acyl derivatives by appropriate condensation reagents. C-Alkylations of amides can also be carried out to modify the carbon skeleton of a compound. *N*-arylamides may be prepared by the Goldberg reaction of an amide with an aryl halide in the presence of potassium carbonate and cuprous iodide.

Since α-haloesters undergo the Reformatskii reaction, α-haloamides may also be subjected to this reaction to give amides.

Another condensation reaction which has been used in the preparation of amides is the Grignard reaction with isocyanates.

Active hydrogen compounds, such as aliphatic nitro-compounds may be added to the double bond of acrylamide in a Michael-type addition to produce saturated amides.

Amides have been prepared by the hydration of nitriles. With secondary or tertiary alcohols or with certain olefins, nitriles react to produce N-substituted amides (Ritter reaction).

A novel oxidation of Schiff bases of arylmethylamines to aromatic amides has been described.

The reduction of aromatic nitro compounds in the presence of molybdenum hexacarbonyl and a carboxylic acid leads to the formation of *N*-aryl-carboxamides.

The Beckmann, Schmidt, and Wolff rearrangements have also been used to form amides.

In the preparation of amides the reported procedures do not always take several factors adequately into account, which may result in difficulties in the isolation and purification of the production. Even the syntheses discussed here do not always take these into consideration.

The first of these is that, contrary to the impression gained from the preparation of amides for identification purposes, not all of them are solids. Thus, for example, such commercially available compounds as formamide, *N*,*N*-dimethylformamide, *N*,*N*-dimethylacetamide, *N*-*tert*-butylformamide, and *N*,*N*-dimethyllauramide are liquids at room temperature. *N*-Acetylethylene-1,2-diamine, several *N*-alkyl acrylamides, and a number of *N*-alkyl amides of perfluorinated acids are also liquids.

The second factor which frequently leads to difficulties in isolation and purification arises from the exceptional solvent properties of amides—both liquid and solid. Many ionic compounds such as salts, water, and a large variety of covalent compounds, including aromatic hydrocarbons, have an appreciable solubility in many amides. The amides, in turn, may exhibit an appreciable solubility in very diversified solvents. Clearly, this situation may bedevil a synthesis with extremely complex solubility distribution coefficient problems. Vapor phase chromatography has been used in our laboratories to advantage in determining whether the amide has been adequately separated from co-products and whether a layer from a phase separation should be retained because it still contains product or whether it should be discarded.

Fortunately, many amides are quite thermally stable. Consequently, fractional distillation can frequently be used to purify those amides which do not melt at elevated temperatures that they seriously clog the condensers.

2. DEHYDRATION OF AMMONIUM SALTS

The reaction of carboxylic acids with ammonia normally produces its ammonium salt. The dehydration of ammonium salts may be used to prepare amides.

The classical dehydration of ammonium acetate to acetamide is accelerated by acetic acid, consequently the usual procedure for the preparation of acetamide involves careful distillation of an ammonium acetate–acetic acid mixture. The distillate is a solution of acetic acid and water. Finally excess acetic acid is distilled out followed by acetamide (yield 87–90% of theory, mp 81°C). The product may be recrystallized from a benzene–ethyl acetate mixture [1].

In the preparation of the required ammonium carboxylates, ammonium carbonate or ammonium bicarbonate is often a more convenient base than aqueous or anhydrous ammonia [1].

Although other methods may often be more convenient, the dehydration of amine salts has wide application, as is illustrated below.

11-1. Preparation of Butyramide [2]

$$CH_3CH_2CH_2\overset{\displaystyle O}{\overset{\|}{C}}-OH + NH_3 \longrightarrow [CH_3CH_2CH_2\overset{\displaystyle O}{\overset{\|}{C}}-O]^-NH_4^+ \qquad (1)$$

$$[CH_3CH_2CH_2\overset{\displaystyle O}{\overset{\|}{C}}-O]^-NH_4^+ \longrightarrow CH_3CH_2CH_2\overset{\displaystyle O}{\overset{\|}{C}}-NH_2 + H_2O \qquad (2)$$

In a hood, to a 250-ml three-necked flask fitted with an empty, electrically heated distillation column maintained between 85° and 90°C and topped with a vacuum distillation head, a gas inlet tube, and a thermometer is placed 88 gm (1 mole) of butyric acid. The butyric acid is heated to 185°C while a steady stream of ammonia is passed directly from a cylinder through the gas inlet tube into the butyric acid for 7 hr. At these temperatures the water formed from the reaction is swept through the apparatus into the distillation head where it may be partially condensed and collected. Then the ammonia flow is stopped, and the contents are distilled under reduced pressure at 130°–145°C (22 mm). Yield of crude product, 86.5 gm (84%). The product may be crystallized from benzene or ether. To separate the mother liquor, centrifugation is recommended. The product is dried in a vacuum desiccator over sulfuric acid, mp of purified product, 114.8°C.

By a similar procedure, a variety of *N,N*-dimethylamides has been prepared. Table I gives preparative details for this procedure.

Free carboxylic acids may also be converted to amides by heating the acid with urea as the source of the amino function [3]. The reactions must be carried out with considerable care since large volumes of gas are evolved during the reaction and since urea tends to sublime into the reflux condenser during the preparation. The preparation of heptamide according to Eq. (3) is described in detail in the literature [4].

$$2 \, CH_3(CH_2)_5CO_2H + H_2NCONH_2 \longrightarrow 2 \, CH_3(CH_2)_5CONH_2 + CO_2 + H_2O \quad (3)$$

TABLE I

PREPARATION OF ALIPHATIC AMIDES AND DIMETHYLAMIDES BY DEHYDRATION [2]

Acid	Flask temp. (°C)	Condenser temp. (°C)	Amine addition time (hr)	Yield (%)	Product bp [°C (mm)]	M P (°C)
Amide of						
Acetic	170°–190°	80°	3–5	96	210°–220° (atm)	81.5°
Propionic	185°	80°	5.5	93	200°–220° (atm)	81.3°
Butyric	185°	85°–90°	7	84	130°–145° (22)	114.8°
Valeric	180°	90°	15	82	100°–130° (6)	105.8°
Caproic	160°	90°	7	75	135°–150° (10)	101.5°
Heptoic	160°–190°	90°	4–7	75	130°–150° (7)	96.5°
Caprylic	180°	90°	11	80	135°–155° (4)	106.0°
Dimethylamide of						
Formic (95%)	95°	60°	3	73	130°–165° (atm)	—
Acetic	150°	80°	3	84	165°–175° (atm)	—
Propionic	155°	80°	3	78	165°–178° (atm)	—
Butyric	155°	85°	2.5	84	180°–194° (atm)	—
Valeric	165°	85°	3	87	205°–215° (atm)	—
Caproic	155°	85°	3	88	220°–230° (atm)	—
Heptoic	160°	85°	3.5	84	165°–175°(95)	—

The dehydration of some amine salts of carboxylic acids may also be carried out by an azeotropic water removal procedure similar to the one commonly used in ester preparations.

While mechanistically, probably quite unrelated to the reactions discussed above, the preparation of amides in the presence of carbodiimides may be looked upon as a reaction of an acid and an amine. While this reaction is of particular value for polypeptide syntheses, it is believed to be quite generally applicable. Of particular interest is the fact that the reaction appears to be specific for amino groups. Hydroxy-containing starting materials do not usually experience *O*-acylation [5, 6]. The reaction is characterized by high yields. The course of the reaction can be followed readily by the precipitation of highly insoluble substituted ureas. While the original work generally refers to the use of dicyclohexylcarbodiimide, other carbodiimides have become commercially available and their applicability should be explored further.

■ **CAUTION:** Carbodiimides must be handled with extreme care since free amino groups in the tissue of the operator may readily be acylated. This may cause physiological effects which are potentially hazardous.

An example of the reaction is given here mainly to illustrate the technique involved. Usually much better yields may be anticipated [7].

11-2. Preparation of *N,N*-Bis(2-chloroethyl)-2-(dichloroacetamido)acetamide [7]

$$
\underset{\substack{\| \\ Cl_2CHC}}{O}\!\!-\!NHCH_2\underset{\substack{\| \\ C}}{O}\!\!-\!OH + (ClCH_2CH_2)_2NH + C_6H_{11}\!-\!N\!=\!C\!=\!N\!-\!C_6H_{11}
$$

$$
\downarrow
$$

$$
\underset{\substack{\| \\ Cl_2CHC}}{O}\!\!-\!NHCH_2\underset{\substack{\| \\ C}}{O}\!\!-\!N\!\!<\!\!\begin{array}{l}CH_2CH_2Cl \\ CH_2CH_2Cl\end{array} + C_6H_{11}NH\underset{\substack{\| \\ C}}{O}NHC_6H_{11} \qquad (3)
$$

■ **CAUTION:** The carbodiimide must be handled so that neither liquid nor vapor comes in contact with personnel by spillage or inhalation.

To a solution of 7.4 gm (0.051 moles) of bis(2-chloroethyl)amine in 60 ml of tetrahydrofuran is added 9.7 gm (0.051 moles of *N*-dichloroacetylglycine. To this solution is added dropwise, with

stirring, over a 30-min period a solution of 10.76 gm (0.051 moles) of
N,N'-dicyclohexylcarbodiimide in 60 ml of tetrahydrofuran. After
an additional stirring period of 30 min, the precipitated N,N'-
dicyclohexylurea is removed by filtration and washed with tetrahydrofuran.

The combined filtrates are evaporated to dryness and the residue
is crystallized from a mixture of ethyl acetate and petroleum ether.
Yield 3.0 gm (18.6%), mp 62.5°–63.5°C. Other preparations of this
compound gave yields up to 38.5%. In other amide or polypeptide
preparations such as those reported by Sheehan and Hess [5, 6], much
higher yields have been obtained.

3. CONDENSATION REACTIONS

A. CONDENSATIONS WITH ACYL HALIDES

The reaction of acyl chlorides with ammonia, primary, or secondary
amines is probably the most generally applicable method of preparing
amides in the laboratory. A large variey of acid chlorides are available
commercially, others are readily prepared from the acids. Not only
halides of carboxylic acids but also those of sulfonic and phosphonic
acids and of picric acid may be converted to amides. Both aliphatic
and aromatic amines may be subjected to acylation with acid halides.
The reactions of aliphatic acid chlorides have been extensively reviewed [8].

The reactions are usually quite rapid, which is desirable in the
preparation of derivatives for the identification of amines or acids
(acid chlorides). The reaction conditions used, however, do not always
produce very high yields.

Since the reaction of an acid chloride with amines is highly
exothermic, often quite violent, reactions must be carefully controlled
from the standpoint of safety. It is customary to carry this reaction out
at ice temperatures. Even so, local overheating may take place as an
acid chloride is being added to the amine. This may cause loss of amine
content, particularly if large-scale laboratory reactions are attempted.
By cooling, the reaction rate is also materially reduced. Therefore,
adequate time for completion of the reaction must be allowed for
isolation of an optimum yield.

The reaction itself is usually carried out in the presence of a base.
This base serves to neutralize the hydrogen chloride formed as a

coproduct. When no base is present, hydrochlorides of the amines may form which may be discarded during the product work-up

Since the base often forms salts in a very finely divided state, considerable product may be occluded, and, in our experience, it is well to attempt to dissolve the salts in water, even though it appears quite dry after its initial separation from the reaction mixtures, and to extract the aqueous solution with solvents. Among the extraction solvents to be considered are aromatic hydrocarbons. Ether should be used only with caution since some simple amine hydrochlorides appear to have significant solubility in this solvent. Obviously, once this solvent is stripped off, a product residue containing both an amide and an amine hydrochloride will be troublesome to separate.

When one of the volatile amines has to be used, a method of measuring the amount of amine utilized is required. We have never found bubbling a gaseous amine directly into an acyl chloride solution to be satisfactory. Aside from the potential violence of reaction, addition rates are difficult to control, amine hydrochlorides may clog the delivery tube, and the reaction mixture may "suck back," even into the gas cylinder.

A technique which is frequently satisfactory consists of bubbling the amine, in an efficient hood, from the gas cylinder, through a suitable gas trap into a tared flask containing the requisite weight of reaction solvent (usually benzene). From time to time the gas flow is shut off, the delivery tube is removed from the flask, and the flask is weighed. This procedure is continued until the desired weight of amine is dissolved in the solvent. As the concentration of amine increases, the solution may have to be cooled in an ice–salt bath. The amine solution may be stored for short periods of time in an ice–salt bath until ready to use.

■ **CAUTION:** This operation, as well as subsequent ones, must be carried out entirely in a hood and the operator should be equipped with a suitable gas mask, gloves, and rubber apron.

If desired, the amine solution may be added gradually to the reaction vessel containing acyl chloride using a pressure-equilizing addition funnel. Usually, however, it is simply transferred to the reaction flask and the acyl chloride is added gradually. This reduces considerably the possibility of forming secondary and tertiary amides. The preparation of *N*-methylacrylamide is an example of this procedure.

11-3. Preparation of *N*-Methylacrylamide [9]

$$2\ CH_3NH_2 + CH_2{=}CHC\overset{\displaystyle O}{\underset{\displaystyle Cl}{\diagdown}}\ \longrightarrow\ CH_2{=}CH{-}C\overset{\displaystyle O}{\underset{\displaystyle NHCH_3}{\diagdown}}\ +\ CH_3NH_2{\cdot}HCl\quad(4)$$

To a well-chilled solution of 122.5 gm (4 moles) of methylamine in 500 ml of dry benzene is added a solution of 176 gm (1.94 moles) of freshy distilled acrylyl chloride in 100 ml of benzene over a 2.75-hr period while a reaction temperature below 5°C is maintained. The solid which forms is filtered off. The filtrate is preserved. The solid is dissolved in water. The aqueous solution is extracted with two portions of benzene. The benzene extracts are combined with the filtrate. The benzene, excess amine, and water are removed from the product solution by distillation. The high-boiling residue is then fractionally distilled. The fraction boiling between 92° and 97°C (4−5 mm) is collected and redistilled at 79°C (0.7 mm). The yield is 107 gm (66% of theory), n_D^{20} 1.4730. The product may be stored after inhibition with 0.1% of *p*-methoxyphenol. By-products are believed to be the addition products of hydrogen chloride or of methylamine to the double bond.

The preparation of amides by reaction of an amine, an acyl chloride, or an acid anhydride in the presence of aqueous alkali is the well-known "Schotten−Baumann" reaction. Since the rate of reaction of the acyl chloride with amines is greater than the rate of hydrolysis of the acyl chloride, amide formation is favored. In some cases, the stability of acyl chlorides to aqueous caustic solution is surprising. For example, benzoyl chloride may be kept in contact with sodium hydroxide solutions for long periods of time. We presume that a thin layer of sodium benzoate forms rapidly and that this acts as a protective coating unless the benzoyl chloride is stirred or shaken to disturb the protective layer. As a matter of fact, the Schotten−Baumann reaction (and the analogous Hinsberg reaction with aromatic sulfonyl chlorides) appears to be most satisfactory when acyl chlorides are used which are relatively insoluble in water. Many years ago, we noted that the addition of a small percentage of a surfactant (e.g., sodium lauryl sulfate as well as cationic surfactants) assisted in the dispersion of the acyl chloride with an increase in reaction rate. These may have been early examples of phase−transfer reactions.

The Schotten–Baumann reaction is widely used, particularly for the preparation of derivatives for the identification of small amounts of amines or acyl chlorides by shaking the reagents together in a glass-stoppered bottle. Sonntag [8] extensively reviewed this reaction.

B. CONDENSATIONS WITH ACID ANHYDRIDES

An example of the use of a mixed anhydride derived from ethyl chlorocarbonate to prepare a rather complex amide is the preparation of 2-phenyl-3-(2′-furyl)propionamide [10]. By the use of such anhydrides it is possible to prepare not only unsubstituted amides, but also various substituted amides.

11-4. Preparation of 2-Phenyl-3-(2′-furyl)propionamide [10]

$$\text{furyl}-CH_2-CH(C_6H_5)-\underset{\underset{\text{O}}{\|}}{C}-OH + Cl-\underset{\underset{\text{O}}{\|}}{C}-OCH_2CH_3 + (C_2H_5)_3N$$

$$\downarrow$$

$$\text{furyl}-CH_2-CH(C_6H_5)-\underset{\underset{\text{O}}{\|}}{C}-O-\underset{\underset{\text{O}}{\|}}{C}-OCH_2CH_3 + (C_2H_5)_3N\cdot HCl \qquad (5)$$

$$\text{furyl}-CH_2-CH(C_6H_5)-\underset{\underset{\text{O}}{\|}}{C}-O-\underset{\underset{\text{O}}{\|}}{C}-OCH_2CH_3 + NH_3 \longrightarrow$$

$$\text{furyl}-CH_2-CH(C_6H_5)-\underset{\underset{\text{O}}{\|}}{C}-NH_2 + \left[CH_3CH_2O\underset{\underset{\text{O}}{\|}}{C}-OH\right] \qquad (6)$$

In a reaction setup, freed and protected from humidity, 50 gm (0.27 moles) of 2-phenyl-3-(2′-furyl)propionic acid and 30 gm

(0.3 moles) of triethylamine in 140 ml of anhydrous chloroform are cooled with stirring in a dry ice–acetone mixture. To this mixture, a solution of 30 gm (0.28 moles) of ethyl chlorocarbonate in 75 ml of anhydrous chloroform is added dropwise. After 1 hr, while maintaining the temperature below 0°C, a stream of anhydrous ammonia is passed through the stirred reaction mixture (hood, suitable gas traps). Within approximately 30 min, the solution is saturated with respect to the ammonia. After the mixture has warmed to room temperature, 500 ml of water are added, and the chloroform layer is separated. The chloroform solution of the product is treated in turn with 4 N hydrochloric acid and water and is then dried with magnesium sulfate. The solvent is then evaporated. The residue is recrystallized from a solution of 80 ml of benzene and 35 ml of petroleum ether. Yield 40 gm (80% of theory), mp 100°–101°C.

C. CONDENSATIONS WITH ESTERS

The aminolysis of esters has been used widely for the preparation of amides. Water and certain solvents, such as glycols, promote the reaction [11], although the presence of glycols may interfere with the isolation of amides at times. In the case of olefinic esters, addition of amine to the double bond also takes place during aminolysis, an example of the Michael condensation.

The aminolysis of esters may be carried out with aqueous ammonia [12], gaseous ammonia, and a variety of amines [13–15]. In the case of malonic esters, the methyl esters appear to react more readily, although ethyl esters have been used. Sodium methoxide is generally used as catalyst in either case [16].

With gaseous amines, the addition techniques mentioned under acyl halide reactions may be applied. With liquid amines, the preparation of N-ethylperfluorobutyramide may serve as a model synthesis. As to final purification procedures, solid amides may be recrystallized (alcohol, alcohol–water, petroleum ether, etc. are typical solvents), or, in many cases, distilled under reduced pressure. Liquid amides are best purified by distillation.

11-5. Preparation of N-Ethylperfluorobutyramide [12]

$$CF_3CF_2CF_2\overset{\overset{\displaystyle O}{\|}}{C}-OCH_3 + C_2H_5NH_2 \longrightarrow CF_3CF_2CF_2\overset{\overset{\displaystyle O}{\|}}{C}-NHC_2H_5 + CH_3OH \quad (7)$$

To an ice-cooled solution of 2736 gm (12 moles) of methyl per-fluorobutyrate in 2 liters of ether (peroxide-free) is added dropwise 750 gm (16.7 moles) of cold ethylamine. After completion of the addition, the reaction mixture is allowed to warm to room temperature and the solvent and excess amine are separated by distillation under reduced pressure and with moderate warming. The residue is fractionally distilled; the fraction boiling quite sharply at 168°C at atmospheric pressure (759 mm) is collected. Yield 2665 gm (92.5%).

Monoacetylated diamines are prepared by using an excess of the diamine over the ester as described below [17].

11-6. Preparation of Monoacetylethylenediamine [17]

$$\text{NH}_2\text{CH}_2\text{CH}_2\text{NH}_2 + \text{C}_2\text{H}_5\text{O}-\overset{\overset{\displaystyle O}{\|}}{\text{C}}-\text{CH}_3 \longrightarrow$$

$$\text{NH}_2\text{CH}_2\text{CH}_2\text{NH}-\overset{\overset{\displaystyle O}{\|}}{\text{C}}-\text{CH}_3 + \text{C}_2\text{H}_5\text{OH} \quad (8)$$

A mixture of 500 gm (6 moles) of ethyl acetate and 1550 gm (18 moles) of commercial 70% aqueous ethylenediamine is prepared and allowed to stand several days after the mixture has become homogeneous. The solution is then distilled, collecting the fraction boiling between 115°–130°C at 5 mm. Upon redistillation, 365 gm (60%) of monoacetylethylenediamine is collected between 125° and 130°C.

The acetylation of amines with isopropenyl acetate appears to be a transition between the highly exothermic reactions of acyl halides and anhydrides with amines on the one hand and the reaction of amines with more conventional esters on the other. While this reagent is of particular value in the preparation of enol acetate, it has been used for the preparation of amides. One interesting aspect of its use is that acetone forms as a coproduct which may distill off as the reaction proceeds. Isopropenyl acetate and other isopropenyl esters may also be used to N-acylate amides and imides. By the judicious selection of starting amides and isopropenyl esters, tertiary amides with three different acyl groups may be synthesized. This may very well be one of very few reaction systems which permits the synthesis of this rare group of tertiary amides.

11-7. Preparation of N-(2-Hydroxyethyl)acetamide [18]

$$HO-CH_2CH_2-NH_2 + CH_2{=}\underset{\underset{CH_3}{|}}{C}-O-\overset{\overset{O}{\|}}{C}-CH_3 \longrightarrow$$

$$HO-CH_2CH_2-NH\overset{\overset{O}{\|}}{C}-CH_3 + CH_3-\overset{\overset{O}{\|}}{C}-CH_3 \quad (8)$$

To 100 gm (1.0 moles) of isopropenyl acetate in a small still is gradually added 63 gm (1.03 moles) of ethanolamine. A vigorous reaction takes place and acetone is distilled off continuously. The residue is distilled at 135°–140°C (1 mm). Yield 85.5 gm (81% of theory).

D. TRANSAMIDATION REACTIONS

Formanilides have been prepared by a base-catalyzed transamidation with dimethylformamide [19]. In this example, DMF serves as the source of the carboxylate-function in the presence of amines.

11-8. Preparation of 2-Iodoformanilide [19]

$$(9)$$

A mixture of 16.2 gm (0.3 mole) of sodium methoxide and 32.9 gm (0.15 mole) of 2-iodoaniline in 150 ml of dimethylformamide is refluxed for 30 min. The product is then isolated by diluting the reaction mixture with water. Yield 25.2 gm (68%), mp 113°–113.5°C.

E. N-ALKYLATION REACTIONS

11-9. Preparation of N-(9-Bromononyl)phthalimide [20]

$$NK + BrCH_2(CH_2)_7CH_2Br \longrightarrow N-(CH_2)_8-CH_2Br + KBr$$

$$(10)$$

A dispersion of 8.45 gm (0.05 mole) of potassium phthalimide, 57.2 gm (0.2 mole) of 1,9-dibromononane, and 3.3 gm of dimethylformamide is heated for $1\frac{1}{2}$ hr at 160°C. The precipitated potassium bromide is separated by filtration, and the excess dibromononane and dimethylformamide are removed by distillation under reduced pressure. The residue is fractionated under reduced pressure. The solid portion of the distillate is recrystallized from ethanol. Yields vary from 40 to 78%, depending on minor variations in procedure; mp 37.5°C.

REFERENCES

1. G. H. Coleman and A. M. Alvarado, *Org. Synth. Collect. Vol.* **1**, 3 (1948).
2. J. A. Mitchell and E. E. Reid, *J. Am. Chem. Soc.* **53**, 1879 (1931).
3. E. Cherbulier and F. Landolt, *Helv. Chim Acta* **29**, 1438 (1946).
4. J. L. Guthrie and N. Rabjohn, *Org. Synth. Collect. Vol.* **4**, 513 (1963).
5. J. C. Sheehan and G. P. Hess, *J. Am. Chem. Soc.* **77**, 1067 (1956).
6. J. C. Sheehan, M. Goodman, and G. P. Hess, *J. Am. Chem. Soc.* **78**, 1367 (1956).
7. I. Levi, H. Blondal, J. W. Weed, A. C. Frosst, H. C. Reilly, F. A. Schmid, K. Sugiura, G. S. Tarnowski, and C. C. Stock., *J. Med. Chem.* **8**, 715 (1965).
8. N. O. V. Sonntag, *Chem. Rev.* **52**, 237 (1953).
9. B. D. Halpern and W. Karo, *U.S. Air Force Syst. Command, Res. Technol. Div. Air Force Mater. Lab., Tech. Rep., WADD* 55–206 (1956).
10. M. Pesson, S. Dupin, M. Antoine, D. Humberto, and M. Joannic, *Bull. Soc. Chim. Fr.* No. **8**, p. 2266 (1965).
11. M. Gordon, J. G. Miller, and A. R. Day, *J. Am. Chem. Soc.* **70**, 1245 (1948).
12. B. B. Corson, R. W. Scott, and C. E. Vose, *Org. Synth. Collect. Vol.* **1**, 179 (1948).
13. B. D. Halpern, W. Karo, L. Laskin, P. Levine, and J. Zomlefer, *U.S. Air Force Syst. Command, Res. Technol. Div., Air Force Mater. Lab., Tech. Rep. WADD* **WADD-TR-54-264** (1954).
14. H. Gilman and R. G. Jones, *J. Am. Chem. Soc.* **65**, 1458 (1943).
15. D. R. Husted and A. N. Ahlbrecht, *116th Am. Chem. Soc. Meet.* Paper 17, Org. Chem. Div. (1949).
16. P. B. Russell, *J. Am. Chem. Soc.* **72**, 1853 (1954).
17. S. R. Aspinall, *J. Am. Chem. Soc.* **63**, 852 (1941).
18. H. J. Hagemeyer, Jr. and D. C. Hull, *Ind. Eng. Chem.* **41**, 2920 (1949).
19. G. R. Petit and E. G. Thomas, *J. Org. Chem.* **24**, 895 (1939).
20. H. B. Donahoe, R. J. Seiwald, M. M. C. Neuman, and K. K. Kimura, *J. Org. Chem.* **22**, 68 (1957).

12

CYANATES, ISOCYANATES, THIOCYANATES, AND ISOTHIOCYANATES

1. INTRODUCTION

Since the cyanates were only first isolated in 1960, satisfactory methods of synthesis are quite limited.

The best available method appears to be the reaction of cyanogen chloride and phenol under such conditions that the level of an added base is carefully limited at all times.

$$ArOH + ClCN \xrightarrow{(C_2H_5)_3N} ArOCN + [(C_2H_5)_3\overset{+}{N}H]\overset{-}{Cl} \qquad (1)$$

From S. R. Sandler and W. Karo, *Organic Functional Group Preparations*, Vol. I, 2d ed. (New York, 1983), 359–378, by permission of Academic Press, Inc.

Isocyanates are prepared by the reaction of amines with phosgene.

$$RNH_2 + COCl_2 \longrightarrow RNCO + 2\,HCl \tag{2}$$

Convenient laboratory methods involve the reaction of alkyl halides or dialkyl sulfate with inorganic cyanates such as silver cyanate.

$$R_2SO_4 + 2\,KOCN \longrightarrow 2\,RNCO + K_2SO_4 \tag{3}$$

$$RX + AgOCN \longrightarrow RNCO + AgX \tag{4}$$

The decomposition of urethanes either thermally or in the presence of phosphorus pentachloride has been reported. An interesting new reaction involves the decomposition of phosphoramidate anions.

Quite recently a more direct preparation has been developed in which aliphatic or aromatic amines are reacted with carbon monoxide in the presence of $PdCl_2$ [1]. The reaction is slow but it has the potential of offering a convenient method of preparing isocyanates in the laboratory.

$$RNH_2 + CO + PdCl_2 \longrightarrow RNCO + Pd + 2\,HCl \tag{5}$$

In some cases the carbonylation of amines gives urethanes instead.

More recently a process has been patented for preparing isocyanates by the reaction of nitro compounds with carbon monoxide in the presence of a rhodium oxide catalyst.

$$ArNO_2 + CO \xrightarrow[\substack{Rh_2O_3 \\ CH_3CN}]{\substack{RhO_2 \\ or}} ArNCO \tag{6}$$

Recently a process has been reported for preparing aromatic isocyanates by the reaction of aromatic nitro compounds with carbon monoxide in alcohol solution to first form the urethanes. Thermal decomposition of the urethane gives the isocyanate.

Furthermore, the reaction of carbon monoxide with a metal salt of carbamic acid has been used to give isocyanuric acid and derivatives or isocyanates.

A recent review on isocyanates is worth consulting for additional details, especially as to commercial uses and manufacturing processes [5].

The classical Curtius, Lossen, Hofmann, and Schmidt rearrangements may be used to prepare isocyanates, although normally isocyanates are considered only intermediates in these reactions.

Since isothiocyanates are generally higher boiling than the corresponding isocyanates, exchange reactions of the type illustrated in Eq. (7) are possible.

$$RNCS + ArNCO \longrightarrow RNCO + ArNCS \qquad (7)$$

The isothiocyanates have been prepared by the condensation of amines with carbon disulfide in the presence of a base to yield the dithiocarbamate which, in turn, is decomposed by reagents such as lead nitrate to the isothiocyanate [see Eqs. (15) and (16)].

Just as amines react with phosgene to give isocyanates, thiophosgene may be used to prepare isothiocyanates.

The thiocyanates are usually prepared by the condensation of alkyl halides (or sulfates) with potassium or ammonium thiocyanate.

$$RBr + KSCN \longrightarrow RSCN + KBr \qquad (8)$$

Aromatic thiocyanates have also been prepared by the action of potassium thiocyanate or cuprous thiocyanate on diazonium salts.

■ **CAUTION:** All compounds considered in this chapter should be handled with great care. Aside from the fact that many have strong, unpleasant odors, many exhibit strong physiological reactions. Many are lachrymators and/or vesicants.

A. NOMENCLATURE

Esters of cyanic acid are referred to as cyanates and have been assigned structure I. The related and better known isocyanates have structure II.

$$R—O—C{\equiv}N \qquad\qquad R—N{=}C{=}O$$
$$\text{I} \qquad\qquad\qquad \text{II}$$
$$\text{Cyanates} \qquad\qquad \text{Isocyanates}$$

Similarly among the sulfur analogs, thiocyanates have structure III while the isothiocyanates have structure IV.

$$R—S—C\equiv N \qquad R—N=C=S$$
$$\text{III} \qquad\qquad \text{IV}$$
$$\text{Thiocyanates} \qquad \text{Isothiocyanates}$$

Other isomers, whose preparation is beyond the scope of this chapter, are the fulminates V and the nitrile oxides VI. Of these, probably because of the extreme explosion hazards, the electronic structure of the fulminates has not been settled completely.

$$R—ONC \qquad R—C\equiv N \rightarrow O$$
$$\text{V} \qquad\qquad \text{VI}$$
$$\text{Fulminates} \qquad \text{Nitrile oxides}$$

2. CYANATES

Up until 1960, the true cyanates, ROCN, had not been isolated. Thus, the reaction of sodium phenolates with cyanogen halides was shown to follow the course given in Eqs.(9) and (10) by both Nef, and Hantzsch and Mai.

$$\text{Ar—ONa + ClCN} \longrightarrow \text{ArO—}\overset{\overset{\displaystyle NNa}{\|}}{C}\text{—Cl} \xrightarrow[\text{2. H}^+]{\text{1. RO}^-} \text{ArO—}\overset{\overset{\displaystyle NH}{\|}}{C}\text{—OR + NaCl} \qquad (9)$$

$$3\ \text{ArO—}\overset{\overset{\displaystyle NH}{\|}}{C}\text{—OR} \longrightarrow \quad + 3\ \text{ROH} \qquad (10)$$

In 1960, however, a few sterically hindered cyanates were produced by essentially the same process. All these procedures involved the addition of cyanogen halides to a reaction medium in which an excess of alcoholates or phenolates was present. Therefore, imido diesters form rapidly in the basic medium followed by trimerization unless steric factors prevent this reaction.

In 1964, Grigat and Pütter began the publication of an extensive series of papers on the chemistry of cyanates in *Chemische Berichte*. In their first paper of the series, Grigat and Pütter point out that, if reaction conditions are such that the base is never present in excess, true cyanates can be prepared from phenolic compounds unless several electron-withdrawing groups are present in the aromatic nucleus. Even some aliphatic cyanates could be prepared by this procedure, provided strongly acidic alcohols such as the trihaloethanols or enols were used as starting materials.

These investigators found that purified cyanates are stable for several weeks. They can be recrystallized and in some cases can even be distilled.

The reactions may be carried out in acetone or other solvents such as ether, carbon tetrachloride, benzene, acetonitrile, or ethyl acetate. The cyanogen halides used may be either cyanogen chloride or, if a higher boiling point or higher reaction temperature is desired, cyanogen bromide.

■ **CAUTION:** These compounds are extremely toxic.

The usual base used in the reaction is triethylamine. With inorganic bases, such as sodium hydroxide, aqueous media have been used, although yields usually are lower. The range of phenols used in the Grigat and Pütter procedure is quite extensive, including phenol, 2-methylphenol, 3-methylphenol, 4-methylphenol, various other mono- and dialkylphenols, naphthols, chlorophenols, nitrophenols, methoxyphenols, trihydroxyquinolines, trihaloethanols, hydroquinones, hydroxybiphenyls, etc.

Depending on the substituents present in the aromatic nucleus, the cyanates show more or less tendency to trimerize in the presence of mineral acids, Lewis acids, bases, as well as phenolic impurities. The reaction proceeds quite smoothly and, therefore, the cyanates become a very useful starting point for the preparation of a large variety of triaryl esters of cyanuric acid.

12-1. Preparation of Phenyl Cyanate [1]

$$\text{C}_6\text{H}_5\text{-OH} + \text{ClCN} \xrightarrow{(\text{C}_2\text{H}_5)_3\text{N}} \text{C}_6\text{H}_5\text{-OCN} + (\text{C}_2\text{H}_5)_3\overset{+}{\text{N}}\text{H}\overset{-}{\text{Cl}} \qquad (11)$$

In a suitable hood and with precautions for handling a toxic material such as cyanogen chloride, a solution of 94.1 gm (1.0 mole) of phenol in 250 ml of acetone is cooled to 0°C. To the cold solution, 65.1 gm (approximately 1.05 moles) of liquid cyanogen chloride is added. While cooling is continued in an ice bath, 101.2 gm (1.0 moles) of triethylamine is added dropwise with vigorous stirring at such a rate that the temperature never exceeds 10°C. After the addition has been completed, stirring is continued for an additional 10 min, and the triethylamine hydrochloride is separated by filtration and extracted three times with 100 ml portions of acetone. The acetone solutions are combined and evaporated under reduced pressure.

■ **CAUTION:** The evaporating solvent may contain cyanogen chloride and due precautions must be taken that the exhaust from the aspirator or pump be properly treated. The residue is then subject to distillation under reduced pressure to yield 112 gm (94%), bp 55°C (0.4 mm).

3. ISOCYANATES

A. CONDENSATION REACTIONS

In view of the fact that the diisocyanates of aromatic diamines are of great industrial importance today, undoubtedly the most common method of preparation of diisocyanates, as well as of monoiso-cyanates, is based on the reaction of amines with phosgene.

■ **CAUTION:** Phosgene is highly toxic and should be handled with great care in well-ventilated hoods.

From the standpoint of laboratory procedures, suitable models for the preparation of many isocyanates are given in the literature. In connection with these preparations, precaution must be taken, since the amines, isocyanates, and phosgene are very toxic. Also, it is important in the laboratory preparations to keep moist air, water, and protic solvents out of contact with the isocyanates.

The reaction of aromatic amines with phosgene to produce isocyanates has wide applicability. The example of the preparation of *p*-nitrophenyl isocyanate illustrates the general procedure. Even complex aromatic amines such as fluorescein amine may be subjected to the reaction to produce fluorescein isocyanate, which has found some application in biochemical research.

In the preparation of aliphatic isocyanates, the volatility of the amines may lead to poor yields when phosgene is bubbled into the reactant. For this reason, thoroughly dried hydrochloride salts of the amines, suspended in high-boiling inert solvents, are substituted for the free amine. Vapor phase preparation of isocyanates has also been reported.

To minimize such side reactions as the formation of substituted ureas, the reaction is generally carried out in an excess of phosgene.

Symmetrical ureas are converted to aliphatic isocyanates in high yield by the reaction of phosgene with the urea at 150°C and higher.

■ **CAUTION:** Phosgene is a war gas and, being extremely toxic, must be handled with extreme care. All work must be carried out in a well-ventilated hood. The operator must wear a gas mask suitable for use with phosgene, rubber gloves, and a rubber apron. It is also recommended that in the general work area warning filter papers which change color on contact with phosgene be prepared.

12-2. Preparation of *p*-Nitrophenyl Isocyanate [2]

$$O_2N-\underset{}{\bigcirc}-NH_2 + Cl-\overset{\overset{\displaystyle O}{\|}}{C}-Cl \longrightarrow O_2N-\underset{}{\bigcirc}-NCO + 2\,HCl \quad (12)$$

■ **CAUTION:** This reaction must be carried out in a well-ventilated hood, behind a shield, and, as a minimum, observing the precautions outlined above.

The reaction system is assembled as follows. The phosgene cylinder is connected through a mercury pressure regulator in turn to a gas wash bottle filled with cottonseed oil to remove chlorine, a wash bottle containing concentrated sulfuric acid, and an empty flask large enough to hold the contents of the reaction flask. To this is connected a gas delivery tube in a 5-liter three-necked flask fitted also with an addition funnel and a gooseneck leading to a condenser. The condenser is connected to a filter flask which serves as a distillate receiver. The vent of this receiver is connected to a gas wash bottle containing 20% sodium hydroxide solution and a safety trap. Finally, this trap is connected to an aspirator which permits drawing the phosgene into the reaction flask by slightly reducing the pressure inside the system.

In the three-necked flask is placed 500 ml of dry, ethanol-free ethyl acetate. This solvent is then saturated with phosgene. While a slow stream of phosgene is passed into the system throughout the reaction, a solution of 150 gm (1.09 moles) of *p*-nitroaniline is 1500 ml of ethyl acetate is added slowly through the addition funnel over a 3-hr period.

The *p*-nitroaniline is added in such a manner that the *p*-nitroaniline hydrochloride which forms initially is allowed to dissolve in the reaction medium. If necessary, the flask is gently warmed.

While the addition is being completed, the solution is gently boiled to assist in breaking up the p-nitroaniline hydrochloride lumps. After completion of the addition, the stream of phosgene is continued for 5 min. The phosgene stream is then turned off. The addition train is removed, boiling chips are added to the flask, the stopper which had carried the gas delivery tube is replaced by a solid stopper, and the ethyl acetate solvent is distilled off by the careful application of heat. To the brown residue is added 800 ml of hot carbon tetrachloride and the insoluble disubstituted urea by-product is separated by filtration.

Then about two-thirds of the solvent is distilled from the filtrate. The solution is then chilled and the crystals of p-nitrophenyl isocyanate are filtered off and stored in a tightly closed container. A further crop of product may be obtained by work-up of the mother liquor. The product may be recrystallized from dry carbon tetrachloride. Yield 160 gm $\pm$ 10 gm (83–95%), mp 56°–57°C, bp 160°–162°C (18 mm), light yellow needles.

4. THIOCYANATES

The sulfur analogs of both cyanates and isocyanates are thermally and hydrolytically more stable than the oxygen analogs. Both series of sulfur analogs have been known for some time from natural and synthetic sources.

A. CONDENSATION REACTIONS

The preparation of alkyl thiocyanates by reaction of cyanogen chloride with a mercaptan according to Eq. (13) has been used.

$$R—SH + ClCN \longrightarrow RSCN + HCl \tag{13}$$

It is believed that this reaction should be reinvestigated in the light of the preparative procedures for cyanates [1].

Perhaps the most widely used preparation of thiocyanates involves the reaction of an alkyl halide with either potassium thiocyanate or ammonium thiocyanate. The reaction is not confined to simple alkyl halides but has also been used for the preparations with dihalides, chlorohydrins, secondary alkyl halides, and acyl halides (in the preparation of acyl thiocyanates). The preparation of undecyl thiocyanate is a typical example of the procedure.

12-3. Preparation of Undecyl Thiocyanate [3]

$$CH_3(CH_2)_9CH_2Br + KSCN \longrightarrow CH_3(CH_2)_9CH_2SCN + KBr \qquad (14)$$

To 145.5 gm (1.5 moles) of potassium thiocyanate dispersed in 340 ml of ethanol heated to reflux, 235 gm (1 mole) of undecyl bromide are added gradually. After addition has been completed, refluxing is continued for 2 hr. The reaction mixture is then cooled to room temperature, diluted with water, and the product extracted with diethyl ether. After the ether solution has been dried with calcium chloride, the ether is evaporated off and the residue fractionally distilled under reduced pressure to yield 184 gm (86.5%), bp 160°–161°C (10 mm).

5. ISOTHIOCYANATES

A. CONDENSATION REACTIONS

Classically, perhaps the most widely used method for the preparation of isothiocyanate involves the reaction of amines with carbon disulfide in the presence of a base such as ethanolic aqueous ammonia or sodium hydroxide to form the appropriate salt of a dithiocarbamate. The conversion of a dithiocarbamate to an isothiocyanate may be carried out by a variety of reagents such as copper sulfate, ferrous sulfate, zinc sulfate, or lead nitrate. A procedure involving the use of lead nitrate for the general preparation of isothiocyanates has been described. This reaction involves the reactions shown in Eqs. (15) and (16).

$$(15)$$

$$(16)$$

Instead of ammonia, aqueous sodium hydroxide or strong organic bases have been used. Alkyl isothiocyanates are prepared by the oxidation of dithiocarbamates with hydrogen peroxide or oxygen.

The dithiocarbamate may be decomposed by the formation of a carboethoxy derivative in the so-called Kaluza reaction, as in Eq. (19). Generally, the reaction has been carried out for both aliphatic and aromatic isothiocyanates, although evidently it cannot be used with aromatic amines containing strong electron-withdrawing groups such as *p*-cyano or *p*-nitro groups. In this reaction, sufficient time has to be allowed for the formation of dithiocarbamate (e.g., 15 min for the reaction of *N*,*N*-dimethylaniline to as long as 7 days for the reaction of *β*-naphthylamine) [4].

12-4. Preparation of *p*-Chlorophenyl Isothiocyanate [4]

$$Cl-\langle\!\!\!\bigcirc\!\!\!\rangle-NH_2 + CS_2 + (C_2H_5)_3N \longrightarrow$$

$$\left[Cl-\langle\!\!\!\bigcirc\!\!\!\rangle-NH-\overset{\overset{S}{\|}}{C}-S\right]^{-} [(C_2H_5)_3NH]^{+} \quad (17)$$

$$\left[Cl-\langle\!\!\!\bigcirc\!\!\!\rangle-NH-\overset{\overset{S}{\|}}{\underset{\underset{S}{|}}{C}}\right]^{-} [(C_2H_5)_3NH]^{+} + Cl-\overset{\overset{O}{\|}}{C}-OC_2H_5 \longrightarrow$$

$$Cl-\langle\!\!\!\bigcirc\!\!\!\rangle-NH-\overset{\overset{S}{\|}}{C}-S-\overset{\overset{O}{\|}}{C}-OC_2H_5 + [(C_2H_5)_3NH]^{+}Cl^{-} \quad (18)$$

$$Cl-\langle\!\!\!\bigcirc\!\!\!\rangle-NH-\overset{\overset{S}{\|}}{C}-S-\overset{\overset{O}{\|}}{C}-OC_2H_5 \xrightarrow{\text{base}}$$

$$Cl-\langle\!\!\!\bigcirc\!\!\!\rangle-NCS + COS + C_2H_5OH \quad (19)$$

In a well ventilated hood, 12.8 gm (0.1 mole) of *p*-chloraniline is dissolved in a minimum amount of benzene and treated with 6.6 ml (0.1 mole) of carbon disulfide and 14 ml (0.1 mole) of triethylamine. The solution is then cooled to 0°C and the low temperature is maintained for 72 hr until the formation of the triethylammonium dithiocarbamate salt has been completed. The solution is then filtered and the solid is washed with anhydrous ether and air dried. Yield 83%.

The salt is then dissolved in approximately 75 ml of chloroform, treated with 14 ml (0.1 mole) of triethylamine, and cooled again to

0°C. To this solution is then added dropwise 10.2 ml (0.1 mole) of ethyl chlorocarbonate over a 15-min period with hand stirring. The resulting solution is stirred at 0°C for 10 min and is then allowed to warm to room temperature during a 1-hr period. The solution is then washed with 3 *M* hydrochloric acid solution, twice with water, and is then dried with sodium sulfate.

The chloroform is then evaporated under reduced pressure and the residual *p*-chlorophenyl isothiocyanate is recrystallized from ethanol. Yield of this step is 70% of theory, mp 46.5°C.

A quite general reaction for the preparation of isothiocyanates involves the use of thiophosgene. Since this material is a liquid, its handling is somewhat simpler than that of phosgene used in the synthesis of isocyanates. Its toxicity is believed to be as great as that of phosgene, if not greater.

REFERENCES

1. E. Grigat and R. Pütter, *Chem. Ber.* **97**, 3012 (1964); **98**, 1359, 2619 (1965).
2. R. L. Shriner and R. F. B. Cox, *J. Am. Chem. Soc.* **53**, 1601 (1931); R. L. Shriner and W. H. Horne, *ibid.* p. 3186; R. L. Shriner, W. H. Horne, and R. F. B. Cox, *Org. Synth. Collect. Vol.* **2**, 453 (1944).
3. P. Allen, Jr., *J. Am. Chem. Soc.* **57**, 198 (1935).
4. J. F. Hodgkins and W. Preston Reeves, *J. Org. Chem.* **29**, 3098 (1964).
5. D. H. Chadwick and T. H. Cleveland, *Kirk-othmer Encycl. Chem. Technol. 3rd ed.*, **13**, 789–818 (1981).

13

AMINES

From S. R. Sandler and W. Karo, *Organic Functional Group Preparations*, Vol. I, 2d ed. (New York, 1983), 377ff., by permission of Academic Press, Inc.

1. INTRODUCTION TO CONDENSATION REACTIONS

A. HOFMANN ALKYLATION OF AMMONIA AND AMINES—TREATMENT OF AMINES WITH HALIDES

The treatment of ammonia and amines with alkyl halides, dialkyl sulfate, or alkyl p-toluenesulfonates we classify as "Hofmann alkylations" to distinguish this fundamental method of preparing amines from other reactions which were discovered by that great organic chemist and which also have his name associated with them.

The treatment of ammonia with alkyl halide normally gives rise to a mixture of primary, secondary, and tertiary amines, and the quaternary ammonium halide salt. The reaction sequence is usually given by Eqs. $(1-4)$

$$RX + NH_3 \longrightarrow RNH_2 + HX \tag{1}$$

$$RX + RNH_2 \longrightarrow R_2NH + HX \tag{2}$$

$$RX + R_2NH \longrightarrow R_3N + HX \tag{3}$$

$$RX + R_3N \longrightarrow R_4N^+ X^- \tag{4}$$

Fortunately, from the preparative standpoint, when R is ethyl or a higher alkyl group, the boiling points of the various amines are sufficiently far apart to permit their separation by fractional distillation. By adjustment of the mole ratio, reaction temperatures, times, and other reaction conditions, it is frequently possible to control the reaction so that any one of these classes of amines becomes the predominant product.

In general, Hofmann alkylations are carried out with appropriate alkyl halides or dialkyl sulfates. These reagents may have to be prepared from the related alcohols, often by rather troublesome methods. A more convenient conversion of alcohols to amines involves the alkylation of amines with toluenesulfonate esters of alcohols. These "tosylates" are generally prepared quite easily and may then be used as alkylating agents [1–14]. If the alcohol which is to be converted to an amine has the proper structural features, the Ritter reaction (see below) is another useful approach to the preparation of primary amines (by hydrolysis of the amide formed in the reaction) and of some secondary amines (by reduction of the amides).

As far as the nitrogen-bearing reaction component is concerned, liquid ammonia, aqueous ammonia, some aqueous amines, and aliphatic and aromatic amines both neat or in one of a host of solvents usually undergo Hofmann alkylations. Since the halides are usually water-insoluble, reactions in aqueous media are sometimes difficult although the addition of alcohols or anionic surfactants and vigorous agitation is helpful. Phase-transfer catalysts in the case of alkylation reactions may be problematical. Many phase-transfer catalysts are cationic surfactants such as quaternary ammonium salts. Consequently there may be a scrambling of alkyl groups in the course of the reaction.

In connection with product work-up, the fact that many amine hydrochlorides are soluble in chloroform while ammonium chloride is insoluble is sometimes useful.

To prepare a primary amine by alkylation of ammonia, the level of ammonia is kept high to reduce the formation of secondary and tertiary amines. The preparation of *n*-butylamine is a typical example of the procedure.

13-1. Preparation of *n*-Butylamine [7]

$$CH_3CH_2CH_2CH_2Br + NH_3 \longrightarrow CH_3CH_2CH_2CH_2NH_2 + HBr \qquad (5)$$

In a 12-liter flask fitted with a stirrer, addition funnel, gas inlet tube, and reflux condenser is placed 8 liters of 90% ethanol. The reaction system is placed in a hood and ammonia is run in with constant stirring until the flask has gained about 300 gm. Then 68.5 gm (0.5 mole) of purified *n*-butyl bromide is added rapidly. Then, while a slow stream of ammonia is passed through the flask, an additional 1438.5 gm (10.5 moles) of *n*-butyl bromide is added continuously at a rate of approximately 17 gm/hr. After the addition has been completed, the flask is stirred for an additional 2 days. Then the reaction mixture is distilled to remove ethanol, and, after approximately 4 liters of ethanol have been distilled off, the flask is cooled and the precipitating ammonium bromide is separated. Another 4 liters of ethanol are then distilled off and more ammonium bromide is filtered off. About 1 liter of solution remains in the flask. To this is added 1 liter of water and the last traces of ethanol are removed by distillation. If necessary, this step is repeated until all of the ethanol has been removed.

To the cooled residue is added a cold solution of 240 gm (6 moles) of sodium hydroxide in 1 liter of water and the mixture is distilled until the lower boiling fraction has come over. This distillate is dried over fused potassium hydroxide.

The pot residue is cooled and the remaining amine layer is separated and combined with the distillate.

The dried amine is then fractionally distilled through a glass-helix-packed distillation column. The fraction boiling at 76.5°C (742 mm) is collected. Yield 191.7 gm (47% of theory), n_D^{20} 1.4008.

B. HOFMANN ALKYLATION OF AMMONIA AND AMINES—REACTION OF TOSYLATES WITH AMINES

An example of the preparation of a secondary amine using a tosylate as the alkylating agent is the preparation of N-methyldihydropyran-2-methylamine [4]. The use of mesylates should give the same results.

13-2. Preparation of N-Methyldihydropyran-2-methylamine [4]

$$(6)$$

To 34.5 gm (0.30 mole) of dihydropyran-2-methanol dissolved in 200 ml of pyridine, is added 75 gm (0.39 mole) of p-toluenesulfonyl chloride. The reaction mixture is warmed to 50°C for $\frac{1}{2}$ hr. After this time, the mixture is cooled to room temperature to produce a white precipitate of pyridinium hydrochloride. This co-product is removed by filtration and the excess pyridine is separated by evaporation under reduced pressure. The residual product is recrystallized from ethanol. Yield of dihydropyran-2-methyl tosylate is 48 gm (59%), mp 47°–48°C.

In a cooled steel pressure bomb are placed 27 gm (0.107 mole) of dihydropyran-2-methyl tosylate and 14 gm (0.45 mole) of methylamine dissolved in 200 ml of absolute methanol. The bomb is sealed and heated with shaking to 125°C for 1 hr. After cooling, the sealed bomb is carefully opened in a hood and the reaction mixture is concentrated in a vacuum evaporator. After removal of the solvent, the semisolid is made basic with 20% solution of sodium hydroxide in water and continuously extracted with ether for 48 hr. The ether layer is dried with potassium carbonate and reduced in volume to yield a crude product. The crude product is purified by distillation under reduced pressure to afford 6.3–8.4 gm (30–40% of theory) of *N*-methyldihydropyran-2-methylamine, bp 60°C (17 mm). During the distillation, the pot temperature should not exceed 150°C to minimize spontaneous decomposition of the crude product.

C. HOFMANN ALKYLATION—MONOALKYL DERIVATIVES OF DIAMINES

The preparation of monoalkyl derivatives of diamines represents something of a synthetic problem. Procedure 13-3 involves acetylation of ethylenediamines under mild conditions to the monoacetyl ethylenediamine, protection of the second amino group by a Hinsberg reaction with benzenesulfonyl chloride, followed by N-alkylation of the acetamido group, and recovery of the *N*-alkylethylenediamine by acid hydrolysis [15].

13-3. Preparation of Monomethylethylenediamine [15]

$$NH_2-CH_2CH_2-NH_2 + CH_3\overset{\overset{\displaystyle O}{\|}}{C}-OC_2H_5 \longrightarrow$$

$$NH_2-CH_2CH_2NH\overset{\overset{\displaystyle O}{\|}}{C}CH_3 + C_2H_5OH \quad (7)$$

$$NH_2-CH_2CH_2-NH\overset{\overset{\displaystyle O}{\|}}{C}-CH_3 + \langle\!\!\bigcirc\!\!\rangle-SO_2Cl \xrightarrow{\text{NaOH}}$$

$$\langle\!\!\bigcirc\!\!\rangle-SO_2NH-CH_2CH_2NH\overset{\overset{\displaystyle O}{\|}}{C}-CH_3 + HCl \quad (8)$$

$$\text{C}_6\text{H}_5\text{—SO}_2\text{—NH—CH}_2\text{CH}_2\text{NH—}\overset{\overset{\displaystyle O}{\|}}{\text{C}}\text{—CH}_3 + \text{CH}_3\text{I} \xrightarrow{\text{KOH}}$$

$$\text{C}_6\text{H}_5\text{—SO}_2\text{NH—CH}_2\text{CH}_2\underset{\underset{\displaystyle \text{CH}_3}{|}}{\text{N}}\overset{\overset{\displaystyle O}{\|}}{\text{C}}\text{—CH}_3 + \text{KI} + \text{H}_2\text{O} \quad (9)$$

$$\text{C}_6\text{H}_5\text{—SO}_2\text{NH—CH}_2\text{CH}_2\text{—}\underset{\underset{\displaystyle \text{CH}_3}{|}}{\text{N}}\text{—}\overset{\overset{\displaystyle O}{\|}}{\text{C}}\text{—CH}_3 + 2\,\text{H}_2\text{O} \xrightarrow{2\,\text{HCl}}$$

$$\text{HCl·NH}_2\text{—CH}_2\text{CH}_2\text{—NHCH}_3\text{·HCl} + \text{C}_6\text{H}_5\text{—SO}_3\text{H} + \text{CH}_3\text{CO}_2\text{H} \quad (10)$$

$$\text{HCl·NH}_2\text{CH}_2\text{CH}_2\text{NHCH}_3\text{·HCl} + 2\,\text{NaOH} \longrightarrow$$

$$\text{NH}_2\text{CH}_2\text{CH}_2\text{NHCH}_3 + 2\,\text{NaCl} + 2\,\text{H}_2\text{O} \quad (11)$$

To 306 gm (3 moles) of monoacetylethylenediamine (for preparation see Chapter 11, Amides) dissolved in 306 gm of water are added slowly and simultaneously 530 gm (3 moles) of benzenesulfonyl chloride and 1200 gm (3 moles) of a 10% aqueous solution of sodium hydroxide. After standing for several hours, this solution is faintly acidified with mineral acid and the product is separated by filtration. The N-benzenesulfonyl-N'-acetylethylenediamine is recrystallized from dilute ethanol, enough of the ethanol being used to prevent oiling out of the amide. A trace of the dibenzenesulfonylethylenediamine arising from traces of ethylenediamine present in the starting material is separated by filtration of the hot recrystallizing mixture. The pure N-benzenesulfonyl-N'-acetylethylenediamine separates on cooling and is collected by filtration. Yield, 500 gm (69%), mp 103°C.

To a boiling solution of 35 gm (0.53 mole) of 85% potassium hydroxide dissolved in 200 ml of absolute alcohol is added 121 gm (0.5 mole) of N-benzenesulfonyl-N'-acetylethylenediamine. To this solution is added 142 gm (1 mole) of methyl iodide dropwise over a 15 min period. The mixture is refluxed for 2 hr after which the precipitated potassium iodide is separated by filtration of the cooled reaction mixture. The filtrate is steam-distilled until the excess methyl iodide and the ethanol are completely removed.

The remaining alkylated sulfonamide is refluxed for 12 hr with 500 ml of concentrated hydrochloric acid which is replenished from time to time with additional amounts of concentrated acid. The hydrolysate is distilled nearly to dryness under reduced pressure and, after addition of an excess of sodium hydroxide, a concentrated water solution of the amine is distilled over. Dry sodium hydroxide is added to the distillate until the amine appears as a separate phase, which is then drawn off and dried over fresh sodium hydroxide. Finally the product may be dried by refluxing over metallic sodium. After cooling and filtration, the amine is obtained as a water-white liquid by fractional distillation of the filtrate under reduced pressure from a fresh piece of sodium; bp 115°–116°C (157 mm). The yield of monomethylethylenediamine is 28 gm (80% of theory).

2. MISCELLANEOUS CONDENSATION REACTIONS

Methods for the preparation of primary amines free of secondary and tertiary amines have occupied the efforts of organic chemists since the days of Hofmann. Several of these have at least a formal resemblance to the Hofmann alkylation.

A. HYDROBORATION

Organoboranes derived from terminal olefins or relatively unhindered olefins are readily converted to the corresponding amine by treatment with chloroamine or hydroxylamine-O-sulfonic acid [16].

Since this early work had been carried out in tetrahydrofuran, in which hydroxylamine-O-sulfonic acid is insoluble, the reaction could not be carried out with relatively hindered olefins. Consequently, it was not possible to take advantage of the highly stereospecific nature of this application of the hydroboration reaction for all types of olefins. More recent work, however, making use of diglyme as the solvent, has extended the reaction [17]. A typical example of the reaction is the preparation of trans-2-methylcyclohexylamine. In Eq. (12), the symbol HB represents the hydroboration reagent. The intermediate hydroborated product is only sketched in for clarity. While the directions of Preparation 13-4 follow closely those of Ref. 17, we suggest that consideration be given to modifying the apparatus. We believe that by the use of serum caps over a standard taper 29/42 joint, nitrogen may be introduced into the flask through a syringe needle. The addition of various reagents may also be done by use of a syringe fitted with an appropriate needle through a serum cap.

13-4. Preparation of *trans*-2-Methylcyclohexylamine [17]

$$\tag{12}$$

A dry 250-ml flask equipped with a dropping funnel, condenser, and magnetic stirrer is flushed with nitrogen. A solution of 0.78 gm (20.6 mmoles) of sodium borohydride and 20 ml of diglyme is introduced, followed by 4.8 gm (50 mmoles) of 1-methylcyclohexene. The flask is immersed in an ice-water bath and hydroboration is carried out by the dropwise addition of 3.90 gm (27.5 mmoles) of boron trifluoride etherate. The solution is then stirred at room temperature for 3 hr. Then a solution of 6.22 gm (55 mmoles) of hydroxylamine-*O*-sulfonic acid in 25 ml of diglyme is added, and the solution is heated to 100°C for 3 hr. The reaction mixture is cooled, cautiously treated with 20 ml of concentrated hydrochloric acid, and then poured into 200 ml of water. The acidic phase is extracted with ether to remove diglyme and the residual boronic acid. The aqueous solution is then made strongly alkaline with sodium hydroxide and the amine is extracted with ether. The ether extract is dried over potassium hydroxide and the dried product solution is fractionally distilled. After removal of the ether, 5 gm (45% of theory) of *trans*-2-methylcyclohexylamine is isolated, bp 148°C (750 mm).

[Reprinted from: M. W. Rathke, N. Inoue, K. R. Varma, and H. C. Brown, *J. Am. Chem. Soc.* **88,** 2870 (1966). Copyright 1966 by the American Chemical Society. Reprinted by permission of the copyright owners.]

B. GABRIEL CONDENSATION AND ING–MANSKE MODIFICATION

Another method for the preparation of primary amines involves the alkylation of potassium phthalimide according to procedures of Gabriel to give *N*-alkylphthalimides, which, on hydrolysis, afford the primary amine and phthalic acid. Since the hydrolysis is sometimes difficult, Ing and Manske [18] developed a modification in which the decomposition of the *N*-alkylphthalimide is carried out in the presence of hydrazine. While the alkylation step has been carried out either without solvent or in the presence of nonpolar high-boiling solvents, the modification [19] suggests the use of dimethylformamide as the solvent. In this solvent the reaction can often be carried out at relatively low temperatures and exhibits a mildly exothermic character. For less reactive alkyl halides, the reaction temperature may be varied.

The use of a phase-transfer catalyst such as hexadecyltributylphosphonium bromide permits the alkylation of potassium phthalimide in toluene in a liquid–solid system [20, 21].

For the sake of applicability, the preparation of α,δ-diaminoadipic acid using the Ing–Manske modification is given here, although the same reference indicates that, at least in this case, the Gabriel condensation and hydrolysis affords a higher yield.

13-5. Preparation of α,δ-Diaminoadipic Acid [19]

A mixture of 69 gm (0.21 mole) of dimethyl α,δ-dibromoadipate, 87 gm (0.47 mole) of potassium phthalimide, and 260 ml of dimethylformamide is gently heated. A mildly exothermic reaction starts at 50°C; however, sufficient heat is applied to maintain the reaction mixture for 40 min at 90°C. Then the reaction mixture is cooled, diluted with 300 ml of chloroform, and poured into 1200 ml of water. The chloroform layer is separated and the aqueous phase is extracted twice with 100-ml portions of chloroform. The combined chloroform extract is washed with 200 ml of 0.1 N sodium hydroxide, and 200 ml of water. Then the solution is dried over sodium sulfate. The chloroform is removed by concentration under reduced pressure to the point of incipient crystallization. The immediate addition of 300 ml of ether induces a rapid crystallization. The product is collected on a filter and washed with ether. Yield 87 gm (90.2%), mp over the range 160°–185°C. After three crystallizations from ethyl acetate and one from benzene, an apparently pure stereoisomer is obtainable which melts at 210.7°–211.4°C.

■ **CAUTION:** Aqueous hydrazine is a toxic reagent.

In a hood, a mixture of 4.64 gm (0.01 mole) of dimethyl α,δ-diphthalamidoadipate (mp 160°–185°C), 50 ml of methanol, and 1.2 ml (0.02 mole) of an 85% aqueous hydrazine solution is heated under reflux for 1 hr. After cooling, 25 ml of water is added and the methanol is removed by concentration under reduced pressure. After 25 ml of concentrated hydrochloric acid has been added to the residual aqueous suspension, the mixture is heated under reflux for 1 hr. After cooling to 0°C, crystalline phthalhydrazide is removed by filtration. The filtrate is concentrated under reduced pressure to remove hydrochloric acid and the moist residue is dissolved in 50 ml of water. A small amount of insoluble matter is removed by filtration and the clear filtrate is neutralized with 2 N sodium hydroxide. After cooling at 0°C for 12 hr, 1.4 gm (79.5%) of α,δ-diaminoadipic acid is obtained.

[Reprinted from J. C. Sheehan and W. A. Bolhofer, *J. Am. Chem. Soc.* **72,** 2786 (1950). Copyright 1950 by the American Chemical Society. Reprinted by permission of the copyright owner.]

C. RITTER REACTION

Nitriles may be reacted with certain types of alcohols or olefins in strongly acidic media to afford an alkylated amide. Since amides may be hydrolyzed to yield a free amine (as, for example, in Ferris *et al.* [22]), a facile method for converting olefins and alcohols to amines is available. The major drawback to the Ritter reaction is the fact that only those alcohols are suitable for the reaction which have branched chains on the carbon adjacent to the hydroxyl group. Thus, 2-propanol is the lowest aliphatic alcohol which undergoes the Ritter reaction. Olefins which are to be subjected to the Ritter reaction also require branching. The example here cited indicates the simplicity of the Ritter reaction as a means of preparing amines.

13-6. Preparation of 1-Methylcyclobutylamine [23]

$$CH_3CN + CH_2{-}C{=}CH_2 \quad \xrightarrow[H_2SO_4]{HOCCH_3} \quad CH_3C{-}NH{-}C{-}CH_2 \qquad (13)$$

$$CH_3C{-}NH{-}C{-}CH_2 \quad \xrightarrow{NaOH} \quad CH_3C{-}ONa + CH_2{-}C{-}NH_2 + H_2O \quad (14)$$

In a 500-ml three-necked flask equipped with a mechanical stirrer, dropping funnel, and reflux condenser, immersed in an ice–salt bath, are placed 9.0 gm (0.22 mole) of acetonitrile, 100 ml of glacial acetic acid, and 20 gm of concentrated sulfuric acid. To the cooled solution, 13.6 gm (0.20 mole) of methylenecyclobutane is slowly added with stirring. After the addition has been completed, stirring is continued for 1 hr at 20°C. The solution is then cooled, diluted with 300 ml of water, and sufficient sodium carbonate is added to render the solution basic. The aqueous phase is then extracted with five 50-ml portions of ether. The etheral solution is dried, filtered, and evaporated to give 17.7 gm (70% of theory) of white crystalline *N*-(1-methyl-cyclobutyl)acetamide.

To 400 ml of a 4 *N* solution of potassium hydroxide in ethylene glycol is added 10.0 gm (0.076 mole) of *N*-(1-methylcyclobutanol)-acetamide. The reaction mixture is heated to reflux for 48 hr. After this reaction period, the materials boiling below 180°C at atmospheric pressure are distilled off. This distillate is extracted continuously with ether. The ether extract is dried over potassium hydroxide. After removal of the ether, the residue is distilled to afford 3.1 gm (46% of theory) of 1-methylcyclobutylamine, bp 85.5°–86.0°C (464 mm).

3. REDUCTION REACTIONS

A. REDUCTION OF NITRILES

Nitriles have been reduced with hydrogen and various catalysts [24] and by chemical means. Among the chemical reducing agents, sodium and alcohol have been used [23]. Deuterated amines have been produced from nitriles by reduction with lithium aluminum deuteride [25]. The prepration of cyclopropylmethylamine with lithium aluminum hydride is a related example of this reduction [26].

13-7. Preparation of Cyclopropylmethylamine [26]

$$\triangleright\!\!-CN + [H] \xrightarrow{\text{LiAlH}_4} \triangleright\!\!-CH_2NH_2 \qquad (15)$$

Using extreme precautions for the reduction of nitriles with lithium aluminum hydride, to a slurry of 52.5 gm (1.4 moles) of lithium aluminum hydride in 1 liter of tetrahydrofuran is added 93.3 gm (1.4 moles) of cyclopropyl cyanide at such a rate as to cause moderate refluxing. The mixture is stirred vigorously for $5\frac{1}{2}$ hr and left overnight. Then the reaction mixture is worked up by the addition of an excess of an aqueous sodium hydroxide solution, the mixture is filtered, and the filter cake is washed well with tetrahydrofuran. Distillation of the filtrates and washings at atmospheric pressure afforded 78.4 gm (79%) of cyclopropylmethylamine, bp 84°–86°C (760 mm).

More recently, diborane–tetrahydrofuran has been suggested as a convenient reducing agent of nitriles [27].

B. REDUCTION OF NITRO COMPOUNDS

Particularly in the aromatic series, many amines have been prepared by the reduction of the corresponding nitro compounds. With the advances in the preparation of aliphatic nitro compounds, the techniques for reducing aliphatic nitro compounds have also been developed to afford aliphatic amines. In the aliphatic series, the techniques permit the preparation of a very large variety of amines. Since aliphatic nitro compounds are active methylene compounds they undergo typical reactions of active methylene groups, such as the various condensation reactions which gives rise to a variety of branched compounds involving the carbon adjacent to the nitro group.

A large number of reducing agents has been used for the reduction of nitro groups. For example, the classical Bechamp method of reducing nitro compounds involves reaction with iron and acetic acid [27a].

a. Reduction with Activated Iron and Water. Both from the laboratory and the industrial standpoints, reduction of aromatic nitro compounds with iron or iron compounds is of considerable importance. By use of a previously "activated" iron, many nitro aromatic compounds have been successfully reduced under very mild conditions (from the point of view of possible hydrolysis of substituents which may be present on the ring).

The preparation of 2-chloroaniline illustrates the method of activating iron for this reaction [28]. In a personal communication, Professor C. F. H. Allen suggested that a mixture of iron powder and rusty iron filings frequently works just as well as specially activated iron [29].

13-8. Preparation of 2-Chloroaniline [28]

$$\tag{16}$$

(a) Activation of Iron. In a hood in an open beaker cooled by an ice-water bath, 56 gm (1.0 mole) of granulated iron (40 mesh, i.e., approximate diameter of 0.64 mm) is rapidly stirred. To the metal, 10 ml of concentrated hydrochloric acid is added very slowly at such a rate that excessive heat is not generated in the course of the ensuing reaction. Stirring is at such a rate that the formation of lumps is prevented. After addition has been completed, the iron powder is allowed to dry thoroughly.

(b) Reduction. In a three-necked flask provided with a reflux condenser and an efficient stirrer, 5 gm (0.032 mole) of 2-chloronitrobenzene is dissolved in 200 ml of benzene. The third neck of the

flask is closed and the benzene solution is heated almost to boiling on a steam bath. Then the iron produced as described in step (a) is introduced while maintaining vigorous stirring. In the subsequent steps, vigorous stirring and refluxing are maintained. After $\frac{1}{2}$ hr, 1 ml of water is added to the reaction mixture. Thereafter, small quantities of water are introduced from time to time at such a rate that at the end of 7 hr 20 ml of water have been added. Refluxing is continued for an additional hour. The hot solution is then filtered, the residual iron is extracted three times with hot benzene, and the benzene extracts are combined with the filtered product solution. Crude amines may be obtained by distilling the benzene from the product, followed by vacuum distillation or crystallization of the amine.

Alternatively the benzene solution is cooled and gaseous hydrogen chloride is passed into the solution to precipitate the amine as the insoluble hydrochloride salt. The salt may then be collected on a filter and recrystallized from alcohol. The hydrochloride salt may then be placed in a suitable distillation apparatus. Concentrated sodium hydroxide solution is added and the product is isolated by steam distillation. In this particular preparation, the product was isolated as the hydrochloride salt, yield 3.7 to 4.8 gm (71–92%).

C. CLASSICAL LEUCKART REACTION

The reaction of formic acid or a variety of formic acid derivatives, such as formate salts and formamides, with ammonia or a variety of amines, as well as various amine derivatives and salts such as ammonium formate salts, and carbonyl compounds, results in the reductive alkylation of the amine in which the entering alkyl group is derived from the carbonyl compound. This reaction is known as the Leuckart reaction [30]. By proper selection of reagents, primary, secondary, and tertiary amines may be prepared. In general this reaction is carried out at elevated temperatures without further solvents. More recent work indicates that magnesium chloride and ammonium sulfate are particularly useful catalysts in the preparation of tertiary amines by the Leuckart reaction [31].

This method appears to be suitable for the preparation of a wide variety of tertiary amines including those derived from α,β-unsaturated carbonyl compounds [31].

13-9. Preparation of 1-(Methyl-3-phenylpropyl)piperidine [31]

In a three-necked flask equipped with a mechanical stirrer, thermometer, and a condenser set downward for distillation, are mixed 19.25 gm (0.13 mole) of 4-phenyl-2-butanone, 58.8 gm (0.52 mole) of formpiperidide, 4.47 gm (0.022 mole) of magnesium chloride hexahydrate, and 7.03 gm (0.13 mole) of 85% formic acid. The mixture is heated gradually and volatile constituents are removed until the pot temperature approximates the boiling point of the formpiperidide. A reflux condenser is then installed in place of the distillation condenser and the mixture is refluxed with stirring for 8 hr. Then the reaction mixture is poured into dilute hydrochloric acid. Unreacted ketone is removed by steam distillation. The ketone-free aqueous residue is then made strongly basic and the amine is distilled out with steam. The steam distillate is saturated with sodium chloride and extracted with ether. The ether extract, after drying over solid potassium hydroxide, is distilled to remove the solvent. The residual amine is then fractionally distilled at 176°–177°C at 25 mm. Yield 15.3 gm (57%).

D. ESCHWEILER–CLARKE MODIFICATION OF THE LEUCKART REACTION

A modification of the Leuckart reaction for the preparation of methylated amines which requires less drastic reaction conditions, known as the Eschweiler–Clarke reaction, involves the use of aqueous formaldehyde and an excess of formic acid as reducing agent, along with the appropriate amine. Usually the reaction conditions are such that only the methylated tertiary amine can be isolated [13]. An example of this reaction is the preparation of 1-methyl-2-(*p*-tolyl)piperidine [32].

13-10. Preparation of 1-Methyl-2-(*p*-tolyl)piperidine [32]

$$ \text{(18)} $$

A mixture of 17.5 gm (0.1 mole) of 2-(*p*-tolyl)piperidine, 30 gm (0.59 mole) of 90% formic acid, and 25 gm (0.24 mole) of 35% formaldehyde is heated under reflux for 12 hr on a steam bath. The reaction mixture is cooled and 15 ml of concentrated hydrochloric acid is added. Heating under reflux is then continued for another 5 hr. Then the reaction mixture is cooled, made strongly basic with sodium hydroxide solution, and extracted with benzene. The benzene extract is dried over anhydrous sodium or potassium hydroxide. The benzene solution is then distilled. After removal of the solvent, the product distills at 117°–118°C (6 mm). Yield 17.8 gm (94%).

By application of methods of experimental design, optimum conditions have been found for the preparation of a secondary amine [33]. We particularly recommend the paper of Meiners *et al.* [33] to the reader, inasmuch as it is one of the few reports describing a statistical design of experiments to a problem in organic synthesis, a procedure which is now finding more extensive application in the laboratory as well as in industry.

4. REARRANGEMENT REACTIONS

A. HOFMANN REARRANGEMENT

The Hofmann rearrangement with hypohalites, the Schmidt reaction, and the Curtius reaction are three closely related methods of preparing amines. While all of them have been used extensively, since they afford a means of converting carboxylic acid derivatives to amines with one less carbon atom, they all are considered somewhat hazardous reactions. Of the three, probably the Hofmann rearrangement is the least hazardous. Both the Schmidt and the Curtius reactions may, under certain circumstances, be extremely hazardous. The Hofmann rearrangement has been reviewed [34, 35]. The overall reaction is represented in Eqs. (19) and (20).

$$\underset{\substack{\| \\ O}}{R C}-NH_2 + NaOX \longrightarrow \left[\underset{\substack{\| \\ O}}{R C}-N\begin{array}{c} \diagup \\ \diagdown \end{array} \right] \longrightarrow R-N=C=O \qquad (19)$$

$$R-N=C=O + R'OH \longrightarrow R-NH-\underset{\substack{\| \\ O}}{C}OR' \xrightarrow{H_2O} RNH_2 + R'\underset{\substack{\| \\ O}}{C}OH \qquad (20)$$

While many preparations involve the use of sodium hypobromite [36], the use of sodium hypochlorite has much to recommend it.

In the case of higher molecular weight fatty acid derivatives, ureas form when some of the free amine product reacts with the isocyanate intermediate of the reaction. Therefore, by the use of inert solvents such as dioxane, the mutual solubilities of isocyanates and the product amine are reduced and the amine may be isolated, as will be shown in the example given below [35].

13-11. Preparation of Nonylamine [35]

$$CH_3(CH_2)_8\underset{\substack{\| \\ O}}{C}-NH_2 + NaOCl \longrightarrow CH_3(CH_2)_7CH_2NH_2 \qquad (21)$$

(a) Preparation of Standard Hypochlorite Solution. In a distilling flask equipped with a dropping funnel is placed 6.7 gm of potassium permanganate. The side arm of the flask is joined by a glass-to-glass connection to a tube dipping below the surface of a solution containing 16 gm (0.4 mole) of sodium hydroxide dissolved in 100 ml of water and cracked ice contained in a graduated cylinder. The cylinder is mounted with an ice bath.

Fifty milliliters of concentrated hydrochloric acid is admitted slowly through the dropping funnel so as to produce a slow stream of chlorine. When all the acid has been added, the dropping funnel is closed and the content of the flask is heated gently with a small flame until the reflux point is a little below the juncture of the side arm. A Pyrex–wool plug below the side arm serves to prevent entrained acid from being carried over. The hypochlorite solution is then made up to 160 ml. The solution contains slightly over 0.1 mole of sodium hypochlorite and 0.2 mole of excess sodium hydroxide.

(b) Hofmann Rearrangement. To a round-bottomed flask, equipped with reflux condenser, magnetic stirrer, thermometer, and addition funnel, containing 17.1 gm (0.1 mole) of capramide in 80 ml of purified dioxane is added 160 ml of the hypochlorite solution prepared as above. The mixture is stirred vigorously and heated to 45°C. The temperature then rises spontaneously to 65°C in 2 min. Stirring is continued without external heating for 2 hr, at which time the temperature has dropped back to 42°C. On cooling, an oily product layer separates. This is removed and the aqueous layer is extracted with benzene. The oily amine product layer and the benzene extract are combined. Then the base is extracted into 100 ml of 1 *N* hydrochloric acid. The resulting acid layer is separated and to it is added a concentrated sodium hydroxide solution. The liberated base is again taken up in benzene, and the benzene extract is dried over potassium hydroxide pellets and filtered. The benzene solution is then saturated with hydrogen chloride gas. The mixture is then concentrated to a volume of 50 ml with an air stream and 50 ml of dry ether is added. The amine hydrochloride is collected on a filter. Yield 11.9 gm (66.4%), mp 185°–186°C.

[Reprinted from: E. Magnien and R. Baltzly, *J. Org. Chem.* **23,** 2029 (1958). Copyright 1958 by the American Chemical Society. Reprinted by permission of the copyright owner.]

REFERENCES

1. V. C. Sekera and C. S. Marvel, *J. Am. Chem. Soc.* **55,** 345 (1933).
2. D. D. Reynolds and W. O. Kenyon, *J. Am. Chem. Soc.* **72,** 1591 (1950).
3. A. Burger, S. E. Zimmerman, and E. J. Ariens, *J. Med. Chem.* **9,** 469 (1966).
4. H. Franke and R. Partch, *J. Med. Chem.* **9,** 643 (1966); R. Partch, private communication (1966).
5. K. Tamaki, S. Yada, and S. Kudo, *Yuki Gosei Kagaku Kyokaishi* **30**(2), 175 (1972); *Chem. Abstr.* **76,** 126065 (1972).
6. I. Ganea and R. Taranu, *Stud. Univ. Babes-Bolyai, Ser. Chem.* **16,** (2), 89 (1971); *Chem. Abstr.* **76,** 112 868 (1972).
7. F. C. Whitmore and D. P. Langlois, *J. Am. Chem. Soc.* **54,** 3441 (1932).
8. N. Bortnick, L. S. Luskin, M. D. Hurwitz, W. E. Craig, L. J. Exner, and J. Mirza, *J. Am. Chem. Soc.* **78,** 4039 (1956).
9. G. F. Hennion and C. V. DiGiovanna, *J. Org. Chem.* **30,** 2645 (1965).
10. I. E. Kopka, Z. A. Fataftah, and M. W. Rathke, *J. Org. Chem.* **45,** 4616 (1980).

11. A. C. Cope and P. H. Towle, *J. Am. Chem. Soc.* **71**, 3423 (1949).

12. G. M. Coppinger, *J. Am. Chem. Soc.* **76**, 1372 (1954).

13. A. C. Cope, E. Ciganek, L. J. Fleckenstein, and M. A. P. Meisinger, *J. Am. Chem. Soc.* **82**, 4651 (1960).

14. L. L. Melhado, *J. Org. Chem.* **46**, 1920 (1981).

15. S. R. Aspinall, *J. Am. Chem. Soc.* **63**, 852 (1941).

16. H. C. Brown, W. R. Heydkamp, E. Breuer, and W. S. Murphy, *J. Am. Chem. Soc.* **86**, 3565 (1964).

17. M. W. Rathke, N. Inoue, K. R. Varma, and H. C. Brown, *J. Am. Chem. Soc.* **88**, 2870 (1966); related reviews are: G. Zweifel and H. C. Brown, *Org. React.* **13**, 1 (1963); H. C. Brown *Tetrahedron* **12**, 117 (1961); "Hydroboration." Benjamin, New York, 1962. L. Verbit and P. J. Heffron, *J. Org. Chem.* **32**, 3199 (1967).

18. H. R. Ing and R. H. F. Manske, *J. Chem. Soc.* p. 2348 (1926).

19. J. C. Sheehan and W. A. Bolhofer, *J. Am. Chem. Soc.* **72**, 2785 (1950).

20. D. Landini and F. Rolla, *Synthesis* p. 389 (1976).

21. E. Ciuffarin, M. Isola, and P. Leoni, *J. Org. Chem.* **46**, 3064 (1981).

22. A. F. Ferris, O. L. Salerni, and B. A. Schultz, *J. Med. Chem.* **9**, 391 (1966).

23. E. F. Cox, M. C. Caserio, M. S. Silver, and J. D. Roberts, *J. Am. Chem. Soc.* **83**, 2719 (1961).

24. M. Freifelder and R. B. Hasbroack, *J. Am. Chem. Soc.* **82**, 696 (1960); F. E. Gould, G. S. Johnson, and A. F. Ferris, *J. Org. Chem.* **25**, 1659 (1960).

25. E. A. Halevi, M. Nussim, and A. Ron, *J. Chem. Soc.* p. 866 (1963).

26. P. M. Carabateas and L. S. Harris, *J. Med. Chem.* **9**, 6 (1966).

27. D. L. Ladd and J. Weinstock. *J. Org. Chem.* **46**, 203 (1981).

27a. H. Koopman, *Recl. Trav. Chim. Pays-Bas* **80**, 1075 (1961).

28. S. E. Hazlet and C. A. Dornfeld, *J. Am. Chem. Soc.* **66**, 1781 (1944).

29. C. F. H. Allen, private communication.

30. M. L. Moore, *Org. React.* **5**, 301 (1959); see also Meiners *et al.* [33] for references to reviews on the mechanism of the Leuckart reaction.

31. J. F. Bunnett and J. L. Marks, *J. Am. Chem. Soc.* **71**, 1587 (1949); J. F. Bunnett, J. L. Marks, and H. Moe, *ibid.* **75**, 985 (1953).

32. L. D. Quin and F. A. Shelburne, *J. Org. Chem.* **30**, 3137 (1965).

33. A. F. Meiners, C. Bolze, H. L. Scherer, and F. V. Morriss, *J. Org. Chem.* **23**, 1122 (1958).

34. E. S. Wallis and J. F. Lane, *Org. React.* **3**, 267 (1946).

35. E. Magnien and R. Baltzly, *J. Org. Chem.* **23**, 2029 (1958).

36. A. C. Cope, T. T. Foster, and P. H. Towle, *J. Am. Chem.* **71**, 3929 (1949).

14

HYDRAZINE DERIVATIVES, HYDRAZONES, AND HYDRAZIDES

1. INTRODUCTION

The simple molecule hydrazine is capable of forming at least six classes of derivatives. Table I gives an overview of the methods of preparing many of these substituted hydrazines. In this connection it must be pointed out that the monomethyl- and monoethylhydrazines are usually prepared by special methods [1, 2]. When a large excess of hydrazine is used the more general alkylation of hydrazine given here will also lead to mono-substituted products.

The six classes of hydrazine derivatives are:

1. Substituted hydrazine with a free amino group, e.g., the simple monoaliphatic and monoaromatic hydrazines and the 1,1-disub-

From S. R. Sandler and W. Karo, *Organic Functional Group Preparations*, Vol. I, 2d ed. (New York, 1983), 434ff., by permission of Academic Press, Inc.

TABLE I

SYNTHESIS OF SUBSTITUTED HYDRAZINES AND HYDRAZINE DERIVATIVES

Product type	Reaction type	Reactant	Reagent	References
Monosubstituted hydrazine	Condensation	Hydrazine (Excess)	Dialkyl sulfate	7, 16
	Condensation	Hydrazine (excess)	Higher alkyl halide	16–18
	Condensation	Hydrazine	Activated aryl halide	19–21
	Condensation	Primary amine	Chloramine	22–26
	Condensation	Primary amine	Hydroxyl-amine-O-sulfonic acid	15–15a
	Reduction	N-Alkyl-N-nitrosoamine	—	27–35
	Reduction	Hydrazone of aldehyde	—	36, 37, 38
	Reduction	Hydrazone of ketone	—	17, 21, 39
	Reduction	Aromatic diazonium salts	—	40–43
1,1-Disubstituted hydrazine	Condensation	Hydrazine	Alkyl halide (2 moles)	18
	Condensation	Secondary amine	Chloramine	23–25
	Reduction	N,N-Dialkyl-N-nitrosoamine	—	27–35, 44
1,2-Disubstituted hydrazine	Condensation	Hydrazine	Activated aryl halide (2 moles)	7
	Condensation	Dihydrazide	Alkyl halide(s)	36, 45, 46
	Reduction	Monosubstituted hydrazone of aldehyde		36, 37, 38
	Reduction	Azo compounds	—	47, 48
	Reduction (bimolecular)	Aromatic nitro compounds	—	49
1,1,2-Trisubstituted hydrazine	Condensation	Hydrazine	Alkyl halide (3 moles)	18
	Reduction	Disubstituted hydrazone of aldehydes	—	37

(*continues*)

TABLE I (*Continued*)

Product type	Reaction type	Reactant	Reagent	References
	Reduction	Monosubstituted hydrazone of ketones	—	*c*
	Reduction	Acyl hydrazone of ketone	—	50
Tetrasubstituted hydrazine	Condensation	Hydrazine	Alkylhalide (4 moles)*a*	18
	Reduction	Disubstituted hydrazone of ketone	—	*c*
	Oxidation	Secondary amine	—	50a
Hydrazone	Condensation	Hydrazine	Aldehyde or ketone	16, 51, 52
	Condensation	Hydrazine	Aldehyde or ketone	50
	Condensation	Substituted hydrazine	Aldehyde or ketone	51, 52
	Exchange reaction	Substituted hydrazone	Hydrazine	53
Hydrazide	Condensation	Hydrazine	Ester	51, 54, 55, 56
	Condensation	Hydrazine	Acyl halide*b*	57
	Condensation	Hydrazine	Amide	58, 59
	Condensation	Substituted hydrazine	Esters	*c*
	Condensation	Substituted hydrazine	Acyl halide	*c*

*a*Excess of alkyl halide may lead to monoquaternized compounds [18].
*b*Frequently leads of 1,2-diacyl hydrazines [36, 45].
*c*Suggested reaction procedures.

stituted hydrazines, sometimes referred to as *N,N*-disubstituted hydrazines or "unsymmetrically disubstituted hydrazines"
2. 1,2-Disubstituted hydrazines, frequently referred to as "hyrdrazo compounds" or "symmetrically disubstituted hydrazines"
3. Trisubstituted hydrazines
4. Tetrasubstituted hydrazines
5. Hydrazones
6. Hydrazides

Hydrazine and its derivatives are well known as reagents used for the identification of carbonyl compounds. Of these, phenylhydrazine, 2,4-dinitrophenyl hydrazine, and Girard's reagent $[(CH_3)_3N^+CH_2CONHNH_2]CL^-$ are of particular importance. Many heterocyclic compounds with two adjacent nitrogens, such as pyrazoles and pyrazolines, may be considered hydrazine derivatives, and methods of preparation of hydrazine derivatives may be applicable to their synthesis. The detailed treatment of synthetic methods for such heterocyclic compounds is beyond the scope of this work, although a few examples of the synthesis of some heterocyclic compounds will be indicated.

References 3–14 are a selection of reviews on the chemistry of hydrazine derivatives.

Hydrazine itself may be subjected to a variety of facile substitution reactions because of its great nucleophilic character. It may react with alkyl halides and activated aryl halides to yield substituted hydrazines. With acyl halides, esters, and amides it may form hydrazides. With carbonyl compounds, hydrazones form. Since hydrazine is bifunctional in nature, both amino groups may undergo reaction. In the case of the reaction of hydrazine with two moles of a carbonyl compound, the products are termed "azines."

Many of the characteristic reactions of hydrazine may also be carried out with substituted hydrazines to afford a wide variety of products. In the case of reactions of substituted hydrazine derivatives, azine formation is ordinarily not possible.

In the case of the reaction of substituted hydrazines, such as phenylhydrazine with glucose or certain other carbohydrates, one molecule of phenylhydrazine reacts with the terminal aldehyde group to form a phenylhydrazone. A second molecule of phenylhydrazine oxidizes the penultimate carbinol grouping to a carbonyl and a third molecule of phenylhydrazine converts this second carbonyl to a phenylhydrazone. This class of di-phenylhydrazones is called an "osazone," a series of compounds not discussed in the present chapter.

Substituted hydrazines can be prepared by the reaction of amines with chloramine and with hydroxylamine-O-sulfonic acid. Of these two reagents, hydroxylamine-O-sulfonic acid has been introduced more recently and the scope of its reaction is still being explored [15]. The present authors believe that this is a more convenient reagent than chloramine. By its use, not only mono- but also 1,1-disubstituted hydrazine should be preparable. The synthesis and utilization of N-substituted chloramines and hydroxylamine-O-sulfonic acids require further exploration.

■ **CAUTION:** In working with hydrazine and many substituted hydrazines, one must bear in mind that many of these compounds are highly toxic. They may also be carcinogenic. Hydrazine has also been suggested as a component in rocket fuel. Consequently, extreme safety precautions must be taken.

Several classes of hydrazines have been prepared by a variety of reductive procedures of *N*-nitrosoamines and hydrazones of various carbonyl compounds. The latter procedure is of particular value, since by the judicious selection of the carbonyl compound and of a substituted hydrazine, hydrazones may be prepared which can lead to new hydrazines with up to three different substituents.

Other functional groups which have been reduced to substituted hydrazines are azides, carbazates (acyl hydrazones), azo compounds, and ketazines.

Since aromatic diazonium salts may be reduced to mono-aryl-substituted hydrazines, a simple route to a vast variety of *N*-arylhydrazines is available.

The use of hydrazine and substituted hydrazines in the preparation of hydrazones by reaction with carbonyl compounds is well known.

Of some interest is the fact that hydrazides also react with carbonyl compounds in an analogous fashion. Since poly(hydrazides) can be prepared from polymeric esters, resins can be prepared which are capable of separating carbonyl compounds from organic mixtures. Such resins might be used in a "carbonyl-exchange" column.

Polyhydrazides can be converted to poly(azides). These polymers find application in affinity chromatography and other biochemical procedures.

The best method of preparing hydrazides is by the reaction of hydrazines with esters. Acyl halides and amides may also be reacted with hydrazines. In the case of the Ing–Manske reaction, N-substituted phythalimides are reacted with hydrazine to generate a primary amine and the cyclic phthalhydrazide.

2. SUBSTITUTED HYDRAZINES

■ **CAUTION:** All reactions involving hydrazine, its hydrates, or its salts are extremely hazardous. Hydrazine, its salts, and its hydrates are toxic and considered cancer suspect agents. Extreme care should be exercised in working with these and related compounds and their derivatives.

The monoethylation of hydrazines given here was found possible when an excess of hydrazine in ethanol was used [16].

14-1. Preparation of Ethylhydrazine Hydrochloride [16]

$$NH_2-NH_2 + (CH_3CH_2)_2SO_4 \xrightarrow[\substack{C_2H_5OH \\ (HCl)}]{KOH} CH_3CH_2NHNH_2 \cdot HCl + C_2H_5OH + KHSO_4 \tag{1}$$

■ **CAUTION:** Both hydrazine and diethyl sulfate are extremely toxic and cancer suspect agents.

In a hood, in a three-necked flask fitted with a mechanical stirrer, an empty distillation column, topped with a total reflux–partial take-off distillation head, an addition funnel, and means to maintain a nitrogen atmosphere are placed 35 gm (0.63 mole) of potassium hydroxide, 30 ml (0.93 mole) of anhydrous hydrazine (95% active), and 60 ml of absolute ethanol. This mixture is cooled in an ice bath while being stirred mechanically.

From the addition funnel, 33 ml (0.25 mole) of acid-free diethyl sulfate is slowly added while maintaining a low temperature. After completion of the addition, the mixture is heated in a bath to 165°C to separate a mixture of hydrazine and ethylhydrazine by distillation. The distillate is cooled and made strongly acid by cautious addition of concentrated hydrochloric acid. The precipitating hydrazine hydrochloride is separated by filtration from the hot solution. The filtrate is then concentrated to half volume. A small amount of hydrochloric acid is added and the solution is allowed to cool. The precipitate that forms is washed in turn with small portions of concentrated hydrochloric acid, alcohol, and ether. The product is then dried in a vacuum desiccator over calcium chloride. A further crop of product may be obtained by concentrating the mother liquors to yield a total of 21 gm (87%) of ethylhydrazine hydrochloride.

The free hydrazine (see warnings above) may be prepared in 80% yields by treatment of the hydrochloride salt with base.

14-2. Preparation of 2-Pentylhydrazine [15]

$$CH_3CH_2-CH_2-\underset{\underset{CH_3}{|}}{CH}-NH_2 + NH_2OSO_3H \xrightarrow{KOH}$$

$$CH_3CH_2CH_2-\underset{\underset{CH_3}{|}}{CH}-NHNH_2 + KHSO_4 \tag{2}$$

To a mixture of 46 gm (0.53 mole) of 2-pentylamine and 9.2 gm (0.164 mole) of potassium hydroxide in 150 ml of water, heated to reflux, is added dropwise with stirring over a $\frac{1}{2}$ hr period, a solution of 9.5 gm (0.084 mole) of hydroxylamine-O-sulfonic acid in 50 ml of water. The reaction mixture is concentrated to half volume under reduced pressure. The solution is then transferred to a separatory funnel, layered with ether, and 10 ml of 10 N sodium hydroxide solution is cautiously added. The ether layer is separated, and the aqueous system is repeatedly extracted with ether. The combined ether extracts are dried with potassium hydroxide. After evaporation of the ether, the residual oil is fractionally distilled under reduced pressure. The product boils between 56°–60°C at 11 mm. The yield is not reported in the patent. However, yields normally run between 30% and 60% by this general procedure [15a].

In the procedure of Gever and Hayes [15a], the isolation of the reaction product involves conversion to the substituted hydrazonē of benzaldehyde, separation of the excess benzaldehyde by steam distillation, hydrolysis of the hydrazone, and the precipitation of the hydrazine as the oxalate salt.

3. HYDRAZONES

The preparation of hydrazones by the reaction of hydrazine derivatives with carbonyl compounds is well known and extensively described in most laboratory manuals. A typical example, used in our laboratory, is the preparation of benzaldehyde p-nitrophenylhydrazine [51] given below.

14-3. Preparation of Benzaldehyde p-Nitrophenylhydrazone [51]

In a 22-liter flask, cooled in an ice bath, is placed 1000 gm (6.45 moles) of p-nitrophenylhydrazine and 10 liters of methanol. The solution is stirred while 800 gm (7.55 moles) of benzaldehyde is added dropwise. During this addition, the temperature within the reaction flask is kept below 0°C by use of an ice–salt bath. After the addition of benzaldehyde is completed, stirring is continued for 16 hr while the reaction flask is allowed to warm up to room temperature. The product is then filtered off and freed of excess benzaldehyde by repeated washing with cold methanol. After drying, the product may be recrystallized from ethanol. Yield 1484 gm (94%), mp 195°C.

A procedure using essentially the same approach but using a small amount of glacial acetic acid as a catalyst for the reaction has also been reported [52, 60, 61].

4. HYDRAZIDES

In general, hydrazides may be prepared by many of the methods analogous to those used in the preparation of amides. For example, hydrazine salts of carboxylic acids and reactions of hydrazine with esters, acyl halides, acyl anhydrides, and amides may be used to produce hydrazides. A reaction analogous to the Hofmann degradation is the formation of hydrazides from ureides (acylureas) [54] (Eq. 4).

$$
\underset{\substack{\| \\ O}}{R-C}-NH\underset{\substack{\| \\ O}}{C}-NH_2 \xrightarrow{\text{NaOX}} \underset{\substack{\| \\ O}}{R-C}-NHNH_2 \tag{4}
$$

The preparation of hydrazides by interaction of esters and hydrazine hydrate is quite straightforward and proceeds in good yields. In many cases, simple addition of hydrazine hydrate to the liquid esters suffices to cause the precipitation of the hydrazide [62]. If the esters are insoluble or solids, a more prolonged treatment is required usually in the presence of an alcohol, as in the preparation of terephthalic hydrazide given below:

14-4. Preparation of Terephthalic Dihydrazide [51, 55]

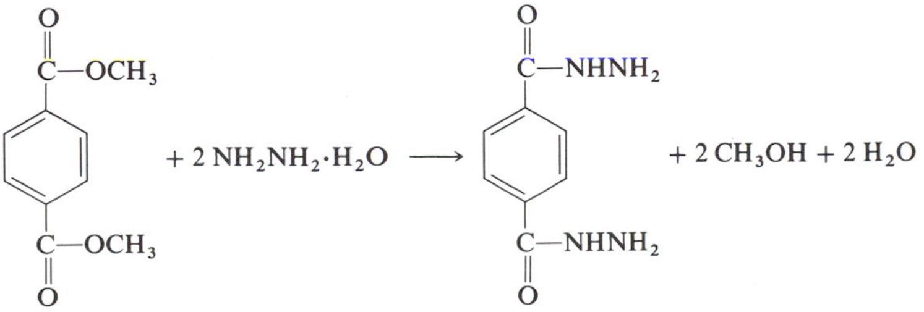

$$
+ 2 NH_2NH_2 \cdot H_2O \longrightarrow \qquad + 2 CH_3OH + 2 H_2O \tag{5}
$$

■ **CAUTION:** Hydrazine hydrate is extremely toxic and a cancer suspect agent.

For convenient handling, a quantity of dimethyl terephthalate is pulverized in a blender. In a flask, 232.8 gm (1.2 moles) of dimethyl terephthalate is slurried with 2760 ml of methanol. To this reaction mixture is added a solution of 504 gm (8.4 moles) hydrazine hydrate (85%) in 240 ml of methanol. The composition is stirred for 16 hr at room temperature. The solid product is then separated by filtration. The solid is repeatedly washed with cold methanol and finally dried in a vacuum oven to yield 222 gm (95%), mp over 330°C.

Since methyl acrylate is difficult to polymerize without some cross-linking, the usual products isolated are somewhat cross-linked. Even the hydrazides prepared from a modest molecular weight polymer of methyl acrylate, which is not cross-linked, tend to cross-link on standing. The preparation below is an example of the preparation of a polymeric hydrazide.

14-5. Preparation of Polyacrylic Hydrazide [56]

$$\left[-CH_2-CH-\right]_n + {}_nNH_2NH_2 \cdot H_2O \longrightarrow$$
$$\begin{array}{c} | \\ CO_2CH_3 \end{array}$$

$$\left[-CH_2-CH-\right]_n + CH_3OH + H_2O$$
$$\begin{array}{c} | \\ CONHNH_2 \end{array} \qquad (6)$$

■ **CAUTION:** Hydrazine hydrate is extremely toxic and a cancer suspect agent.

In an Erlenmeyer flask, 5 gm of polymethyl acrylate (molecular weight approx. 80,000), which has been pulverized, and 50 gm of hydrazine hydrate are warmed on a water bath until a homogeneous solution is formed. To this solution is added 500 ml of methanol containing 1 ml of glacial acetic acid. The product thereupon precipitates.

To purify the product, the polyacrylic hydrazide is dissolved in 50 ml of water and again precipitated with methanol. This procedure is repeated several times. The product is finally dried in a vacuum desiccator over sulfuric acid at room temperature. The product is stored in the cold in a vacuum desiccator with the exclusion of light. The stability of this polymer, as a non-cross-linked raw material, is poor. In general, this preparation must be carried out quite rapidly, with a minimum amount of exposure to heat during the preparation and with a minimum exposure to methanol.

REFERENCES

1. J. Thiele, *Justus Leibigs Ann. Chem.* **376,** 244 (1910).
2. H. H. Hatt, *Org. Synth. Collect. Vol.* **2,** 395 (1943).
3. H. O. Wieland, "Die Hydrazine" *in* "Chemie in Einzeldarstellungen" (J. Schmidt, ed.), Enke, Stuttgart, 1913.
4. L. F. Audrieth and B. A. Ogg, "The Chemistry of Hydrazine." Wiley, New York, 1951.
5. C. C. Clark, "Hydrazine." Mathieson Chem. Corp., Baltimore, 1953.
6. R. A. Reed, *R. Inst. Chem., Rep.* No. **5** (1957).
7. N. V. Sidgwick, *in* "The Organic Chemistry of Nitrogen" (T. W. J. Taylor and W. Baker, eds.), pp. 378ff. Oxford Univ. Press, London and New York, 1942. "N. V. Sidgwick's Organic Chemistry of Nitrogen." 3 ed., Chapter 15. Oxford Univ. Press, London and New York, 1966, and Japan Hydrazine Co. in U.S. Patent 4,310, 696 (1/12/82).
8. L. I. Smith, *Chem. Rev.* **23,** 193 (1938).
9. U. V. Solmssen, *Chem. Rev.* **37,** 490 (1945).
10. H. H. Sisler, G. M. Omietanski, and B. Rudner, *Chem. Rev.* **57,** 1021 (1957).
11. E. Enders, *in* "Houben-Weyl's Methoden der organischen Chemie," 4th ed., Vol. X/2, pp. 169, 750. Thieme Verlag, Stuttgart, 1967.
12. E. Müller, *in* "Houben-Weyl's Methoden der organischen Chemie," 4th ed., Vol. X/2, p. 121. Thieme Verlag, Stuttgart, 1967.
13. S. Patai, ed., "Chemistry of Hydrazo-, Azo-, and Azoxy Groups. Wiley, New York, 1975.
14. H. W. Schiessl, *Aldrichimica Acta* **13**(2), 33 (1980).
15. J. Druey, P. Schmidt, K. Eichenberger, and M. Wilhelm, Swiss Patent 372,685 (1963); *Chem. Abstr.* **16,** 5517e (1964).
15a. G. Gever and K. Hayes, *J. Org. Chem.* **14,** 813 (1949).
16. R. D. Brown and R. A. Kearley, *J. Am. Chem. Soc.* **72,** 2762 (1950).
17. E. F. Elslager, E. A. Weinstein, and D. F. Worth, *J. Med. Chem.* **7,** 493 (1964).
18. C. Westphal, *Ber. Dtsch. Chem. Ges. B.* **74B,** 739, 1365 (1941).
19. C. F. H. Allen, *Org. Synth. Collect. Vol.* **2,** 228 (1943).
20. A. Ault, *J. Chem. Educ.* **42,** 267 (1965).
21. D. G. Holland, G. J. Moore, and C. Tamborski, *J. Org. Chem.,* **29,** 1562, 3042 (1964); D. G. Holland and C. Tamborski, *ibid.,* **31,** 280 (1966).
22. F. Raschig, *Ber. Dtsch. Chem. Ges.* **40,** 4587 (1907).
23. L. F. Audrieth and L. H. Diamond, *J. Am. Chem. Soc.* **76,** 4869 (1954).
24. L. H. Diamond and L. F. Audreith, *J. Am. Chem. Soc.* **77,** 3131 (1955).
25. G. M. Omietanski and H. H. Sisler, *J. Am. Chem. Soc.* **78,** 1211 (1956).
26. P. R. Steyermark and J. L. McMclanahan, *J. Org. Chem.* **30,** 935 (1965).
27. W. W. Hartman and L. J. Roll, *Org. Synth. Collect. Vol.* **2,** 418 (1943).
28. H. H. Hatt. *Org. Synth. Collect. Vol.* **2,** 211 (1943).
29. G. W. Smith and D. N. Thatcher, *Ind. Eng. Chem., Prod. Res. Dev.* **1,** 117 (1962).

30. P. Besson, A. Nallet, and G. Luiset, French Patent 1,364,573 (1964); *Chem Abstr.* **61,** 11892b (1964).
31. C. G. Overberger, L. C. Palmer, B. S. Marks, and N. R. Byrd, *J. Am. Chem. Soc.* **77,** 4100 (1955).
32. G. Neurath, B. Pirmann, and M. Dünger, *Chem. Ber.* **97,** 1631 (1964).
33. J. Neurath and M. Dünger, *Chem. Ber.* **97,** 2713 (1964).
34. R. J. Hedrich and R. T. Major, *J. Org. Chem.* **29,** 2486 (1964).
35. H. Zimmer, L. F. Audrieth, M. Zimmer, and R. A. Rowe, *J. Am. Chem. Soc.* **77,** 790 (1955).
36. H. L. Lochte, W. A. Noyes, and J. R. Bailey, *J. Am. Chem. Soc.* **44,** 2556 (1922).
37. H. W. Stewart, Belgian Patent 630,723 (1963); *Chem. Abstr.* **61,** 1487c (1964).
38. J. D. Benigni and D. E. Dickson, *J. Med. Chem.* **9,** 439 (1966).
39. N. I. Ghali, D. L. Venton, S. C. Hung, and G. C. LeBreton, *J. Org. Chem.* **46,** 5413 (1981).
40. F. E. Condon and G. L. Mayers, *J. Org. Chem.* **30,** 3946 (1965).
41. N. Kornblum, *Org. React.* **2,** 287 (1944).
42. G. H. Coleman, *Org. Synth. Collect. Vol.* **1,** 442 (1948).
43. D. S. Tarbell, C. W. Todd, M. C. Paulson, E. G. Lindstrom, and V. P. Wystrach, *J. Am. Chem. Soc.* **70,** 1381 (1948).
44. D. M. Lemal, F. Menger, and E. Coats, *J. Am. Chem. Soc.* **86,** 2395 (1964).
45. H. H. Hatt, *Org. Synth, Collect. Vol.* **2,** 208 (1943).
46. H. C. Ramsperger, *J. Am. Chem. Soc.* **51,** 918 (1929).
47. W. E. Bachman, *J. Am. Chem. Soc.* **53,** 1524 (1931).
48. T. G. Back, S. Collins, and R. G. Kerr, *J. Org. Chem.* **46,** 1564 (1981).
49. H. R. Snyder, C. Weaver, and C. D. Marshall, *J. Am. Chem. Soc.* **71,** 289 (1949).
50. L. Spialter, D. H. O'Brien, G. L. Untereiner, and W. A. Rush, *J. Org. Chem.* **30,** 3278 (1965).
50a. F. A. Neugebauer and P. H. H. Fischer, *Chem. Ber.* **98,** 844 (1965).
51. Authors' Laboratory.
52. H. C. Yoa and P. Resnick, *J. Org. Chem.* **30,** 2832 (1965).
53. G. R. Newkome and D. L. Fishel, *J. Org. Chem.* **31,** 677 (1966).
54. P. Schestakov, *Ber. Dtsch. Chem. Ges* **45,** 32 73 (1912).
55. W. Sweeney, private communication to W. R. Sorenson and T. W. Campbell; see "Preparative Methods of Polymer Chemistry," p. 103. Wiley (Interscience), New York, 1961.
56. W. Kern, T. Hucke, R. Holländer, and R. Schneider, *Makromol. Chem.* **22,** 31 (1957).
57. F. K. Velichko, B. I. Keda, and S. D. Polikarpova, *Zh. Obshch. Khim.* **34,** 2356 (1964); *Chem. Abstr.* **61,** 9400b (1964).
58. T. A. Geissman, M. J. Schlatter, I. D. Webb, and J. D. Roberts, *J. Org. Chem.* **11,** 741 (1946).
59. Y. Ohno and H. A. Stahmann, *Macromolecules* **4,** 350 (1971).
60. R. Fusco and F. Sannicolo, *J. Org. Chem.* **46,** 90 (1981).
61. T. Iida and F. C. Chang, *J. Org. Chem.* **46,** 2786 (1981).
62. P. A. S. Smith, *Org. React.* **3,** 337 (1946).

DIAZO AND DIAZONIUM COMPOUNDS

1. INTRODUCTION

A number of reviews on the aliphatic diazo compounds have appeared from time to time (for representative references, see [1–13a]). The chemistry, properties, and uses of aromatic diazonium compounds have been reviewed extensively (refs. [14–22] are representative of these).

The preparation of diazo hydrocarbons is generally carried out by the alkaline decomposition of *N*-nitroso-*N*-alkyl derivatives such as the *N*-nitroso derivatives of sulfonamides, amides, and phthalamides, urethanes, ureas, nitroguanidines and β-alkylaminoisobutyl ketones. Of these, the decompositions of *N*-alkyl-*N*-nitroso-*p*-toluenesulfonamide and of bis (*N*-methyl-*N*-nitroso)terephthalamide

From S. R. Sandler and W. Karo, *Organic Functional Group Preparations*, Vol. I, 2d ed.(New York, 1983), 466ff., by permission of Academic Press, Inc.

are perhaps the most convenient laboratory methods for the preparation of diazomethane.

Also of interest in the preparation of diazoalkanes are the diazotization of primary amines with activating substituents in the α-position, reaction of hydrazine or hydrazides with dichlorocarbene, diazo-group transfer reactions, the oxidation of hydrazones, and condensation reactions of active methylene compounds.

Aromatic diazonium salts are generally prepared by diazotization of aromatic amines in aqueous systems with nitrous acid. In nonaqueous systems, diazotizations have been carried out with isoamyl nitrite.

2. ALIPHATIC DIAZO COMPOUNDS

The aliphatic diazo compounds find application as intermediates in a variety of organic reactions such as the Arndt–Eistert synthesis.

■ **CAUTION:** Because several explosions have been reported in the preparation of such materials as diazomethane, great care must be taken during the preparation of all diazo compounds, and the isolation of the pure compounds should normally be avoided. The scale of reaction should be kept small and extreme precautions against explosion hazards must be taken. Furthermore, diazomethane and presumably many other diazo compounds as well as many of the intermediates used in the preparation of diazo compounds are toxic. Some of the compounds used as intermediates may initially cause sensitization so that upon further contact severe physiological reactions may take place. Intermediates may also be carcinogenic. For example, nitrosomethylurea is considered a potent carcinogen [23].

In preparations involving the use of aliphatic diazo compounds, due caution must be exercised in the transfer of the diazoalkane solution to subsequent reaction systems and in the isolation of the final product. Particular attention must also be paid to the disposal of by-products, to the handling of residues in the reaction flasks, and to traces of diazo compounds in the apparatus. Extensive notes on safety and health considerations for handling of diazomethane are givn by Gutsche [6], Moore and Reed [24], and De Boer and Backer [25]. The recommendations made for diazomethane should be applied to all the diazoalkanes and related compounds. To be kept in mind are the following points:

1. The starting materials, particularly the *N*-nitroso derivatives used as starting materials, may be toxic, may cause skin irritations, and other serious allergic reactions on contact. They may also be carcinogenic. Therefore protection against contact, inhalation, and spillage must be provided.
2. The diazo compounds may be both toxic and explosive; therefore, hoods and other provision for protection against explosive hazards must be provided, such as heavy shields, heavy gloves, protective goggles, and helmets.
3. The explosions due to diazo compounds may be initiated by a variety of factors such as exposure to sunlight or other strong light, contact with sharp edges, corners, ground surfaces, chipped glass surfaces, sticks of potassium hydroxide, or crystalline side products. For this reason it is generally recommended that ground glass equipment not be used. Only clean, new glassware should be used. The flame-polishing of ground glass joints has also been suggested [26].

In one report of a serious diazomethane explosion, an investigation into the accident concluded that a static charge had built up on the chemist's polyester-and-cotton lab coat when the chemist had walked about in the laboratory just prior to pouring a 3% diazomethane solution. When the solution was poured, a spark may have discharged and initiated the explosion [27]. The use of all-cotton lab coats was recommended at that time. However, today polyester and other synthetics are so generally worn that other measures to prevent static discharge will have to be found.

The early directions for the preparation of diazo compounds used to call for shaking the intermediates with alkalies by hand. In view of the explosion hazards involved, Moore and Reed [24] recommend that Teflon-coated magnetic stirrers be used instead.

Recently, new designs for equipment have been suggested for the codistillation of diazomethane with ether using a special dry ice condenser to assure reasonably complete condensation of the diazomethane–ether azeotrope [26].

This equipment is commercially available [28]. An apparatus which allows the generation of diazomethane and its isolation without codistillation with ether using a hot bath has also been described. This equipment is probably most useful for the preparation of derivatives on a millimolar scale [29].

A. DECOMPOSITION OF N-ALKYL-N-NITROSO-p-TOLUENESULFONAMIDES

The base-catalyzed decompositions of a large variety of N-nitroso compounds to aliphatic diazo compounds are well known. Primary emphasis has been on methods for the generation of diazomethane. A number of these will be mentioned below.

The method of De Boer and Backer makes use of N-methyl-N-nitroso-p-toluenesulfonamide as the source of diazomethane [25, 30–32]. The starting material is available from suppliers of specialty organic chemicals, it is reasonably stable at room temperature, and it seems to give fewer allergic reactions than some of the other nitroso compounds used in the preparation of diazomethane. It is believed to have a lower degree of carcinogenicity than many of the other proposed intermediates [33].

De Boer and Backer described their procedure in *Organic Syntheses* [25]. Despite the fact that these authors demonstrated that the maximum yield of diazomethane is obtained when the molar ratio of potassium hydroxide to N-methyl-N-nitroso-p-toluenesulfonamide is 0.8 to 1 [31], the usual practice appears to be to use equimolar quantities.

Preparation 15-1 is a recent adaptation of the De Boer–Backer method using the Hudlicky apparatus with its special dry ice condenser [26].

15-1. Preparation of Diazomethane from N-Methyl-N-nitroso-p-toluenesulfonamide [26]

$$H_3C-\langle\hspace{-0.3em}\bigcirc\hspace{-0.3em}\rangle-SO_2N-CH_3 + ROH \xrightarrow{KOH} CH_2N_2 + H_3C-\langle\hspace{-0.3em}\bigcirc\hspace{-0.3em}\rangle-SO_2OR + H_2O$$
$$|$$
$$NO$$

$$(1)$$

In a hood, behind a safety shield, in an apparatus (illustrated in Refs. [26] and [28]) consisting of a 500-ml three-necked round bottom flask immersed in a water bath and equipped with a Teflon-coated magnetic stirring bar, a thermometer, a 500-ml addition funnel with pressure equilizer, and a dry ice reflux condenser, arranged like a distillation head with a Teflon take-off stopcock and a liquid overflow trap connected by a ground glass joint to a receiver for the product and also fitted with a vent which is connected to a dry ice trap, are placed 10 gm (0.32 mole) of potassium hydroxide dissolved in 30 ml of water, 105 ml of Carbitol (diethylene glycol monoethyl ether), and 30 ml of ether. After filling the condenser with dry ice and acetone, to the addition funnel is charged a solution of 64.2 gm (0.3 mole) N-methyl-N-nitroso-p-toluenesulfonamide in 375 ml of ether. After attaching a 500-ml flask as a receiver and cooling it in an ice–water bath, the magnetic stirrer is started and the water bath is heated to 60°C. Then the addition of the nitroso compound is begun at such a rate that all of the yellow vapor formed is completely condensed when the take-off stopcock is closed. Upon opening the stopcock, the first portion of the diazomethane–ether distillate is allowed to flow into the receiver at such a rate that no vapor is permitted to pass the over-flow trap. The bath temperature is gradually raised to approximately 80°C until the addition has been completed. Distillation is continued until only colorless ether collects in the overflow trap. The process is complete within approximately 100 min. and dry-ice consumption in this time is on the order of 2 kg. The yield is said to be approx. 0.27 mole (90% of theory).

3. AROMATIC DIAZONIUM SALTS

A. NOTES ON THE PREPARATION OF DIAZONIUM SALT SOLUTIONS IN AQUEOUS MEDIA

In normal laboratory practice, diazonium salts are used as interme-diates in the preparation of a variety of aromatic compounds. Since many diazonium salts may detonate when warmed or when dry, they are usually used in solution without isolation. Obviously, since diazonium compounds are used widely in the dye industry, dry di-azonium salts of considerable stability can be prepared. Particularly those diazonium salts which contain electron-withdrawing groups may be converted to relatively stable salts. Even in those cases, it would be well if materials were handled with considerable care. In partic-ular, the compounds should be stored in a cool dark place. Further-more, since many diazonium salts are quite sensitive to ultraviolet light, preparation, handling, and storage in a cool, shaded area are recommended.

In general, the most common method of preparing diazonium salts involves the treatment of a soluble aromatic amine salt in aqueous mineral acid at low temperature with sodium nitrite. In this connection it should be kept in mind that many aromatic amines require considerable purification prior to use since they are subject to air oxidation. Frequently, purification of the amine in the presence of traces of sodium hydrosulfite in a recrystallizing solvent is helpful in overcoming discolorations due to oxidation. Also to be kept in mind in preparing diazonium salts is the fact that the hydrochloride salts of many aromatic amines are more soluble at low temperatures than at high temperatures.

The purity even of reagent grades of sodium nitrite is sometimes questionable, making it difficult to weigh out an equivalent amount of the reagent for a given preparation. It is recommended, therefore, that the diazotization be carried out with an excess of sodium nitrite solution on hand from the beginning of the reaction. The reaction is carried out in an excess of mineral acid with an excess of sodium nitrite. Starch−iodide paper is used to control the addition of the nitrite. When the paper turns blue, addition is considered completed.

Since the usual reactions of diazonium salts do not proceed satisfactorily with an excess of nitrous acid present, the addition of either crystalline sulfamic acid or urea assists in destroying the excess nitrous acid.

While it is useful to surround the reaction flask with an ice bath during diazotization and storage of the diazonium salt for subsequent reactions, the use of ice cubes or cracked ice *in* the reaction flask is strongly recommended to minimize problems of local overheating.

B. STABILIZED DIAZONIUM SALTS

Many fluoroborate salts of aromatic diazonium compounds have a high degree of stability.

The stabilization of some diazonium salts may be carried out by the addition of naphthalene-1,5-disulfonic acid or 2-naphthol-1-sulfonic acid to the hydrochloric salt solution produced after diazotization [33]. The usual procedure simply involves addition of a slurry of acids to the diazonium salt solutions and, if necessary, assisting the precipitation of the product by the addition of sodium chloride. Some substituted amines may be diazotized in concentrated sulfuric acid or in glacial acetic acid at relatively high temperatures [34].

In our own laboratory we have even prepared 1-anthraquinone-diazonium chloride at 45°–50°C and found the product to be reasonably stable.

15-2. Preparation of 1-Anthraquinonediazonium Chloride [35]

$$+ NaNO_2 + 2\,HCl \longrightarrow$$

$$Cl^- + NaCl + 2\,H_2O \qquad (2)$$

In a 12-liter flask, 273 gm (1.25 moles) of 1-aminoathraquinone and 685 ml of concentrated hydrochloric acid are mixed at 30°C. Then a solution of 86.5 gm (1.25 moles) of sodium nitrite in 450 gm of distilled water is added slowly and the temperature of the reaction product is raised to 50°C. After diazotization has been completed, 8 liters of warm distilled water is added and the solution is filtered while hot. The insoluble material is washed with 500 ml of warm water and discarded. To the warm filtrate is added approximately 2400 gm of sodium chloride. The solution is stirred until all of the sodium chloride has gone into solution. Upon cooling the 1-anthraquinone diazonium chloride precipitates. The product is collected by filtration and washed with saturated sodium chloride solution. The product may then be air-dried. Yield 285 gm (85.5% of theory), melting point range 117°–118°C with violent explosion.

In the field of immunochemistry, the fractionation of proteins is of importance. One technique involves the conjugation of an antigen protein to a polymer. This conjugated resin is then used as a column packing which may be used to fractionate antibodies. Many systems have been described. Among these has been the use of poly(styrenediazonium chloride). The preparation of this material is given here.

15-3. Preparation of Poly(styrenediazonium chloride) [36]

$$
\left[\!\!\begin{array}{c} -CH-CH_2- \\ \bigcirc \\ NH_2 \end{array}\!\!\right]_x + x\,NaNO_2 + 2x\,HCl \longrightarrow \left[\!\!\begin{array}{c} -CH-CH_2- \\ \bigcirc \\ {}^{+}N_2 + Cl^{-} \end{array}\!\!\right]_x + x\,NaCl + 2x\,H_2O
$$

$$(3)$$

A 2% solution (0.002 mole) of poly(aminostyrene) in 10 ml of 2 N hydrochloric acid is cooled to between $0°$ and $-5°C$. The solution is then diazotized by the slow addition of 1.5 ml of 14% (0.0038 mole) solution of sodium nitrite. The excess of nitrous acid is destroyed by the addition of 2 ml of a 6% solution (0.002 mole) of urea previously cooled to $0°C$. The reaction mixture is maintained between $-5°$ and $0°C$ with stirring for about 1 hr until a test with starch–iodide paper no longer gives a positive test for nitrous acid. This solution may then be used to carry out reactions typical of diazonium salts. It is our belief that if a cross-linked poly(aminostyrene) were used in this preparation, diazotization would proceed on the slurry, resulting in a diazotized resin which may be particularly useful as a column packing in immunochemistry.

REFERENCES

1. L. I. Smith, *Chem. Rev.* **23,** 193 (1938).
2. B. Eistert, *Z. Angew. Chem.* **54,** 99 (1941).
3. W. E. Bachman and W. S. Struve, *Org. React.* **1,** 38 (1942).
4. B. Eistert, *in* "Newer Methods of Preparative Organic Chemistry," p. 513. Wiley, New York (translation) (see B. Eistert, *in* "Neuere Methoden der Preparativen organischen Chemie," p. 361. Verlag Chemie, Weinheim, 1949).
5. A. F. McKay *et al., Chem. Rev.* **51,** 331 (1952).
6. C. D. Gutsche, *Org. React.* **8,** 391 (1954).
7. M. Regitz, *Synthesis* **7,** 351 (1972).
8. M. Regitz, *Angew. Chem. Int. Ed. Engl.* **6,** 733 (1967).
9. M. Regitz, I. K. Korobitsyna, and L. C. Redina, *Method. Chim.* **6,** 205 (1975).
10. A. L. Fridman, F. A. Gabitov, O. B. Kremeleva, and V. S. Zalesov, *Nauk. Tr. Perm. Formatsevot.* **8,** 3 (1975); *Chem. Abstr.* **85,** 5524 (1976).

11. R. F. Muraca, "Diazo Group" *in Treatise Anal. Chem. Vol.* **15,** p. 347 (1976); I. M. Kolthoff and P. J. Elving, ed., Wiley, New York, 1971.

12. I. D. R. Stevens, *Org. React. Mech.* p. 327 (1976); *Chem. Abstr.* **85,** 62264 (1976).

13. R. Fields, M. S. Gibson, and G. Holt, *in* "Rodd's Chemistry of Carbon Compounds" (M. F. Ansell, ed.), 2nd ed., Vol. 1, Parts A–B, Suppl. 125 Elsevier, Amsterdam, 1975.

13a. T. H. Black, *Aldrichimica Acta.* **16** (1), 3 (1983).

14. P. H. Groggins, "Unit Processes in Organic Synthesis," p. 129. McGraw-Hill, New York, 1947.

15. K. H. Saunders, "The Aromatic Diazo-Compounds and Their Technical Applications," p. 1. Longmans, Green, New York, 1947.

16. W. E. Bachman and R. A. Hoffman, *Org. React.* **2,** 224 (1944).

17. N. Kornblum, *Org. React.* **2,** 262 (1944).

18. H. Zollinger, "Azo and Diazo Chemistry, Aliphatic and Aromatic Compounds," Wiley (Interscience), New York, 1961.

19. J. H. Ridd, *Q. Rev. Chem. Soc.* **15,** 418 (1961).

20. K. Schank, *Method. Chim.* **6,** 159 (1975).

21. M. Yamamoto, *Senvyo To Yakuhin* **21** (2), 30; (3), 58 (1976); *Chem. Abstr.* **85,** 178948 (1976).

22. S. Patai, ed., "The Chemistry of Diazonium and Diazo Compounds." Wiley, New York, 1978.

23. A. Graffi and F. Hoffman, *Acta Biol. Med. Ger.* **16,** K-1 (1966); *Org. Synth. Collect. Vol.* **2,** 165 (1943).

24. J. A. Moore and D. E. Reed, *Org. Synth. Collect. Vol.* **5,** 351 (1973).

25. T. J. De Boer and H. J. Backer, *Org. Synth. Collect. Vol.* **4,** 250 (1963).

26. M. Hudlicky, *J. Org. Chem.* **45,** 5377 (1980).

27. Anonymous, *Catalyst* **53,** 202 (1968).

28. Aldrich Chemical Co., *Aldrichimica Acta* **14**(3), 59 (1981).

29. H. M. Fales and T. M. Jaouni, *Anal. Chem.* **45,** 2302 (1973).

30. T. J. De Boer and H. J. Backer, *Recl. Trav. Chim. Pays-Bas* **73,** 229 (1954).

31. T. J. De Boer and H. J. Backer, *Recl. Trav. Chim. Pays-Bas* **73,** 582 (1954).

32. C. G. Overberger and J.-P. Anselme, *J. Org. Chem.* **28,** 592 (1963).

33. H. H. Hodgson and E. Marsden. *J. Chem. Soc.* p. 207 (1940).

34. R. Howe, *J. Chem. Soc. C* p. 478 (1966).

35. W. Karo, unpublished procedure.

36. L. H. Kent and J. H. R. Slade, *Biochem. J.* **77,** 12 (1960).

16

NITRO COMPOUNDS

1. INTRODUCTION

The present chapter is divided into two sections, the first section deals with aliphatic nitro compounds and the second section deals with aromatic nitro compounds.

While the direct nitration of aliphatic hydrocarbons enjoys considerable importance in the industrial sphere, in the laboratory it has only limited value since complex mixtures of products are usually formed. Treatment of olefins with dinitrogen tetroxide may lead to dinitroparaffins and mixtures of nitro nitrites. The latter may be oxidized to nitro nitrates or hydrolyzed to nitro alcohols or thermally degraded to olefinic nitro compounds under the reaction conditions.

From S. R. Sandler and W. Karo, *Organic Functional Group Preparations*, Vol. I, 2d ed. (New York, 1983), 497ff., by permission of Academic Press, Inc.

With solutions of acetyl nitrate, olefins may be converted to acetate esters of β-nitro alochols produced from the olefins.

Active methylene compounds have been nitrated directly with nitric acid. In this synthesis one or more labile hydrogens are replaced by a nitro group.

Olefins may be chloronitrated with nitryl chloride. In this reaction, the nitro group appears to add to the olefinic carbon atom bearing the larger number of hydrogens.

Under alkaline conditions, active methylene compounds have been nitrated with alkyl nitrates. Under the conditions of the reaction, considerable cleavage of the reaction product takes place.

Among the indirect methods of preparing aliphatic nitro compounds are the reaction of various alkyl halides with silver nitrite or sodium nitrite, the oxidation of oximes and amines with peroxytrifluoroacetic acid, and the potassium permanganate oxidation of tertiary amines.

Since aliphatic nitroalkanes are active methylene compounds, these may be used as starting materials for the preparation of more complex products by typical reactions of methylene compounds such as alkylations, aldol condensations, Michael condensations, and Mannich reactions.

Aromatic compounds are readily nitrated with nitric acid–sulfuric acid mixtures, acyl nitrates, nitronium tetrafluoroborate, and the oxides of nitrogen.

Among the indirect methods of preparing aromatic nitro compounds are the displacement of sulfonic acid groups with nitro groups and the replacement of diazonium groups.

By oxidation, aromatic amines and nitroso compounds may be converted to nitro compounds.

■ **CAUTION:** It is generally known that trinitrotoluene is a high explosive and that the treatment of glycerol, cellulose, and other carbohydrates with nitric acid/sulfuric acid leads to other explosives. The nitration of many other compounds may form products (or intermediates) which may be explosives, and precautions against detonations must be taken. Alkali salts of nitromethane or nitromethane derivatives are sensitive to impact when dry [1]. In fact, the addition of bases or acids to nitromethane is said to render the compound susceptible to detonation according to a recent publication [2]. An explosion during the nitrolysis of an acetyl compound may have been caused by the formation of acetyl nitrate in the reaction mixture. In general, mixtures of concentrated nitric acid and acetic anhydride or trifluoroacetic anhydride may form potentially explosive compositions in the reaction such as, for example, acetyl nitrate, as well as traces of tetranitromethane [3]. These safety notes [3] indicate that explosion hazards are quite prevalent in reaction systems designed to produce nitro compounds and due precautions must be taken quite generally.

Animal studies have implicated 2-nitropropane as a possible human carcinogen. Therefore, the use of approved respirators and full-body protective clothing is indicated for handling this compound and perhaps for related compounds [3].

In our discussion below, the hazards may not always be pointed out, but it should be understood that safety shields, even barricades, respirators, and full-body protective clothing, should always be used when working with nitro compounds.

2. ALIPHATIC NITRO COMPOUNDS

A. DIRECT NITRATION OF ACTIVE METHYLENE COMPOUNDS WITH NITRIC ACID

Active methylene compounds have been nitrated directly with nitric acid. Since the nitro group thus introduced also activates the carbon adjacent to it, the product resulting from the direct nitration of an active methylene compound is frequently highly reactive and consequently may be quite unstable—in fact, explosive. Therefore, as with all aliphatic nitro compounds, precaution must be taken to reduce the hazards of explosion to personnel and equipment. A typical example of the reaction is given in the following preparation.

16-1. Preparation of Diethyl Nitromalonate [4]

$$
\begin{array}{ccc}
\overset{O}{\underset{}{\underset{\diagdown}{C}}}-OC_2H_5 & & \overset{O}{\underset{}{\underset{\diagdown}{C}}}-OC_2H_5 \\
CH_2 & + \ HNO_3 \longrightarrow & CHNO_2 \\
\underset{\diagup}{\underset{O}{C}}-OC_2H_5 & & \underset{\diagup}{\underset{O}{C}}-OC_2H_5
\end{array}
\tag{1}
$$

With all due precautions against possible explosion hazards, in a 500-ml three-necked flask fitted with a dropping funnel, stirrer, thermometer, and an outlet protected by a drying tube is placed 80.0 gm (0.5 mole) of diethyl malonate. The flask is cooled by tap water at 12°C and 184 ml of fuming nitric acid (density: 1.5 gm/ml) is added at a rate sufficient to maintain the temperature between 15° and 20°C. The addition requires approximately 1 hr. After this period, the mixture is allowed to stir for $3\frac{1}{2}$ hr at 15°C. The solution is then poured into 1 liter of ice and water and the ester is extracted with two portions of toluene, the first being 200 ml, the second 100 ml. The combined toluene extracts are washed twice with water and then with 200 ml portions of 5% aqueous urea solution until a starch–potassium iodide test for the oxide of nitrogen is negative.

To isolate the product from its toluene solution, the toluene solution is extracted with several portions of a 10% solution of sodium carbonate in water until the acidification of a test portion of the aqueous extract shows that no more nitro ester is being extracted by the sodium carbonate solution.

The sodium carbonate extracts are combined and washed once with 200 ml of toluene. The aqueous solution is then carefully acidified to Congo Red paper with concentrated hydrochloric acid while cooling by the occasional addition of ice. The ester is collected by extracting in turn with 500-, 200-, and 100-ml portions of toluene. The toluene solutions are washed twice with 200 ml portions of water and then again with a 5% aqueous urea solution, checking again with starch–potassium iodide test papers for complete absence of oxides of nitrogen. The toluene solution is then dried over magnesium sulfate. For many purposes this solution may be used as is. To assay the solution, an aliquot is added to an equal volume of ethanol and titrated to a phenolphthalein end point with 1 N sodium hydroxide.

In this case, the assay depends on the acidic nature of the nitro esters. The assay indicates a yield of 94.1 gm (91.7%) of product. To isolate the pure ester, the toluene is evaporated under reduced pressure, and the residue is distilled at 81°–83°C (0.3 mm).

[Reprinted from: D. I. Weisblat and D. A. Lyttle, *J. Am. Chem. Soc.* **71**, 3079 (1949). Copyright 1949 by the American Chemical Society. Reprinted by permission of the copyright owner.]

B. INDIRECT NITRATION OF SODIUM NITRITE CONDENSATIONS

The reaction of alkyl halides or dialkyl sulfates with both silver nitrite and sodium nitrite has been extensively studied by Kornblum and co-workers (see review by Kornblum [5]). These indirect methods of nitration offer procedures for the preparation of primary, secondary, and tertiary nitro compounds.

The reaction of alkyl halides with silver nitrite (Victor Meyer reaction) is of value in the preparation of primary nitroalkanes. In the case of secondary nitroalkanes, reactions are slow and yields are low. The reaction is of little value for the preparation of tertiary nitroalkanes.

Whereas the older literature indicated that the reaction of alkyl halides with sodium nitrite afforded primarily nitrite esters rather than nitro compounds, more recent investigations have shown that, by use of dimethylformamide [6] or dimethyl sulfoxide [7], good yields of either primary or secondary nitro compounds may be obtained. In these solvents, sodium nitrite is soluble, at least to some extent, which promotes the desired course of the reaction. Since more concentrated solutions are possible in dimethyl sulfoxides, the latter solvent offers some advantage in terms of shorter reaction times. The solubility of sodium nitrite in dimethylformamide is enhanced by the addition of urea. With open-chain secondary bromides, somewhat higher yields are obtained in dimethylformamide than in dimethyl sulfoxide. By the silver nitrite methods, yields are somewhat higher; however, for commercial operations the cost of the reagent outweighs this advantage. Table 1 indicates the situations in which silver nitrite may be the reagent choice.

■ **CAUTION:** The reactions with silver nitrite may be particularly hazardous.

Phloroglucinol, which is indispensable in the preparation of α-nitro esters, as a scavenger of nitrite esters, has a strong retarding influence on the reaction in dimethyl sulfoxide. However, the reaction in this solvent is intrinsically so rapid that the reaction still proceeds at a reasonable rate even in the presence of phloroglucinol [7]. With secondary bromides [6] and also cyclopentyl and cycloheptyl iodides, urea in dimethylformamide is used along with a scavenger for nitrite esters. Suitable scavengers are phloroglucinol, catechol, and resorcinol. Of these, phloroglucinol appears to be the most satisfactory. The use of such scavengers is mandatory in the preparation of α-nitro esters. In preparations involving the use of dimethyl sulfoxide, the use of urea is omitted.

TABLE I

NITRO COMPOUND TYPES FORMED BY INDIRECT PROCEDURES

Starting material	Reagent	Product type
RCH_2X	$AgNO_2$	Primary nitroalkanes (particularly suitable where R contains electron-withdrawing groups)
ICH—CO$_2$R' | R	$AgNO_2$	α-Nitro esters (limited to iodo compounds) (method of choice for ethyl α-nitroacetate)
RCH_2X	$NaNO_2$/DMF or DMSO	Primary nitroalkanes (less useful when R contains electron-withdrawing groups)
R \CHX R'	$NaNO_2$/DMF or DMSO	Secondary nitroalkanes (reagent of choice) (fails for cyclohexyl halides)
Br—CH—CO$_2$R' | R	$NaNO_2$/DMF or DMSO	α-Nitro esters (generally useful process) (not suitable for ethyl α-nitroacetate)
R—CH=NOH	$\overset{O}{\overset{\|\|}{CF_3C}}$—OOH	Primary nitroalkanes
R—C=NOH | R'	$\overset{O}{\overset{\|\|}{CF_3C}}$—OOH	Secondary nitroalkanes
R—C=NOH | R'	1. N-Bromosuccinimide 2. NaBH$_4$	Secondary nitroalkanes
R—CH—NH$_2$ | R'	$\overset{O}{\overset{\|\|}{CH_3C}}$—OOH	Secondary nitroalkanes
R | R'—C—NH$_2$ | R''	$\overset{O}{\overset{\|\|}{CH_3C}}$—OOH	Tertiary nitroalkanes
R | R'—C—NH$_2$ | R''	$KMnO_4$	Tertiary nitroalkanes

Phloroglucinol is added in the preparation of α-nitro esters in dimethyl sulfoxide [7, 8]. While alkyl bromides or alkyl iodides are generally used in this preparation, tosylate derivatives of alcohol may also be used [6, 9].

16-2. Preparation of 2-Nitrooctane (in DMF) [6]

$$CH_3(CH_2)_5\text{—}\underset{\underset{Br}{|}}{CH}\text{—}CH_3 + NaNO_2 \xrightarrow[\text{2. } H_2O]{\text{1. DMF}} CH_3(CH_2)_5\text{—}\underset{\underset{NO_2}{|}}{CH}\text{—}CH_3 + NaBr$$

$$+ CH_3(CH_2)_5\text{—}\underset{\underset{OH}{|}}{CH}\text{—}CH_3 \quad (2)$$

In a hood, behind a shield, to a stirred mixture of 600 ml of dimethylformamide, 36 gm (0.52 mole) of sodium nitrite, 40 gm (0.67 mole) of urea, and 40 gm (0.3 mole) of anhydrous phloroglucinol cooled in a water bath is added with vigorous stirring 58 gm (0.30 mole) of 2-bromooctane while maintaining a temperature no higher than 25°C in the reaction flask. The stirring at room temperature is continued for 45 hr. Then the reaction mixture is poured into 1.5 liters of ice water, layered over with 100 ml of petroleum ether (bp 35°–37°C). The aqueous phase is extracted four times with 100-ml portions of petroleum ether, after which the extracts are washed with four 75-ml portions of water and dried over anhydrous magnesium sulfate. The petroleum ether is then removed by distillation under reduced pressure through a column, heat being applied to the distillation flask with a bath which is gradually raised to approximately 65°C. Distillation of the residue through a 60 × 1 cm externally heated distillation column packed with ⅛ inch glass helices and equipped with a total reflux-variable take-off head yielded 15.0 gm (35%) of 2-octanol at 45°C (1 mm) and 27.6 gm (58%) of 2-nitrooctane at 67°C (3 mm).

3. AROMATIC NITRO COMPOUNDS

A. DIRECT NITRATION WITH MIXED ACID OR NITRIC ACID

In view of the hazardous reagents frequently used in these nitration reactions and the fact that many aromatic nitro compounds are explosive in nature, great care must be exercised in handling these reactions. There also was a report that 4-nitrodiphenyl was suspected of being carcinogenic [10]. It is prudent, therefore, to consider all nitro compounds circumspectly.

By far the most generally useful and most commonly used nitrating procedure involves the use of solutions of nitric acid in sulfuric acid, commonly referred to as "mixed acid" [11, 12].

Extensive studies of the nitration in mixed acid have indicated that the nitrating agent is the nitronium ion NO_2^+ formed according to Eq. (3) [13].

$$HNO_3 + 2H_2SO_4 \rightleftharpoons NO_2^+ + 2HSO_4^- + H_3O^+ \tag{3}$$

Theoretically and experimentally, the most rapid nitrations take place in systems which are above 90% in sulfuric acid.

Preparative procedures usually use larger concentrations of nitric acid than required by the theoretical requirements in an attempt to compromise the rate of reaction with the isolation of practical quantities of product. The use of sulfuric acid has the following advantages. (1) It will combine with water generated in the reaction and thus prevent the dilution of nitric acid. Since dehydration may also be accomplished with phosphorus pentoxide, the fact that phosphorus pentoxide in the presence of nitric acid does not increase the rate indicates that chemical dehydration alone is not a particularly important factor as far as obtaining the maximum rate of reaction is concerned. (2) Many organic compounds are soluble in concentrated sulfuric acid. (3) It also diminishes the oxidizing action of nitric acid. (4) In some cases, it is believed that the intermediate formation of a sulfonic acid takes place which subsequently is readily replaced by a nitro group.

16-3. Preparation of *p*-Nitrobromobenzene [11]

$$(4)$$

In a 200-ml flask, cooled with an ice–water bath, to 28 gm of concentrated nitric acid is cautiously added 37 gm of concentrated sulfuric acid. While maintaining the flask at room temperature, there is added, in 2–3 ml portions, 16 gm of bromobenzene with vigorous shaking of the flask (rubber gloves, rubber apron, and face shield must be worn throughout the preparation). During the addition, the reaction temperature is maintained between 50° and 60°C by cooling the flask in running water as necessary.

After all the bromobenzene has been added and the temperature no longer tends to rise, the flask is warmed on a steam bath for $\frac{1}{2}$ hr. The reaction mixture is then cooled to room temperature and poured into approximately 100 ml of cold water. The crude product (which contains some *ortho* isomer) is filtered off, washed with water, and pressed dry.

To purify the product, the crude material is recrystallized from 100–125 ml of hot ethanol. The *ortho* isomer, being more soluble, remains in solution while the *para* isomer crystallizes out on cooling. Yield 10–14 gm (50–70%) mp 126°–127°C.

Under more vigorous reaction conditions, dinitration takes place with mixed acids. The semimicro preparation of 2,4-dinitrobromobenzene illustrates this procedure.

16-4. Preparation of 2,4-Dinitrobromobenzene [14]

$$\text{(C}_6\text{H}_5\text{Br)} + 2\ HNO_3 \xrightarrow{H_2SO_4} \text{(2,4-dinitrobromobenzene)} + 2\ H_2O \tag{5}$$

In a hood, the mixed acid is prepared as follows. In a 50-ml Erlenmeyer flask cooled in an ice–water bath, to 5 ml of concentrated nitric acid is added cautiously 15 ml of concentrated sulfuric acid. The flask is then heated to 85°–90°C, and 2.0 ml (3.0 gm) (0.19 mole) of bromobenzene is added in three or four portions during 1 min. The reaction mixture is swirled well after each addition. The temperature rises to 130°–135°C. The reaction is allowed to stand with occasional swirling for 5 min and is then cooled to nearly room temperature. Then the reaction mixture is poured over 100 gm of ice. The resulting mixture is stirred until the product solidifies. The lumps are crushed and the crude product is collected by filtration. The crude product is washed in turn with cold water, sodium bicarbonate solution, and again with cold water. To recrystallize the product, it is dissolved in 50 ml of hot 95% ethanol. The solution is then allowed to cool. Since the product tends to separate as an oil, crystallization is promoted by swirling the reaction mixture vigorously as soon as the oil appears. After crystals have started to form, the solution is cooled in cold water and finally in an ice bath. Crystallization is completed in about 5 min. The product is collected by suction filtration. The final product is then washed with a small quantity of cold ethanol. Yield 3.3 gm (70%), mp 69°–71°C.

REFERENCES

1. H. G. Adolph, R. E. Oesterling, and M. E. Sitzman, *J. Org. Chem.* **33,** 4296 (1968).
2. F. Cooke, *Chem. Eng. News* **59,** 3 (August 24, 1981), and Editor's note.
3. Anonymous, *Chem. Eng. News* **58,** 23 (Oct. 6, 1980).
4. D. I. Weisblat and D. A. Lyttle, *J. Am. Chem. Soc.* **71,** 3079 (1949).
5. N. Kornblum, *Org. React.* **12,** 101 (1962).
6. N. Kornblum, H. O. Larson, R. K. Blackwood, D. D. Mooberry, E. P. Oliveto, and G. E. Graham, *J. Am. Chem. Soc.* **78,** 1497 (1956).
7. N. Kornblum and J. W. Powers, *J. Org. Chem.* **22,** 455 (1957).
8. N. Kornblum, R. K. Blackwood, and J. W. Powers, *J. Am. Chem. Soc.* **79,** 2507 (1957).
9. F. T. Williams, Jr., P. W. K. Flanagan, W. J. Taylor, and H. Shechter, *J. Org. Chem.* **30,** 2674 (1965).
10. Anonymous, *Chem. Eng. News* **45,** 41 (1967).
11. R. Adams and J. R. Johnson, "Laboratory Experiments in Organic Chemistry," 4th ed., pp. 299, 301, 303. Macmillan, New York, 1949.
12. H. H. Hodgson and F. Heyworth, *J. Chem. Soc.* p. 1624 (1949).
13. C. K. Ingold, "Structure and Mechanism in Organic Chemistry," 2nd ed. Cornell Univ. Press, Ithaca, New York, 1969.
14. A. Ault, *J. Chem. Educ.* **42,** 267 (1965).

17

NITRILES
(CYANIDES)

1. INTRODUCTION

The best known reactions for the preparation of nitriles are the dehydration of amides and oximes, the condensation of halides with cyanides, and the Sandmeyer reaction. Other methods of preparation involve the addition of hydrogen cyanide to double bonds, oxidation and reduction reactions, and nitrile interchange reactions. A variety of condensation reactions such as the aldol reaction and alkylation of active methylene compounds are also used for the synthesis of nitriles.

From S. R. Sandler and W. Karo, *Organic Functional Group Preparations*, Vol. I, 2d ed. (New York, 1983), 549ff., by permission of Academic Press, Inc.

TABLE I

REPRESENTATIVE CYANIDES PREPARED FROM THEIR CORRESPONDING HALIDE,
AMIDE, OR OXIME

Compound	Yield (%)	Method	Reference
n-Butyl cyanide	93	NaCN–DMSO	1
tert-Butyl cyanide	80	Amide	2
tert-Butyl cyanide	73	Amide	3
n-Hexyl cyanide	72	NaCN–H_2O	4
Cyclohexyl cyanide	93	Amide	5
Benzyl cyanide	90	NaCN–H_2O	6
Benzyl cyanide	87	Amide	7
4-Cyanobiphenyl	50	Sandmeyer	8
o-Chlorophenylacetonitrile	64	Oxime	9

In these preparations the carbon skeleton associated with an existing nitrile-containing structure is changed in a controlled manner.

By the dehydration of amides, primary, secondary, and tertiary nitriles can be obtained in good yields.

The method is applicable to aliphatic and aromatic amides. For example, isobutyronitrile is formed in 86% yield by heating iso-butyramide and phosphorus pentoxide in the absence of a solvent at 100°–200°C and distilling the product as it is formed. Table I illustrates a few of the products prepared by the most common methods.

The most common example of the addition of hydrogen cyanide to a double bond is the formation of cyanohydrins from aldehydes and ketones with this reagent. The addition of hydrogen cyanide to olefins and acetylenes is important industrially but requires special equipment and precautions.

A. HAZARDS AND SAFE HANDLING PRACTICES

■ **CAUTION:** It is of prime importance to recognize and become familiar with the extreme toxicity of cyanide salts and HCN gas.

The alkali cyanides are readily soluble in water and sparingly soluble in the lower alcohols. These solutions are extremely toxic if taken internally and can be absorbed through the skin. The alkali cyanides are alkaline and rapidly liberate HCN on contact with acids or slowly with moist air. Hydrogen cyanide is a volatile liquid, bp 25°–26°C, f.p. −13°C, flammable and explosive with air at 5.6°C at 40% concentration. It decomposes with great violence when exposed to bases and therefore is sold with an acid-stabilizer. When cyanides are being used, each person should work in a well-ventilated hood and wear rubber gloves, an apron, and a gas mask with an HCN canister good for HCN up to a 2% concentration. Antidotes such as sodium thiosulfate and amyl nitrite must be used promptly, but they can be administered only by a qualified physician. One should never work alone and the other party should also be familiar with the necessary precautions. Signs should be posted to alert people entering the area.

It should also be understood that the organic cyanides and nitriles are toxic and are easily absorbed through the skin.

In our own experience while working with sodium cyanide, the question has arisen as to whether exposure to a fatal dose has taken place. If it has, immediate and drastic action by a physician is mandatory; if it has not, administration of antidotes for prophylactic purposes is hazardous, since some of them may be quite toxic.

The safe disposal of excess cyanides and of waste by-products, as well as the proper clean-up of all equipment, is also a problem. Thorough soaking with potassium permanganate solutions is believed to be of some value in the decontamination of inorganic cyanides. Never dispose of cyanides by pouring down the drain. Pour solutions into special waste containers set aside for work with cyanides.

From the legal standpoint, in many, if not all, states in the United States, a laboratory may *not* provide any first aid equipment or antidotes which must be administered orally, by injection, or by inhalation, unless this be done by a physician. Self-administration is done at the risk of the individual, administration of antidotes by a lab partner or other nonmedical individual might subject that person to the most serious legal difficulties under both civil and criminal law.

2. ELIMINATION REACTIONS

A. DEHYDRATION OF AMIDES

The dehydration of amides is probably the best known of the elimination reactions for the preparation of nitriles. One method, the dehydration with phosphorus pentoxide, has been used for many years although it is undoubtedly one of the most messy synthetic procedures available. The procedure normally involves a dry distillation of a mixture of a primary amide and phosphorus pentoxide. Under the reaction conditions, this composition usually melts, chars, and, because a gaseous product is formed, quickly foams up to fill the available space in the cooler parts of the reactor.

Aliphatic and aromatic amides, N-methyl aromatic amides, and benzaldoxime have been dehydrated with silazanes and related compounds [10]. Procedure 17-1 outlines the procedure which we believe to have wide applications.

17-1. Preparation of Benzonitrile [10]

$$3 \; C_6H_5-\overset{O}{\overset{\|}{C}}-NH_2 + \left(HN\overset{CH_3}{\underset{CH_3}{\overset{|}{\underset{|}{Si}}}} \right)_3 \longrightarrow 3 \; C_6H_5-C\equiv N$$

$$+ \; 3\,NH_3 + \left(O\overset{CH_3}{\underset{CH_3}{\overset{|}{\underset{|}{Si}}}} \right)_x \quad (1)$$

In a reflux apparatus topped with a drying tube, equipped with a flask thermometer, and a stirrer, a mixture of 12.1 gm (0.1 mole) of benzamide and 7.8 gm (0.102 equivalents) of hexamethylcyclotrisilazane is heated for 20 hr at 220°C. When the evolution of ammonia has become negligible, the reaction mixture is distilled. The product is isolated at 70°C (approx. 10 mm). Yield: 9.6 gm (91%). By gas–liquid chromatography analysis this product was essentially pure.

3. CONDENSATION REACTIONS

A. THE USE OF CYANIDE SALTS

The nucleophilic displacement reaction of both alkyl and aryl halides by cyanide ions in dipolar aprotic solvents has been studied extensively [1, 6, 11]. This reaction can be visualized as a condensation of NaCN with RX to give RCN. In the alkyl series, sodium cyanide is usually used with dimethyl sulfoxide or dimethylformamide as the solvent [12–20]. Some examples from Friedman and Shechter [1] are reproduced in Table II.

Perhaps the earliest example of a nucleophilic displacement reaction for the synthesis of aliphatic nitriles was the Pelouze reaction

TABLE II

REACTIONS OF ALKYL AND CYCLOALKYL HALIDES WITH SODIUM OR POTASSIUM
CYANIDE IN DIMETHYL SULFOXIDE[a]

Halide	Cyanide[b]	Reaction temp (°C)	Reaction time (hr)[c]	Yield of nitrile (%)
1-Chlorobutane	NaCN	140°	0.25	93
1-Chlorobutane	KCN	120°–140°	10	69
1-Chloro-2-methylpropane	NaCN	140°[d]	0.5	88[e]
1-Chloro-3-methylbutane	NaCN	100°–140°	2	85[f]
1-Chloro-2-methyl-2-phenylpropane	NaCN	120°	24	26
Benzyl chloride	NaCN	35°–40°[g]	2.5	92[h]
p-Nitrobenzyl chloride	NaCN	35°–40°	1	0[i]
2-Chlorobutane	NaCN	120°–145°	3	64[j]
2-Chlorobutane	KCN	120°–138°[k]	24	42
Chlorocyclopentane	NaCN	125°–130°	3	70
Chlorocyclohexane[n]	NaCN	130°–80°	4	0[l]
2-Chloro-2-methylpropane	NaCN	130°–105°	4	0[m]
1-Bromobutane[o]	NaCN	60°–90°	0.6	92[o]
1-Bromo-2-methylpropane	NaCN	70°	2	62
2-Bromobutane	NaCN	70°	6	41[p]

[a] Reprinted from L. Friedman and H. Shechter, *J. Org. Chem.* **25**, 877 (1960). Copyright 1960 by the American Chemical Society. Reprinted by permission of the copyright owner.

[b] The ratio of halide (moles), cyanide (moles), and DMSO (ml) usually used was 1:1.2:250.

[c] The reaction time listed is the sum of that for addition of the halide and subsequent reaction at the given temperature.

[d] The halide was added in 10 min to the initial mixture at 80°C. The reaction is mildly exothermic and was completed by heating to 140°C until refluxing ceased.

[e] Bp 128°C, n_D^{20} 1.3926.

[f] Bp 151°–155°C, n_D^{20} 1.4047–1.4051; lit. b.p. 150°–155°C [H. Rupe and K. Glenz, *Helv. Chim. Acta* **5**, 939 (1922)].

[g] The reaction mixture was cooled externally.

[h] Bp 90.5°–91°C (5 mm), n_D^{20} 1.5237–1.5238; lit. bp 115°–120°C (10 mm), n_D^{20} 1.5242 [J. W. Bruhl, *Z. Phys. Chem.* **16**, 218 (1895)].

[i] 4,4'-Dinitrostilbene is formed in 78% crude yield, m.p. 286°–288°C; lit. 288°C [P. Ruggli and F. Lang, *Helv. Chim. Acta* **21**, 42 (1938); R. Walden and A. Kernbaum, *Ber. Dtsch. Chem. Ges.* **23**, 1959 (1890)].

[j] Bp 123.5°–124°C (742 mm), n_D^{20} 1.3898–1.3900; lit. bp 125°C [M. Hanriot and L. Bouveault, *Bull. Soc. Chim. Fr.* [3] **1**, 172 (1889)].

[k] The halide was added dropwise in 6 hr to the mixture at 120°–125°C; the mixture was then heated for 18 hr until it reached 138°C.

[l] The reaction mixture became dark and gave cyclohexene, gases, and a black intractable product.

[m] Upon initiating reaction at 130°C, gases (2-methylpropene, hydrogen cyanide, and formaldehyde) were evolved, the temperature dropped to 105°C, and black resinous materials were formed. The desired product was not obtained at lower reaction temperatures.

[n] The bromide was added in 30 min to the cyanide mixture at 60°C while effecting cooling of the reaction; the mixture was then heated for 15 min at 90°C.

[o] Bp 138°–139°C (742 mm), n_D^{20} 1.3970.

[p] Upon addition of the 2-bromobutane, the temperature was maintained at 55°–60°C by intermittent cooling. In subsequent reaction gases were evolved and the mixture became progressively darker and malodorous.

of dimethyl sulfates or potassium monoalkysulfates with potassium cyanide [21].

The reaction is also applicable to both primary and secondary halides and to α-halo ethers [16]. Iodide ions catalyze the replacement of chlorine. Tertiary halides and alicyclic halides tend to form olefins or decomposition products. The preparation of pentanonitrile using dimethyl sulfoxide is representative of the procedure although dimethylformamide has also been used [17].

17-2. Preparation of Pentanonitrile (Valeronitrile) [1]

$$CH_3CH_2CH_2CH_2Cl + NaCN \xrightarrow{\text{DMSO}} CH_3CH_2CH_2CH_2CN + NaCl \qquad (2)$$

■ **CAUTION:** See notes at the beginning of this chapter on the hazards in handling sodium cyanide. Dimethyl sulfoxide should also be handled with caution. This solvent is readily absorbed through the skin, carrying with it some impurities which may be present on the skin (e.g., residual soap), leading to physiological reactions. The solvent is also said to reduce the sensation of pain. Test animals have exhibited reactions in their eyes upon treatment with DMSO.

In a hood, with due precautions, to a rapidly stirred mixture of 53 gm (1.08 moles) of reagent grade sodium cyanide in 240 ml of dimethyl sulfoxide at 80°C is added 93 gm (1 mole) of 1-chlorobutane over a 15-min period. During this period the reaction temperature rises rapidly and has to be controlled at 140° $\pm$ 5°C by water cooling. The reaction mixture cools rapidly after the addition is completed.

The reaction mixture is cooled, diluted with water to approximately 1 liter, and extracted with three 150-ml portions of ether.

In the hood, the ether extracts are washed in turn with 6 N hydrochloric acid (to hydrolyze a small amount of isocyanide) and with water, and are then dried over calcium chloride. The ether is evaporated and the residue is fractionally distilled from phosphorus pentoxide. The pentanonitrile passes over at 138°–139°C (747 mm). Yield 77 gm (93%).

A paper by Regen and co-workers reports on a convenient synthesis of nitriles which we believe to have application to other displacement reactions. In this procedure, alumina is impregnated with sodium cyanide. This solid is stirred with a solution of an alkyl halide at 90°C. The insoluble reagent is filtered off and the filtrate is freed of solvent to leave a high purity residue of the nitrile in substantial yield [22]. While the use of impregnated ion exchange resins has been suggested before, this particular procedure seems particularly simple [cf Refs. 23 and 24]. Procedure 17-3 illustrates the newer technique.

17-3. Preparation of 1-Cyanododecane [22]

$$CH_3(CH_2)_{10}CH_2Br \xrightarrow{\text{NaCN on } Al_2O_3} CH_3(CH_2)_{10}CH_2CN \tag{3}$$

(a) Preparation of Alumina Impregnated with Sodium Cyanide. In a 200-ml round-bottom flask is dissolved 8.0 gm (0.1232 mole) of sodium cyanide in 20 ml of distilled water. Then 16.0 gm of neutral alumina (Bio-Rad Laboratories, Alumina AG-7, 100–200 mesh) is added. The flask is then attached to a rotary evaporator in a hood and freed of water under reduced pressure while warming with a water bath maintained below 65°C. The impregnated alumina is then dried at 110°C at 0.05 mm for 4 hr.

(b) Synthesis of 1-Cyanododecane. In a 100-ml round-bottom flask containing a Teflon-coated magnetic stirrer and equipped with a thermometer and a reflux condenser topped with a drying tube are placed 2.5 gm (0.01 mole) of 1-bromododecane, 15 gm of the impregnated alumina, and 30 ml of toluene. With stirring, this mixture is heated at 90°C for 45 hr.

At the end of this period, the reaction mixture is cooled to room temperature. The spent impregnated alumina is washed with 100 ml of toluene. The toluene extract is combined with the filtrate. The combined toluene solution is freed of solvent, under reduced pressure. The product is a colorless liquid identical to an authentic sample as determined by infrared and NMR spectra. Yield: 1.96 gm (100%).

B. CONVERSION OF CARBONYL DERIVATIVES TO NITRILES BY CYANOETHYLATION OF ACTIVE METHYLENE COMPOUNDS

An example of the cyanoethylation reaction is the preparation of 1,1,1-*tris*-(2-cyanoethyl)acetone [25]. In this case, alcoholic potassium hydroxide is used as a catalyst, while benzyltrimethylammonium hydroxide is frequently used in other preparations.

17-4. Preparation of 1,1,1-*Tris*(2-cyanoethyl)acetone [25]

$$CH_3-\overset{\overset{\displaystyle O}{\|}}{C}-CH_3 + 3\ CH_2{=}CHCN\ \longrightarrow\ CH_3-\overset{\overset{\displaystyle O}{\|}}{C}-C(CH_2CH_2CN)_3 \qquad (4)$$

To a stirred solution of 29 gm (0.5 mole) of acetone, 30 gm of *tert*-butyl alcohol, and 2.5 gm of 30% ethanolic potassium hydroxide solution maintained between 0° and 5°C, a solution of 80 gm (1.5 moles) of acrylonitrile in 37 gm of *tert*-butyl alcohol is added dropwise over a period of $1\frac{1}{2}$ hr. Stirring is continued for 2 hr at 5°C. Then the product is filtered off. Yield 84 gm (79.5%), mp (after crystallization from water) 154°C.

C. DEHYDRATION OF OXIMES

The dehydration of oximes by such reagents as phosphorus pentoxide, thionyl chloride, acetic anhydride, acyl chlorides, and phosphorus pentachloride is well known [26]. In effect, this dehydration procedure permits the conversion of aldehydes to nitriles with the same number of carbon atoms. A modification applicable only to the aromatic series makes use of boiling acetic acid as a dehydrating agent [27]. With other dehydrating agents, aliphatic aldehydes also may be converted to nitriles. Oximes may be converted readily to nitriles by an acid-catalyzed reaction with ortho esters [28].

17-5. Preparation of Heptanonitrile [28]

$$CH_2CH_2CH_2CH_2CH_2CH_2CH{=}NOH + HC(OC_2H_5)_3$$

$$\downarrow{\scriptstyle H^+}$$

$$CH_2CH_2CH_2CH_2CH_2CH_2C{\equiv}N + HCO_2C_2H_5 + 2\ C_2H_5OH \qquad (5)$$

In a 100-ml flask attached to a short distillation column are placed 15.0 ml (0.1 mole) of *n*-heptaldehyde oxime and 20.0 ml (0.12 mole) of triethyl orthoformate containing one drop of methanesulfonic acid. The mixture is heated gently to distil out ethyl formate and ethanol as they form. After the coproducts have been removed, the residue is distilled under reduced pressure at 70°–72°C (10 mm). Yield: 95%.

REFERENCES

1. L. Friedman and H. Schechter, *J. Org. Chem.* **25,** 877 (1960).
2. S. M. McElvain, R. L. Clarke, and G. D. Jones, *J. Am. Chem. Soc.* **64,** 1968 (1942).
3. F. C. Whitmore, C. I. Moll, and V. C. Meunier, *J. Am. Chem. Soc.* **61,** 683 (1939).
4. G. H. Jeffery and A. I. Vogel, *J. Chem. Soc.* p. 674 (1948).
5. C. H. Tilford, M. G. Van Campen, Jr., and R. S. Shelton, *J. Am. Chem. Soc.* **69,** 2902 (1947).
6. R. Adams and A. F. Thal, *Org. Synth. Collect. Vol.* **1,** 107 (1941).
7. J. A. Mitchell and E. E. Reid, *J. Am. Chem. Soc.* **53,** 321 (1931).
8. L. Bauer and J. Cymerman. *J. Chem. Soc.* p. 2078 (1950).
9. N. Campbell and J. E. McKail, *J. Chem. Soc.* p. 1251 (1948).
10. W. E. Dennis, *J. Org. Chem.* **35,** 3253 (1970).
11. L. Friedman and H. Schechter, *J. Org. Chem.* **26,** 2522 (1961).
12. R. A. Smiley and C. Arnold, *J. Org. Chem.* **25,** 257 (1960).
13. M. S. Newman and S. Otsuka, *J. Org. Chem.* **23,** 797 (1958).
14. M. Hirakura, M. Yanargita, and S. Inayama, *J. Org. Chem.* **26,** 3061 (1961).
15. A. J. Parker, *Q. Rev. Chem. Soc.* **16,** 163 (1962).
16. P. A. Argabright and D. W. Hall, *Chem. Ind. (London)* p. 1365 (1964).
17. Y. Oshiro, H. Tanisake, and S. Komori, *Yuki Gosei Kagaku Kyokaishi* **24,** 950 (1966).
18. J. J. Bloomfield and A. Mitra, *Chem. Ind. (London)* p. 2012 (1966).
19. D. Brett, I. M. Downie, and J. B. Lee, *J. Org. Chem.* **32,** 855 (1967).
20. M. A. Schwartz, M. Zoda, B. Vishnuvajjala, and I. Mami, *J. Org. Chem.* **41,** 2502 (1976).
21. J. Pelouze, *Justus Liebigs Ann. Chemie* **10,** 249 (1834).
22. S. L. Regen, S. Quici, and S.-J. Liaw, *J. Org. Chem.* **44,** 2029 (1979).
23. M. Gordon and C. E. Griffin, *Chem. Ind. (London)* p. 1091 (1962).
24. Y. Urata, *Nippon Kagaku Zasshi* **83,** (10), 1105 (1962); *Chem. Abstr.* **59,** 11240c (1963).
25. H. A. Bruson and T. W. Riener, *J. Am. Chem. Soc.* **64,** 2850 (1942).
26. Houben-Weyl, *in* "Methoden der organischen Chemie" (E. Mueller, ed.), 4th ed., Vol. VII, p. 325. Thieme, Stuttgart, 1952.
27. J. H. Hunt, *Chem. Ind. (London)* p. 1873 (1961).
28. M. M. Rogié, J. F. Van Peppen, K. P. Klein, and T. R. Demmin, *J. Org. Chem.* **39,** 3424 (1974).

18

MERCAPTANS, SULFIDES, AND DISULFIDES

1. INTRODUCTION

The reaction of sodium hydrosulfide with activated aliphatic or aromatic halides yields mercaptans. If aromatic nitro substituents are present, they are reduced to amino groups.

$$RX + NaSH \longrightarrow RSH + NaX \tag{1}$$

Dihalides yield dithiols. Tertiary aliphatic halides usually give olefins.

From S. R. Sandler and W. Karo, *Organic Functional Group Preparations*, Vol. I, 2d ed. (New York, 1983), 585–602, by permission of Academic Press, Inc.

Two other methods which have found general applicability in the laboratory are the reaction of alkyl halides with thiourea with subsequent alkaline hydrolysis, and the reaction of free sulfur with aryl lithium or Grignard reagents as shown below.

$$(NH)_2CS + RX \longrightarrow NH_2{-}C(SR){=}NH_2{}^+ + X^- \xrightarrow{\text{NaOH}} RSH + (NH_2CN)_x \quad (2)$$

$$ArM + S \longrightarrow ArSM \xrightarrow{\text{H}_2\text{O}} ArSH + MOH \qquad\qquad (3)$$
$$M = MgX \text{ or } Li$$

A widely used laboratory method for the preparation of sulfides is the action of halides on metallic sulfides.

$$Na_2S + 2\,RX \longrightarrow R_2S + 2\,NaX \qquad\qquad\qquad (4)$$

The monohydrate of sodium sulfide is a satisfactory reagent to afford high yields of symmetrical sulfides.

The conversion of mercaptans to sulfides is effected by converting to the sodium mercaptide and reacting with active halides or dialkyl sulfates.

$$RSH \longrightarrow RSNa \xrightarrow[\text{or } R_2SO_4]{R'X} RSR' \qquad\qquad\qquad (5)$$

Polar solvents such as amides favor this reaction.

The addition of hydrogen sulfide or mercaptans to olefins is another important method for the preparation of mercaptans and sulfides via ionic (6a) or free radical means (6b).

$$RCH{=}CH_2 + R'SH \xrightarrow{\text{H}^+} \underset{\underset{SR'}{|}}{RCH{-}CH_3} \qquad\qquad (6a)$$

$$\xrightarrow[\text{or } h\nu]{H_2O_2} RCH_2CH_2SR' \qquad\qquad (6b)$$
$$R = H, \text{ alkyl, or aryl group}$$

The reaction of activated alkyl or aryl halides with sodium sulfide yields disulfides.

$$2\,RX + Na_2S_2 \longrightarrow RSSR + 2\,NaX \qquad\qquad\qquad (7)$$

The oxidation of mercaptans by hydrogen peroxide is the best method of preparing disulfides in good yields if the corresponding mercaptan is readily available. Recently it has been reported that dimethyl sulfoxide oxidizes thiols at 80°–90°C to give disulfides in good yields. Oxidation by free sulfur has also been reported.

2. MERCAPTANS (THIOLS)

A. CONDENSATION REACTIONS OF METAL SULFIDES OR HYDROSULFIDES WITH ALKYL HALIDES

Sodium of potassium hydrosulfide react with activated halides to give mercaptans. Alkyl sulfates and primary or secondary halides act as alkylating agents. Such hydrosulfides are reducing agents some structural features of the alkylating agent may be reduced. Thus, for example, the nitro group in *p*-chloronitrobenzene is reduced to give *p*-amino thiophenol. Sodium sulfide reacts with 1 mole of an activated halide to give the sodium salt of the thiol. The addition of acid liberates the free thiol. In addition hydrogen sulfide may be capable of reacting with an intermediate carbonium ion to give a thiol.

Recently it has been reported that alkyl thiols and silyl-substituted alkyl thiols have been synthesized from the corresponding alkyl halides using a combination of hydrogen sulfide and ammonia or alkylamines.

18-1. Preparation of *p*-Aminothiophenol [1]

$$Na_2S + Cl\text{—}\langle\bigcirc\rangle\text{—}NO_2 \longrightarrow HS\text{—}\langle\bigcirc\rangle\text{—}NH_2 \tag{8}$$

To a flask containing 192 gm (2 moles) of sodium sulfide monohydrate dissolved in 2 liters of water is added 128 gm (0.81 mole) of *p*-chloronitrobenzene and the mixture is refluxed for 8 hr. A small amount of an orange colored oil separates which is ether-extracted and discarded. The remaining aqueous layer is saturated with sodium chloride and 240 gm (4 moles) of glacial acetic acid is added. The liberated oil is extracted several times with ether and the ether extract is dried. Evaporation of the solvent leaves a residue which upon distillation under reduced pressure yields 70 gm (69%) of product, bp 143°–146°C (17 mm), mp 43°–45°C (literature 46°C).

18-2. Preparation of Triphenylmethyl Mercaptan [2]

$$(C_6H_5)_3-C-Cl + H_2S \xrightarrow{Al_2O_3} (C_6H_5)_3-C-SH + HCl \qquad (9)$$

To 400 ml of dry dioxane (dried by refluxing with 4% of its weight of sodium for 4 hr) are added 100 gm of activated alumina (Alcoa F-20) and 100 gm of triphenylmethyl chloride. Dry hydrogen sulfide is then passed into the mixture below the level of the suspended alumina at such a rate as to agitate the solution and to keep the solution saturated. After 15 hr the alumina is filtered off and washed with two 50-ml portions of dioxane. The filtrate and dioxane washings are poured into 2 liters of ice water and the contents are stirred until a granular product precipitates. The product is filtered, and then dissolved in 500 ml of boiling isopropanol. Slow cooling yields 75–80 gm (75–80%) of pale yellow crystals of triphenyl mercaptan, mp 106°–107°C. The mother liquor yields no appreciable amount of product but a small amount of triphenylcarbinol is isolated.

B. ADDITION OF HYDROGEN SULFIDE TO OLEFINS AND OTHER UNSATURATED COMPOUNDS

Hydrogen sulfide adds to olefins and unsaturated compounds to give thiols in good to excellent yields. The reaction is usually very rapid and is catalyzed either by free radical initiators (peroxides or ultraviolet light) or ionic types of catalysts such as transition metals, sulfuric acid, or even free sulfur. The free radical initiated reaction addition gives anti-Markovnikoff addition products [Eq. (10)] whereas the ionic catalyzed reaction gives Markovnikoff addition products [Eq. (11)]. Both types of reactions work well in the laboratory and are also practiced commercially.

$$RCH{=}CH_2 + H_2S \xrightarrow[\text{or } h\nu]{\text{ROOH}} RCH_2-CH_2SH \qquad (10)$$

$$RCH{=}CH_2 + H_2S \xrightarrow[\substack{\text{or metals and} \\ N \text{ compounds}}]{H_2SO_4} \underset{\underset{SH}{|}}{RCH}-CH_3 \qquad (11)$$

18-3. Preparation of *n*-Butylmercaptan [3, 4]

(a) Sealed Tube Experiment [3]. 1-Butene (0.044 mole) and hydrogen sulfide (0.088 mole) are sealed in a 10-mm I.D. quartz tube.

The contents are cooled to 0°C and illuminated with a ultraviolet light source (quartz mercury arc having a wavelength below about 2900–3000 angstroms) for 4 min. The tube is cooled and opened. The unreacted (approx. 20%) butene and hydrogen sulfide are recovered and the product (3.8 ml) is distilled to give 85% yield of *n*-butylmercaptan (bp 97°–98°C) and 15% yield of di-*n*-butyl sulfide (bp 188°–189°C.)

(b) Pressure Autoclave [4]. A mixture of hydrogen sulfide (2.0 moles) and 1-butene (1.0 mole) is charged to a 3-gal autoclave equipped with a mechanical stirrer, cooling coil, and a quartz immersion well containing a 200-watt high pressure ultraviolet light. The pressure of the reaction vessel at the start is 200 psi and the reaction is maintained at 90°–130°F. The ultraviolet lamp is started and samples from the reaction are analyzed by gas chromatography. Olefin conversion is 7–22% per pass and mercaptan yield is about 90–97%. The unreacted olefin and hydrogen sulfide are recycled to the reaction mixture.

3. SULFIDES

A. CONDENSATION REACTION OF SODIUM OR OTHER METAL MERCAPTIDES WITH ACTIVE ALKYL OR ARYL HALIDES

Sodium mercaptides are prepared from the mercaptans and aqueous or alcoholic solutions of sodium hydroxide or alcoholic sodium ethoxide. The sodium mercaptide reacts with halides, chlorohydrins, esters of sulfonic acid, or alkyl sulfonates [6] to give sulfides in yields of 70% or more. A recent report describes a general procedure for synthesizing aryl thioesters by a nucleophilic displacement of aryl halide with thiolate ion in amide solvents. No copper catalysis is necessary as in an Ullmann-type reaction.

$$ArSK + Ar'X \xrightarrow{\text{DMF}} ArSAr' + KX \qquad (12)$$

■ **CAUTION:** Lithium *p*-nitrophenylthiolate (prepared from free thiol and *n*-butyllithium) has been reported to detonate fairly violently after drying and being exposed to the atmosphere.

B. CONDENSATION OF METALLIC SULFIDES WITH HALIDES

Sodium sulfide reacts with aqueous alcoholic solutions of the halides to give good yields of the symmetrical sulfide. Halides containing β-carboxy, hydroxyl, ethoxyl, or diethylamino groups are effective in this reaction. Long chain halides give cyclic sulfides.

C. MERCAPTYLATION OF THE DOUBLE BOND

Mercaptans add to olefins in good yields according to Markovnikoff's rule in the presence of sulfuric acid, boron trifluoride, or sulfur and also in an anti-Markovnikoff fashion in the presence of peroxides or via photochemical means. Vinyl chloride and allyl alcohol give lower yields than conjugated olefinic ketones, aldehydes, esters, and cyanides. Cupric acetate or triethylamine can be used as a catalyst for the reaction of methyl mercaptan with acrolein to give β-mercaptopropionaldehyde in 84% yield. Allene reacts homolytically with methanethiol to give allyl sulfide and the 1,3- and 1,2-dimethylthiopropanes.

18-4. Preparation of *tert*-Butyl Sulfide [5]

$$(CH_3)_2C=CH_2 + (CH_3)_3C-SH \longrightarrow [(CH_3)_3C-]_2S \qquad (13)$$

To an ice cold mixture of 225 gm (2.3 mole) of concentrated sulfuric acid and 65 gm of water is added 50.4 gm (0.60 mole) of *tert*-butyl alcohol at such a rate to keep the temperature at 10°C. After the addition of the alcohol, 27 gm (0.30 mole) of *tert*-butylmercaptan is added over a 30-min period. The ice bath is removed and the mixture is warmed to room temperature. The mixture is poured into 500 gm of ice, extracted with ether, the ether extract dried over magnesium sulfate, and concentrated. Distillation of the residue yields 27.9 gm (87% based on the reacted mercaptan) of *t*-butyl sulfide, bp 148°–149°C.

4. DISULFIDES

A. CONDENSATION REACTIONS

The reaction of alkyl halides or activated aryl halides with sodium disulfide produces disulfides in good yields. 1,3-Dihalides yield cyclic sulfides and polysulfides are produced with two to five equivalents of sodium sulfide. The hydroxyl or nitro groups do not interfere with the reaction.

Recently the preparation of unsymmetrical disulfides in fair to moderate yields has been reported to be feasible by the reaction of alkylthiosulfates with sodium alkylthiolates.

$$RBr + Na_2S_2O_3 5H_2O \longrightarrow RS-SO_3Na + NaBr \qquad (14)$$

$$RS-SO_3Na + R'SNa \longrightarrow RS-SR' + Na_2SO_3 \qquad (15)$$

18-5. Preparation of Di-*n*-amyl Disulfide [6]

$$2 CH_3(CH_2)_3CH_2Br + Na_2S_2 \longrightarrow [CH_3(CH_2)_3CH_2]_2S_2 \qquad (16)$$

To 58.5 gm (0.75 mole) of sodium sulfide dissolved in 750 ml of 95% ethanol are added while refluxing 24 gm (0.75 mole) of sulfur. The mixture is stirred until the sulfur dissolves. This hot solution is added to 151 gm (1.0 mole) of *n*-amyl bromide in 250 ml of 95% ethanol at such a rate as to maintain gentle refluxing (20 min). The mixture is refluxed for 3 hr and then allowed to stand overnight. One-third of the alcohol is removed under reduced pressure and the remaining solution is extracted with 500 ml of benzene. The extract is washed several times with water, dried, and distilled to yield 62 gm (69%), bp 90°–92°C (1 mm), n_D^{25} 1.4875.

REFERENCES

1. J. E. Bittell and J. L. Speier, *J. Org. Chem.* **43**, 1687 (1978).
2. N. Kharasch and H. R. Williams, *J. Am. Chem. Soc.* **72**, 1843 (1950).
3. W. E. Vaughan and F. F. Rust, *J. Org. Chem.* **7**, 472 (1942).
4. J. R. Edwards, U.S. Patent 3,412,001 (1968).
5. E. A. Fehnel and M. Carmack, *J. Am. Chem. Soc.* **71**, 92 (1949).
6. D. S. Tarbell and D. K. Fukushima, *J. Am. Chem. Soc.* **68**, 1458 (1946).

19

SULFOXIDES

1. INTRODUCTION

The first reported synthesis of a sulfoxide was by Märcker in 1865. The methods generally involved the controlled oxidation of a sulfide by oxidizing agents such as hydrogen peroxide, ozone, peracids, hydroperoxides, manganese dioxide, selenium dioxide, nitric acid, chromic acid, dinitrogen tetroxide, iodosobenzene, and others.

$$R_2S \xrightarrow{\text{[O]}} R_2S{=}O \tag{1}$$

2. OXIDATION METHODS

One reported method describes the convenient use of sodium metaperiodate as an oxidizing agent to form sulfoxides from sulfides free of

From S. R. Sandler and W. Karo, *Organic Functional Group Preparations*, Vol. I, 2d ed. (New York, 1983), 603–609, by permission of Academic Press, Inc.

sulfone contaminants. The method finds use in preparing linear and cyclic aliphatic or aryl sulfoxides as shown in Table I.

The reaction is carried out by adding the sulfide in a methanol–water mixture to a slight excess of 0.5 M aqueous sodium metaperiodate at ice bath temperatures. Higher temperatures lead to sulfone formation. The reaction is complete in 3 to 12 hr, and yields of 90% or better of the sulfoxide are obtained.

■ **CAUTION:** Dimethyl sulfoxide has been reported to be easily absorbed through the skin and to then pass into the blood stream. Others have given some indication that skin irritation or burns and eye injuries may result from prolonged exposure to sulfoxides. Therefore, great caution should be exercised in handling and preparing sulfoxides since these compounds possess this great tendency of skin penetration.

A laboratory explosion has been reported in the preparation of methyl sulfinyl carbanion with sodium hydride and dimethyl sulfoxide.

$$(CH_3)_2SO + NaH \longrightarrow CH_3SOCH_2^- Na^+ + H_2 \qquad (2)$$

Whether this occurs with other sulfoxides is unknown.

19-1. General Method of Oxidation of Alkyl Sulfides to Sulfoxides Using Sodium Metaperiodate [1]

$$R_2S + NaIO_4 \longrightarrow R_2SO + NaIO_3 \qquad (3)$$

To 210 ml (0.105 mole) of a 0.5 M solution of sodium metaperiodate at 0°C is added 0.1 mole of sulfide. The mixture is stirred at ice bath temperature, usually overnight. The precipitated sodium iodate is removed by filtration, and the filtrate is extracted with chloroform. The extract is dried over anydrous magnesium sulfate, and the solvent is removed under reduced pressure. The sulfoxide is purified by distillation, crystallization, or sublimation.

Typical uses for this procedure are found in Table I where 13 sulfoxides are shown to be prepared by this method. The overall yield ranges from 65 to 99 per cent.

REFERENCES

1. N. J. Leonard and C. R. Johnson, *J. Org. Chem.* **27**, 284 (1962).

TABLE I
Sulfoxides Produced by Sodium Metaperiodate Oxidation of the Corresponding Sulfides[a,u]

Name	Structure	Yield (%)	Mp (Bp) (°C)	
			Found	Reptd.
1-Thiacyclooctan-5-one 1-oxide	OC⟨(CH$_2$)$_3$ / (CH$_2$)$_3$⟩SO	91	91–92	—
1-Thiacyclohexan-4-one 1-oxide	OC⟨(CH$_2$)$_2$ / (CH$_2$)$_2$⟩SO	97	109–110	113[c]
Methyl 4-ketopentyl sulfoxide	$CH_3SO(CH_2)_3COCH_3$	98	22.5–23.5 (99–101/ 0.12 mm)	—
Phenyl sulfoxide	$(C_6H_5)_2SO$	98	69–71	69–71[e,f]
Methyl phenyl sulfoxide	$CH_3SOC_6H_5$	99	29–30 (83–85/0.1 mm)	29.5 (104.5/0.7 mm)[h,i]
Thian 1-oxide	$(CH_2)_5SO$	99	67–68.2	60–61.5[j]
1,4-Oxathian 4-oxide	O⟨(CH$_2$)$_2$ / (CH$_2$)$_2$⟩SO	83[k]	46–47.2	44.5–45[l]
Bis(2-diethylaminoethyl) sulfoxide	$[(C_2H_5)_2N(CH_2)_2]_2SO$	85[m]	Dipicrate 146–148	—
1-Benzylsulfinyl-2-propanone	$C_6H_5CH_2SOCH_2COCH_3$	89	126–126.5	125[n]
Acetoxymethyl methyl sulfoxide	$CH_3COOCH_2SOCH_3$	72	(85–90/0.1 mm)	—
Phenylsulfinylacetic acid[o]	$C_6H_5SOCH_2COOH$	99	118–119.5	113–115[p]
Benzyl sulfoxide[r]	$(C_6H_5CH_2)_2SO$	96	135–136	132–133[f]
Ethyl sulfoxide	$(C_2H_5)_2SO$	65[s]	(45–47/0.15 mm)	88–89/15 mm)[t]

[a] Carbon tetrachloride solution.

[b] Chloroform solution.

[c] G. M. Bennett and W. B. Waddington [*J. Chem. Soc.* p. 2829 (1929)] reported mp 113°C, but were unable to repeat their preparation.

[d] Repeated purification procedures did not improve analysis.

[e] H. H. Szmant and R. L. Lapinski, *J. Am. Chem. Soc.* **78**, 458 (1956).

[f] R. L. Shriner, H. C. Struck, and W. J. Jorison. *J. Am. Chem. Soc.* **52**, 2060 (1930).

[g] $\lambda_{max}^{C_4H_5OH}$ 274 mμ (log of 3.3), 233 (4.2) [H. P. Koch. *J. Chem. Soc.* p. 2892 (1950)].

[h] L. Horner and F. Hübenett, *Justus Liebigs Ann. Chem.* **579**, 193 (1953).

[i] C. C. Price and J. J. Hydock, *J. Am. Chem. Soc.* **74**, 1943 (1952).

[j] M. Tamres and S. Searles, Jr., *J. Am. Chem. Soc.* **81**, 2100 (1959).

[k] Yield based on technical thioxane.

TABLE I (*Continued*)

$v_{max}S{=}O$ (cm^{-1})	Formula	C (%) Calcd.	C (%) Found	H (%) Calcd.	H (%) Found	Other identifying properties, Remarks
1049[a]	$C_7H_{12}O_2S$	52.48	52.46	7.55	7.86	$v_{C=O}$[a] 1710 cm^{-1} extremely hygroscopic
1055[a]	$C_5H_8O_2S$	45.43	44.72[d]	6.10	6.19	$v_{C=O}$[a] 1725 cm^{-1}
1058[b]	$C_6H_{12}O_2S$	48.66	48.86	8.17	8.25	n_D^{25} 1.4873, $v_{C=O}$[b] 1718 cm^{-1}
1033[b]	$C_{12}H_{10}OS$	—	—	—	—	$\lambda_{max}^{C_2H_5OH}$ 274 mμ (log ϵ 3.2) 233 mμ (log ϵ 4.1)[g]
1050[a]	C_7H_8OS	59.90	59.75	5.71	6.18	Very hygroscopic
1045[a]	$C_5H_{10}OS$	—	—	—	—	Liquifies immediately on exposure to the atmosphere
1026[a]	$C_4H_8O_2S$	—	—	—	—	Hygroscopic
—	Dipricate $C_{24}H_{34}N_8O_{15}S$	40.80	40.98	4.86	5.16	—
1046[b]	$C_{10}H_{12}O_2S$	61.17	61.33	6.16	6.22	$v_{C=O}$[b] 1705 cm^{-1}
1044[a]	$C_4H_8O_3S$	35.28	35.17	5.92	5.80	n_D^{25} 1.4798; $v_{C=O}$[a] 1762 cm^{-1}
1015[q]	$C_8H_8O_3S$	—	—	—	—	$v_{C=O}$[q] 1732 cm^{-1}
1025[b]	$C_{14}H_{14}OS$	73.00	73.12	6.13	6.06	Analytically pure after single crystallization from ethanol
1047[a]	$C_4H_{10}OS$	—	—	—	—	n_D^{25} 1.4676

[l] French Patent 859, 886 (1940), *Chem. Abstr.* **42**, 3783 (1948).

[m] Crude yield. No formal purification of free base was made. Characterized as the dipricate.

[n] C. Wahl, *Ber. Dtsch. Chem. Ges.* **55**, 1449 (1922).

[o] Isolated by lyophilization of reaction mixture, followed by extraction with hot ethyl acetate.

[p] A. Tananger, *Ark. Kemi. Mineral. Geol.* **24A**, No. 10 (1947).

[q] Nujol mull.

[r] Oxidation by 0.25 M sodium metaperiodate in 50% methanol.

[s] Lower yield due to incomplete extraction. Ethyl sulfone (5%) was formed during heating used to concentrate reaction mixture prior to extraction.

[t] R. Pummerer, *Ber. Dtsch. Chem. Ges.* **43**, 1401 (1910).

[u] Reprinted from N. J. Leonard and C. R. Johnson, *J. Org. Chem.* **27**, 283 (1962). Copyright 1962 by the American Chemical Society. Reprinted by permission of the copyright owner.

20
SULFONES

1. INTRODUCTION

The oxidation of either sulfides or sulfoxides yields the corresponding sulfone. Some of the oxidizing agents that are described in the literature are hydrogen peroxide, peracids, oxygen, ozone, organic peroxides, potassium permanganate, potassium persulfate, sodium hypochlorite, hypochlorous acid, ruthenium tetroxide, oxides of nitrogen, nitric acid, and anodic oxidation.

Salts of sulfinic acids, especially benzene sulfinites, are easily alkylated by primary and secondary benzyl halides and by alkyl sulfates to sulfones.

$$ArSO_2Na + RX \longrightarrow ArSO_2R + NaX \tag{1}$$

Aryl halides also undergo this reaction, provided the halogen is activated by nitro groups in the *ortho* or *para* position.

From S. R. Sandler and W. Karo, *Organic Functional Group Preparations*, Vol. I, 2d ed. (New York, 1983), 610–618, by permission of Academic Press, Inc.

The Friedel–Crafts condensation reaction of aromatic hydrocarbons with sulfonyl chlorides yields sulfones.

The reaction of Grignard reagents with sulfonyl chlorides also yields sulfones.

Sulfones are produced as by-products in the sulfonation of aromatic hydrocarbons, probably as a result of the condensation of the sulfonic acid with unreacted hydrocarbon. A more recent modification of preparative value is the preparation of aromatic sulfones by the condensation of aromatic sulfonic acids and aromatic hydrocarbons in a polyphosphoric acid medium.

$$\text{ArSO}_3\text{H} + \text{Ar}'\text{H} \xrightarrow{-\text{H}_2\text{O}} \text{ArSO}_2\text{Ar}' \tag{2}$$

2. CONDENSATION REACTIONS

20-1. Preparation of Phenyl Benzyl Sulfone [1]

$$\text{C}_6\text{H}_5\text{SO}_2\text{Na} + \text{C}_6\text{H}_5\text{CH}_2\text{Cl} \xrightarrow{\text{C}_2\text{H}_5\text{OH}} \text{C}_6\text{H}_5\text{SO}_2\text{CH}_2\text{C}_6\text{H}_5 + \text{NaCl} \tag{3}$$

To a flask containing 178 gm (1 mole) of sodium benzenesulfinate is added 127 gm (1 mole) of benzyl chloride in 500 ml of absolute alcohol. The mixture is refluxed for 8 hr and the hot mixture is poured into 1 liter of ice water. The crude product is filtered, dried, and recrystallized from ethanol to give 120 gm (52%), mp 146°–146.5°C.

NOTE: The sulfinate does not completely dissolve during the reaction and sodium chloride is precipitated at the same time. Dimethylformamide may be a more useful solvent.

3. OXIDATION METHODS

The most useful reagents for the laboratory preparation of sulfones are 30% hydrogen peroxide or nitric acid. Other reagents have also been described.

More recently benzyl triethylammonium permanganate has been reported to be a mild, one-phase oxidizer of sulfides and sulfoxides to sulfones in 63–98% yield.

Oxidation of sulfoxides to sulfones with oxygen in the presence of soluble iridium and rhodium complexes has also been reported. Metachloroperbenzoic acid has also been reported to be a good laboratory oxidizing agent for converting sulfides to sulfones.

20-2. Preparation of Tetramethylene Sulfone [2]

$$\begin{array}{c} H_2C\text{------}CH_2 \\ | \quad\quad | \\ H_2C_{\diagdown S \diagup}CH_2 \end{array} + H_2O_2 \longrightarrow \begin{array}{c} H_2C\text{------}CH_2 \\ | \quad\quad | \\ H_2C_{\diagdown S \diagup}CH_2 \\ O_2 \end{array} \tag{4}$$

To a flask containing 8.8 gm (0.1 mole) of tetramethylene sulfide is added in one portion 22.8 gm (0.2 mole) of 30% hydrogen peroxide. The reaction is exothermic and after 1 hr the solution becomes homogeneous. The solution is refluxed for 4 hr and then water is distilled off over a 1-hr period. The remaining solvent is stripped off under reduced pressure to leave 11.7 gm (97%) of colorless tetra-methylene sulfone, mp $10°-10.5°C$.

The reaction can also be carried out using glacial acetic acid [1] as a solvent and the extent of the reaction can be followed by G.C.

REFERENCES

1. R. L. Heath and A. Lambert, *J. Chem. Soc.* p. 1477 (1947).
2. D. S. Tarbell and C. Weaver, *J. Am. Chem. Soc.* **63,** 2941 (1941); V. G. Kulkarni and
   G. V. Jadhov, *J. Indian Chem. Soc.* **34,** 245 (1957).

SULFONIC ACIDS

1. INTRODUCTION

The literature on the sulfonation reactions is voluminous and several earlier reviews are worth consulting. The industrial aspects of sulfonation with SO_3 to produce dodecylbenzenesulfonate salts and lauryl sulfate salts (sulfation reaction) have been reported in detail in an article also discussing plant design. The use of sulfonation reaction to produce surfactants has been described.

The industrial chemist will find several references to the direct sulfonation of polymers to produce ion exchange resins and water-soluble materials. For example, chlorosulfonic acid is used to treat polystyrene to give an ion exchange resin. In addition phenol formaldehyde resins can also be sulfonated.

The most direct preparation of aromatic sulfonic acids is by the replacement of the hydrogen atom by one of the reagents shown in Eq. (1).

From S. R. Sandler and W. Karo, *Organic Functional Group Preparations*, Vol. I, 2d ed. (New York, 1983), 619–639, by permission of Academic Press, Inc.

$$\text{(benzene)} \begin{array}{l} H_2SO_4 \\ SO_3\cdot H_2SO_4 \\ SO_3 \text{ in dioxane} \\ SO_3 \text{ in pyridine} \\ SO_3 \text{ in } SO_2 \\ + ClSO_3H \\ NaHSO_3 \\ H_2SO_4 + P_2O_5 \\ (ClSO_2)_2O + AlCl_3 \\ SO_2Cl_2 + AlCl_3 \\ SO_2Cl_2 + ClSO_3H \end{array} \longrightarrow \text{(benzenesulfonic acid)} \qquad (1)$$

These sulfonations are more commonly used in the laboratory than the indirect methods which involve the oxidation of thiols, sulfinic acids, disulfides, or the conversion of the diazonium group into a sulfonic acid group.

Olefins have been found to react by a free radical addition with thiolacetic acid to form thiolacetates, which yield sulfonic acids upon oxidation by hydrogen peroxide–acetic acid.

$$RCH{=}CH_2 + CH_3COSH \rightarrow RCH_2{-}CH_2SCOCH_3 \xrightarrow{H_2O_2-CH_3COOH}$$
$$RCH_2CH_2SO_3H \quad (2)$$

where R represents straight and branched groups; the olefin can be terminal and internal.

In addition sodium bisulfite reacts with olefins by a free radical mechanism to give sodium sulfonates.

$$RCH{=}CH_2 + NaHSO_3 \longrightarrow RCH_2{-}CH_2SO_3Na \qquad (3)$$

$$\text{(maleic anhydride)} + NaHSO_3 \longrightarrow \text{(sulfosuccinic anhydride)} \qquad (4)$$

The Strecker synthesis involves the reaction of an active halogen compound with alkali or ammonium sulfites to give good yields of sulfonic acid salts.

$$RX + Na_2SO_3 \longrightarrow RSO_3Na + NaX \qquad (5)$$

The sulfoxidation and chlorosulfoxidation of hydrocarbons yields sulfonic acids and alkane sulfonyl chlorides, respectively.

The reaction of alkyl mercaptans and dialkyl disulfides with chlorine in an aqueous system yields alkane sulfonyl chlorides in good yield.

A. THE REACTION OF SULFURIC ACID AND ITS DERIVATIVES WITH AROMATIC HYDROCARBONS

Sulfuric acid is satisfactory for the sulfonation of the more reactive aromatic hydrocarbons. However, a large excess of reagent is required to give a good yield since the reaction is reversible.

$$\text{(benzene)} + H_2SO_4 \rightleftharpoons \text{(benzene-SO}_3\text{H)} + H_2O \qquad (6)$$

Removal of the water as it is formed will drive the reaction to completion and will allow one to use the stoichiometric amount of sulfuric acid. Aromatic sulfonic acids hydrolyze easily when heated in the presence of water and dilute acids.

Solvents are employed to moderate the sulfonation reaction as in the case of biphenyl, where chlorosulfonic acid in chloroform or tetrachloroethane is employed to give monosulfonation. A solvent also minimizes the formation of sulfonyl chlorides.

The sulfonic acids are usually isolated as their sodium salt and then hydrochloric acid is added to give the sulfonic acid. The sulfonic acids are hygroscopic solids or liquids which are difficult to purify.

■ **CAUTION:** Some sulfonations are exothermic and may take place with explosive violence when the reaction is conducted at an elevated temperature.

Heating a solution of *p*-nitrotoluene and sulfuric acid at 160°C initiates an exothermic reaction which results in an explosion.

21-1. Preparation of *p*-Toluenesulfonic Acid [1]

$$C_6H_5CH_3 + H_2SO_4 \rightarrow p\text{---}HSO_3\text{---}C_6H_4\text{---}CH_3 + H_2O + \text{some} \qquad (7)$$

$$\textit{ortho} \text{ and } \textit{meta} \text{ isomers}$$

To an Erlenmeyer flask are added 53 ml (0.5 mole) of distilled toluene, a boiling chip, and 29 ml (52 gm, 0.52 mole) of concentrated sulfuric acid. The mixture is gently refluxed by heating in an oil bath. The flask is gently swirled so that the toluene and sulfuric acid can react. (A stirring hot plate would be more convenient.) After about 1 hr the toluene layer is almost gone and there is very little reflux of toluene. The flask is cooled and the crystals are filtered to give 79 gm (83%) of crude *p*-toluenesulfonic acid monohydrate. The acid is crystallized from chloroform and dried to give 71 gm, mp 104°–106°C. The product is predominantly the *para* isomer but contains small amounts of the *ortho* and *meta* isomers. Azeotropic removal of water has been reported to aid this reaction and increase the yield [2].

B. THE STRECKER SYNTHESIS

The halogen groups of halogen compounds, especially reactive halogens, is easily replaced by the $^-SO_3Na$ groups to give sodium salts of sulfonic acids. Aliphatic compounds give high yields but branched chains give lower yields. *tert*-Butyl bromide yields only 23% of the sulfonate salt. Higher temperatures are required for high molecular weight halogen compounds. Phase transfer catalysts such as $R_4N^+Cl^-$ (R = Bu, Et) have been reported to facilitate this reaction. The Strecker method is not preferred in preparing sodium trifluoroethanesulfonate (10% yield) via the reaction of trifluoroethyl bromide with sodium sulfite.

21-2. Preparation of β-Phenoxyethanesulfonic Acid [3]

$$C_6H_5OCH_2CH_2Cl + NaHSO_3 \longrightarrow C_6H_5OCH_2CH_2SO_3Na \xrightarrow{H^+}$$

$$C_6H_5OCH_2CH_2SO_3H \quad (8)$$

To a flask containing 469.5 gm (3.0 moles) of β-chloroethyl phenyl ether is added a solution of 390 gm (3.0 moles) of sodium bisulfite in 1380 ml of water. The mixture is stirred vigorously and refluxed for 21 hr. Upon cooling to 20°C a crystalline precipitate of sodium β-phenoxyethanesulfonate appears. The product is washed with ether, and dried at 125°C to give 289 gm (43%).

An aqueous solution of 288 gm (1.29 moles) of the sodium salt is passed through an ion exchange column containing 454 gm of

Dowex–50X, a cation exchange resin in the hydrogen form. Evaporation of the effluent gives an almost quantitative recovery of β-phen-
oxyethanesulfonic acid which after drying over phosphorous pentoxide *in vacuo* melts below 100°C.

REFERENCES

1. Authors' Laboratory.
2. Chemetron Corp., British Patent 968,874 (1964).
3. R. M. Beringer and R. A. Falk, *J. Am. Chem. Soc.* **81,** 2997 (1959).

22

ALLENES

1. INTRODUCTION
    22-1. Preparation of 3-Methyl-1,2-butadiene

1. INTRODUCTION

A llenes have the 1,2-diene structure I, where R^1, R^2, R^3, $R^4 = H$, alkyl, aryl, halogen, heterocyclic, ether, etc. Since the terminal methylene groups lie in mutually perpendicular planes, optical isomers are possible [1].

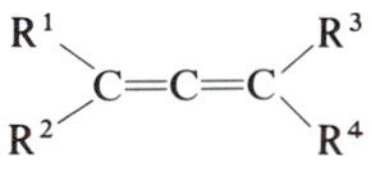

$$\begin{array}{c} R^1 \qquad\qquad R^3 \\ \diagdown \qquad\qquad \diagup \\ C{=}C{=}C \\ \diagup \qquad\qquad \diagdown \\ R^2 \qquad\qquad R^4 \end{array}$$

I

Allene is the generally accepted class name for all such compounds. However, the systematic name of 1,2-diene is also used as described for II and III.

$$CH_2{=}C{=}CH_2$$
Allene (1,2 Propadiene)
II

$$CH_3{-}CH{=}C{=}CH{-}CH_3$$
1,3-Dimethylallene (2,3-Pentadiene)
III

From S. R. Sandler and W. Karo, *Organic Functional Group Preparations*, Vol. II, 2d ed. (Orlando, Florida, 1986), 1ff., by permission of Academic Press, Inc.

Ward [2a] and then Skattebøl [3] reported that methyllithium or *n*-butyllithium reacts with *gem*-dibromocyclopropanes to give allenes in high yield. The related dichloro compounds were found to be inert to methyllithium but reacted slowly with *n*-butyllithium. Several examples of the preparation of allenes from *gem*-dibromocyclopropanes are shown in Table I.

The *gem*-dibromocyclopropanes are treated with an etheral solution of either methyllithium or *n*-butyllithium at 0° to −80°C. Methyllithium is preferable to *n*-butyllithium because occasionally difficulties are encountered in completely separating the *n*-butyl bromide from the allene product.

TABLE I

PREPARATION OF ALLENES FROM *gem*-DIBROMOCYCLOPROPANES

Starting olefin	Product allene	Dehalogenating reagent	Yield (%)	Ref.
trans-2-Butene	2,3-Pentadiene	Na–alumina	44	4ab
1-Pentene	1,2-Hexadiene	Na–alumina	64	4ab
2-Methyl-2-butene	2-Methyl-2,3-pentadiene	Mg	34	4ab
		CH_3Li	69	3
Isobutylene	3-Methyl-1,2-butadiene	CH_3Li	92	3
1-Ethoxy-2-methyl 2-propene	1-Ethoxy-3-methyl-1,2-butadiene	CH_3Li	71	3
1-Pentene	1,2-Hexadiene	Mg	45	4ab
2-Hexene	2,3-Heptadiene	CH_3Li	88	2abc
		C_4H_9Li	83	2abc
1-Octene	1,2-Nonadiene	CH_3Li	81	2abc
		C_4H_9Li	49	2abc
1-Decene	1,2-Undecadiene	CH_3Li	68	2abc
		C_4H_9Li	43	2abc
cis-Cyclooctene	1,2-Cyclononadiene	CH_3Li	81	2abc
cis-Cyclononene	1,2-Cyclodecadiene	C_4H_9Li	78	2abc
cis-Cyclodecene	1,2-Cycloundecadiene	C_4H_9Li	89	2abc
Styrene	1-Phenylpropadiene	CH_3Li	82	3
1,1-Diphenylethylene	1,1-Diphenylpropadiene	CH_3Li	43	3
Cycloocta-1,5-diene	1,2,6-Cyclononatriene	CH_3Li	80	3
1-Dodecene	Decylallene	Mg	45	5
Bicyclobutylidene	1,3-Bis(trimethylene) propradiene	CH_3Li	83	6
4,4,9,9-Tretramethoxy cyclododeca-1,6-diene	5,5,11,11-Tetramethoxy-1,2,7,8-cyclododeca-tetraenes	CH_3Li	—	7
tert-Butylethylene	*tert*-Butylallene	CH_3Li	56	8
Cyclotetradecene	Cyclopentadeca-1,2-diene	*n*-BuLi	—	9

22-1. Preparation of 3-Methyl-1,2,-Butadiene [3]

$$\text{(structure) } Br, Br + CH_3Li \xrightarrow[\substack{-30° \text{ to } -40°C \\ -CH_3Br \\ -LiBr}]{ether} \quad \overset{H}{\underset{H}{>}}C=C=C\overset{CH_3}{\underset{CH_3}{<}} \tag{1}$$

To a flask cooled with dry ice–acetone ($-30°$ to $-40°C$) and containing 22.8 gm (0.10 mole) of 1,1-dibromo-2,2-dimethylcyclopropane and 25 ml of dry ether is added 2.2 gm (0.10 mole) of methyllithium in 80 ml of ether over a $\frac{1}{2}$ hr period. The reaction mixture is stirred for $\frac{1}{2}$ hr, hydrolyzed with water, the ether separated, washed with water, dried, and distilled to afford 6.2 gm (92%), b.p. 40°C, n_D^{24} 1.4152.

1-Phenyl-1,2-propadiene, b.p. 64°–65°C (11 mm), n_D^{24} 1.5809, is obtained in a similar manner in 82% yield by the reaction of 1,1-dibromo-2-phenyl-cyclopropane with methyllithium at $-60°C$.

REFERENCES

1. A. Rank, A. F. Drake, and S. F. Mason, *J. Amer. Chem. Soc.* **101,** 2284 (1979); M. Nakazaki, K. Yamamoto, M. Maeda, P. Sato, and T. Tsutsui, *J. Org. Chem.* **47,** 1435 (1982).
2. W. R. Moore and H. R. Ward, *J. Org. Chem.* **27,** 4179 (1962).
2b. E. T. Marquis and P. D. Gardner, *Tetrahedron Lett.* 2793 (1966).
2c. L. Skattebøl and S. Solomon, *Org. Synth., Collect. Vol.* **5.** 306 (1973).
3. L. Skattebøl, *Acta Chem. Scand.* **17,** 1683 (1963); J. Leland, J. Boucher, and K. Anderson, *J. Polym. Sci. Chem. Ed.* **15,** 2785 (1977).
4a. S. R. Sandler, *J. Org. Chem.* **32,** 3876 (1967).
4b. W. von E. Doering and P. M. LaFlamme, *Tetrahedron* **2,** 75 (1958).
5. T. J. Logan, U.S. Pat. 3,096,384 (1963).
6. L. K. Bee, J. Beeby, J. W. Everett, and P. T. Garratt, *J. Org. Chem.* **40,** 2212 (1975).
7. P. J. Garratt, K. C. Nicolaou, and F. Sondheimeir, *J. Am. Chem. Soc.* **95,** 4582 (1973).
8. K. C. L. Lje and R. S. Macomber, *J. Org. Chem.* **39,** 3600 (1974).
9. P. A. Verbrugge and W. Brunmayer-Schilt, U.S. Pat. 4,025,562 (1977).

ORTHO
ESTERS

1. INTRODUCTION

1. INTRODUCTION

Carboxylic ortho esters can be regarded as ester–acetal derivatives of the hydrates of carboxylic acids (ortho acids).

$$\underset{\text{O}}{\overset{\text{O}}{\underset{\|}{RC}}}\text{—OH} + H_2O \;\rightleftharpoons\; RC\underset{\text{OH}}{\overset{\text{OH}}{|}}\text{—OH} \qquad (1)$$

The free ortho acids have not been isolated because of their thermodynamic instability, and early attempts aimed at their detection were not successful [1]. However, the ortho esters are stable derivatives which have the general structure I. Related to the ortho esters are derivatives in which the R′, R″, R‴ = —OR, —OH, —COR. The thio ortho esters have sulfur in place of oxygen in the structures above. The orthocarbonates $ROC(OR)_3$ and $C(OR)_4$ can also be considered ortho esters.

From S. R. Sandler and W. Karo, *Organic Functional Group Preparations*, Vol. II, 2d ed. (Orlando, Florida, 1986), 48ff., by permission of Academic Press, Inc.

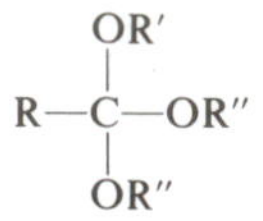

$$\begin{array}{c} OR' \\ | \\ R-C-OR'' \\ | \\ OR'' \end{array}$$

R, R', R'', R''' = alkyl or aryl groups

I

The most important synthetic methods for preparing ortho esters are shown in Eqs. (2)–(7).

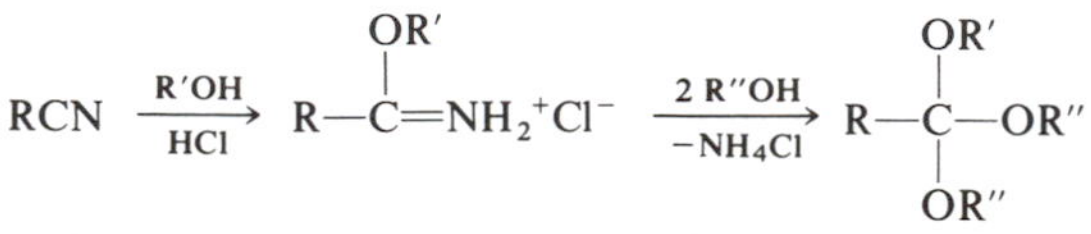

$$RCN \xrightarrow[HCl]{R'OH} R-\underset{\underset{}{\overset{OR'}{|}}}{C}=NH_2^+Cl^- \xrightarrow[-NH_4Cl]{2\ R''OH} R-\underset{\underset{OR''}{|}}{\overset{OR'}{\overset{|}{C}}}-OR'' \qquad (2)$$

R' = or ≠ R''

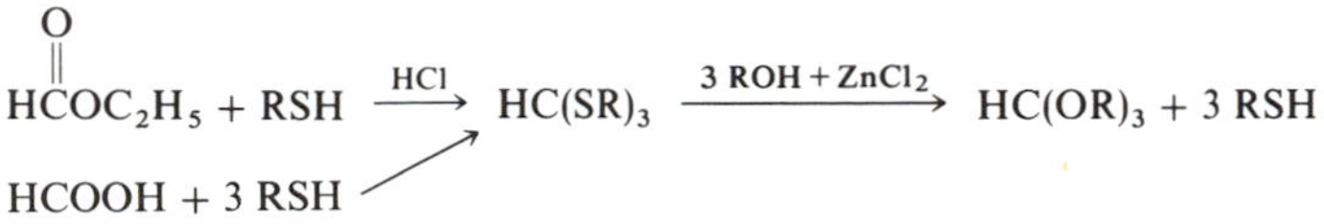

$$\begin{array}{c} \overset{O}{\overset{\|}{H}COC_2H_5} + RSH \xrightarrow{HCl} HC(SR)_3 \xrightarrow{3\ ROH + ZnCl_2} HC(OR)_3 + 3\ RSH \qquad (3) \\ HCOOH + 3\ RSH \nearrow \end{array}$$

$$RC(OR)_3 + 3\ R'OH \underset{}{\overset{H^+}{\rightleftharpoons}} RC(OR')_3 + 3\ ROH \qquad (4)$$

$$R-CX_3 + 3NaOR' \longrightarrow R-C(OR')_3 + 3NaCl \qquad (5)$$
$$R = H, \text{aryl, halogen, } -NO_2, -SCl$$

$$RO-CH-X_2 + 2NaOR' \longrightarrow ROCH(OR')_2 + 2NaX \qquad (6)$$

$$(RO)_2CHX + NaOR' \longrightarrow (RO)_2CHOR' + NaX \qquad (7)$$

Ortho esters are either colorless liquids or solids, depending on their molecular weight and structure. They are slightly soluble or very slightly soluble in neutral to basic water. They are soluble in many organic solvents and decompose under acidic conditions as shown in Eqs. (8) and (9). The ortho ester functional group has characteristic absorption in the infrared spectra at 1100 cm^{-1} for the C—O stretching band, and the NMR spectrum shows no unusual effects. The ortho ester group does not give any characteristic ultraviolet absorption.

$$RC(OR')_3 + HCl \longrightarrow RCOOR' + R'OH + R'Cl \qquad (8)$$

$$RC(OR')_3 + H_2O \longrightarrow RCOOR' + 2R'OH \qquad (9)$$

Alcoholysis of nitriles, of ortho and thio ortho esters (trans-esterification), and of halides is the most common method of preparing the ortho ester functional group (see Eqs. 2–7).

23-1. Preparation of Ethyl Orthoacetate (One-Step Synthesis) [2]

$$CH_3CN + 3\,C_2H_5OH \xrightarrow[CHCl_3]{HCl} CH_3C(OC_2H_5)_3 + NH_4Cl \qquad (10)$$

Into a water-cooled mixture of 1025 gm (25.0 moles) of aceto-nitrile, 1150 gm (25.0 moles) of absolute ethanol and 900 ml of chloroform is slowly bubbled 913 gm (25.0 moles) of anhydrous hydrogen chloride. The temperature rises to 35°–40°C and then is recooled to 20°C. After 48 hr at 20°–25°C, 5 liters of absolute ethanol is added and then the mixture is left for 2 days. The precipitated ammonium chloride is filtered, washed with ethanol, and the filtrate and washings added to 20 liters of 5% sodium hydroxide solution. The product is extracted with chloroform, concentrated, and fractionated under reduced pressure to afford 2430–3250 gm (60–80%), b.p. 70°–80°C (60 mm).

23-2. Preparation of 2-Butyl Orthoformate [3]

$$HC(OC_2H_5)_3 + 3\,CH_3{-}\underset{\underset{OH}{|}}{CH}{-}CH_2CH_3 \longrightarrow HC\left(OCH\underset{C_2H_5}{\overset{CH_3}{\diagdown}}\right)_3 + 3\,C_2H_5OH \quad (11)$$

To a 500-ml flask equipped with 14-inch glass-packed distillation column and distillation head are added 74.1 gm (0.5 mole) of ethyl orthoformate and 128.2 gm (2.0 moles) of 2-butanol. The flask is heated for 24 hr or until the ethanol (1.5 moles) is removed. The resulting mixture is distilled under reduced pressure to afford 101.5 gm (87.5%), b.p. 115°C (23 mm), n_D^{20} 1.4141.

Using (+)-(S)-2-butanol having $[\alpha]_D^{25}$ +8.402° (optical purity 61.64%) and letting it react with methyl orthoformate affords a 67% yield of 2-butyl orthoformate having $[\alpha]_D^{25}$ + 28.40° (61.58% optical purity) [4].

TABLE I

Reaction of Trihalomethyl Compounds and Halogenated Ethers with Alkoxides or Carboxylates to Give Ortho Esters or Carboxylate Derivatives

RCCl₃	RCCl₂OR	RCCl(OR)₂	RCOONa	RONa	Product	Yield (%)	B.p., °C (mm Hg)	n_D (°C)	Ref.
HCClF₂	—	—	—	CH₃ONa	(CH₃O)₃CH	—	98–99	1.377 (25)	a
HCCl₃	—	—	—	C₂H₅ONa	(C₂H₅O)₃CH	45	140–146	1.391 (25)	b
HCCl₃	—	—	—	CH₃ONa	(CH₃O)₃CH	84	103–105	—	c–e
HCCl₃	—	—	—	o-CH₃C₆H₄ONa	(o-CH₃C₆H₄O)₃CH	8.1	m.p. 96°C	—	f
C₆H₅CCl₃	—	—	—	C₂H₅ONa	C₆H₅C(OC₂H₅)₃	22	108–112 (13)	1.4930 (25)	g
	CHCl₂OCH₃	—	—	CH₃ONa	(CH₃O)₃CH	43	100.5–101	1.3787 (20)	h
	CHCl₂OCH₃	—	—	C₂H₅ONa	(C₂H₅O)₂(CH₃O)CH	56	133–134	1.3868 (20)	h
	CHCl₂OCH₃	—	—	n-C₃H₇ONa	(C₃H₇O)₂(CH₃O)CH	58	61.5–62 (11)	1.4010 (20)	h
	CHCl₂OCH₃	—	—	cycl-C₆H₁₁ONa	(C₆H₁₁O)₂(CH₃O)CH	36	107–109 (0.4)	1.4671 (20)	h
	CHCl₂OCH₃	—	—	C₆H₅ONa	(C₆H₅O)₂(CH₃O)CH	41.5	114 (0.05)	1.5517 (20)	h
	CHCl₂OCH₃	—	CH₃COONa	—	(CH₃COO)₂CHOCH₃	61	85 (7)	1.4052 (20)	h
	CHCl₂OCH₃	—	C₂H₅COONa	—	(C₂H₅COO)₂CHOCH₃	61	100–101.5 (12)	1.4136 (20)	h
	CHCl₂O-n-C₄H₉	—	CH₃COONa	—	(CH₃COO)₂CHO-n-C₄H₉	53	111–112 (12)	1.4153 (20)	h
(benzodioxole with Cl Cl)		—	—	C₂H₅ONa	(benzodioxole)C(OC₂H₅)₂	60	123 (15)	1.4943 (20)	i
NO₂CCl₃	—	—	—	C₂H₅ONa	(C₂H₅O)₄C	46–49	158–161	1.3905 (25)	j
Cl₃C—CCl₃	—	—	—	CH₃ONa	(CH₃O)₄C	48	113.5	1.3858 (20)	k
CCl₄	—	—	—	{ H(CF₂)₆CH₂OH / FeCl₃ }	[H(CF₂)₆CH₂O]₄C	35	170 (0.008)	—	l
		(C₆H₅O)₂CHCl	—	{ Et₃N / CH₃OH }	(C₆H₅O)₂CH—OCH₃	86	145 (3)	—	m
		(C₆H₅O)₂CHCl	—	{ Et₃N / C₆H₅OH }	(C₆H₅O)₂CHOC₂H₅	84	150 (3)	—	m
		(C₆H₅O)₂CHCl	—	{ Pyridine / C₆H₅OH }	(C₆H₅O)₃CH	96	m.p. 77°C	—	m

[a] J. Hine and J. J. Porter, J. Amer. Chem. Soc. 79, 5493 (1957).
[b] W. E. Kaufmann and E. E. Dreger, Org. Syn. Coll. Vol. 1, 258 (1932).
[c] C. Lenz, K. Hass, and H. Epler, Ger. Pat. 1,217,943 (1966).
[d] Feldmuehle Papier and Zellstoffwerke A. G., Belg. Pat. 613,988 (1962).
[e] T. A. Weidlich and W. Schulz, Ger. Pat. 919,465 (1954).
[f] J. E. Driver, J. Amer. Chem. Soc. 46, 2090 (1924).
[g] S. M. McElvain, H. I. Anthes, and S. H. Shapiro, J. Amer. Chem.

[h] H. Gross and A. Rieche, Chem. Ber. 94, 538 (1961).
[i] H. Gross, J. Rusche, and H. Bornowski, Ann. Chem. 675, 146 (1964).
[j] J. D. Roberts and R. E. McMahon, Org. Syn. Coll. Vol. 4, 457 (1963).
[k] H. Tieckelmann and H. W. Post, J. Org. Chem. 13, 265 (1948).
[l] M. E. Hill, D. T. Carty, D. Tegg, J. C. Butler, and A. F. Strong, J. Org. Chem. 30, 411 (1965).
[m] H. Scheibler and M. Depner, J. Prakt. Chem. 7, 60 (1958); Chem. Ber. 68B, 2151 (1935).

23-3. Preparation of Methyl Orthoformate [5]

$$3 \text{ NaOCH}_3 + \text{CHCl}_3 \xrightarrow{\text{solvent}} \text{HC(OCH}_3)_3 + 3 \text{ NaCl} \qquad (12)$$

A mixture consisting of 100 gm (1.85 moles) of powdered sodium methoxide suspended in 120 gm of benzene (or crude methyl orthoformate) is heated to 50°C and then 74 gm (0.62 mole) of chloroform is added dropwise over a 1 hr period, the reaction temperature being kept between 60°–80°C. The precipitated sodium chloride is removed by filtration and the filtrate distilled to afford 552 gm (84%), b.p. 103°–105°C, n_D^{25} 1.3770.

Ethyl orthoformate is prepared in 45% yield by adding sodium metal portionwise to a mixture of excess absolute alcohol and chloroform [6]. (See Kaufmann and Dreger [6] for earlier references to this reaction.) A recent references describes this preparation from chloroform with 3 moles of EtONa in excess EtOH [7]. See Table I for additional and related examples with references for the preparation of orthoesters.

REFERENCES

1. W. Colles, *J. Chem. Soc.* **89**, (1906).
2. VEB Filmfabrik Agfa Wolfen, Belg. Pat 617,666 (1962).
3. E. R. Alexander and H. M. Busch, *J. Amer. Chem. Soc.* **74,** 554 (1952).
4. F. Piacenti, M. Bianchi, and P. Pino, *J. Org. Chem.* **33,** 3653 (1968).
5. A. Lenz, K. Hass, and H. Epler, Ger. Pat. 1,217,943 (1966).
6. W. E. Kaufmann and E. E. Dreger, *Org. Syn. Coll. Vol.* **1,** 258 (1932).
7. W. Grabowicz and Z. Cybulska, Pol. Pat. 125,872 (1984); *Chem. Abstr* **102**, 45490a (1985).

24

SULFITES

1. INTRODUCTION

T he first report on sulfites appeared in 1846, and several synthetic procedures were later published in 1858–1859 [1]. The literature contains numerous reports on sulfites from 1909 to the present, yet sulfites have not even been mentioned briefly in some well-known texts on sulfur chemistry. However, the first comprehensive review appeared in 1963 in *Chemical Reviews*. Sulfites are characterized by the structure I where R = aryl or alkyl groups. The sulfites may have two

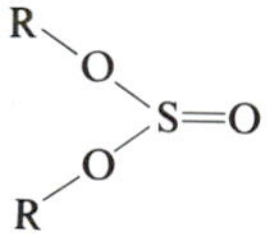

$$
\begin{array}{c}
R \\
\diagdown O \\
 \diagdown S{=}O \\
\diagup O \\
R
\end{array}
$$

I

similar groups, or two different groups. Recently the use of sulfites as insecticides, pesticides, plant-growth regulators, and plasticizers

From S. R. Sandler and W. Karo, *Organic Functional Group Preparations*, Vol. II, 2d. ed. (Orlando, Florida, 1986), 78–94, by permission of Academic Press, Inc.

and their ability to form polyesters when dicarboxylic acids react with ethylene sulfite have revived great interest in this functional group. For example, 2-(*p-t*-butylphenoxy)-1-methylethyl-2-chloroethyl sulfite has been reported for insect control on currants.

Chemical Abstracts refers to sulfites as sulfurous acid esters. Simple esters (benzyl, phenyl, ethyl, etc.) are listed as sulfurous acid esters under the names of the corresponding hydroxy compound. All mixed esters are indexed separately under the heading "sulfurous acid esters."

The best methods of preparing sulfites involve the reactions outlined in Eq. (1).

$$\tag{1}$$

2. CONDENSATION REACTIONS

A. REACTION OF THIONYL CHLORIDE WITH ALCOHOLS

Thionyl chloride is known to react with alcohols in the absence of hydrogen chloride acceptors to give reactions (2)–(7). Optimization of reaction (4) is of primary concern in this chapter.

$$SOCl_2 + ROH \longrightarrow ROSOCl + HCl \tag{2}$$

$$ROSOCl + ROH \longrightarrow (RO)_2SO + HCl \tag{3}$$

$$SOCl_2 + 2ROH \longrightarrow (RO)_2SO + 2HCl \tag{4}$$

$$(RO)_2SO + SOCl_2 \longrightarrow 2ROSOCl \tag{5}$$

$$ROSOCl \longrightarrow RCl + SO_2 \tag{6}$$

$$SOCl_2 + ROH \longrightarrow olefin + 2HCl + SO_2 \tag{7}$$

TABLE I

Preparation of Cyclic Sulfites

1,2-Diol	(moles)	SOCl$_2$ (moles)	R$_3$N (moles)	Solvent (ml)	Yield (%)	B.p., °C (mm Hg) or m.p., °C	n_D (°C)
			Pyridine	CH$_2$Cl$_2$			
Cycloheptane [3]							
cis-	0.1	0.1	0.24	250	41	90 (0.5)	1.4860 (24)
trans-	0.1	0.1	0.24	250	38	92 (0.5)	1.4865 (24)
cis-Indane [3]	0.1	0.1	0.24	250	72	70	—
1,4-Anhydroerythritol [3]	0.1	0.1	0.24	250	55	106–108	—
Ethane [15]	1.35	1.35	—	—	90	86–88 (38)	—
				Ether			
1,1-Dimethylol-3-cyclopentene [4]	0.274	0.823	0.55	150	59	117–119 (8) m.p. 49–50 (ether)	—
				C$_6$H$_6$			
2,3-Diphenyl-2,3-butanediol [5]	0.06	0.15	0.12	120	62	163 (1.3)	1.5752 (24)
3,3-Diphenyl-2-methyl-2,3-propanediol [5]	0.06	0.15	0.12	120	57	83–84	—

				Ether			
Hydrobenzoin [6]							
meso-	0.094	0.093	0.19	540 C_6H_6	83	126–128	—
dl-	0.023	0.024	0.051	420 Dioxane	83	84–86	—
Cyclohexane [6]							
cis-1,2-	0.086	0.087	0.18	100	76	90 (2.0) m.p. 6–8.0	1.4832 (20)
trans-1,2-	0.17	0.17	0.34	200 CS_2	76	94–96 (2.0) m.p. −15	1.4847 (20)
Catechol [7]	1.0	1.0	2.0	230 $ClCH_2CH_2Cl$	26	137–138 (105 mm)	—
2,2-Dinitro-1,3-propanediol [7a]	0.5	0.5	12 drops	300	81	m.p. 37–38	—
Trichloropropylene glycol [7b]	1.0	1.0	—	—	100	68 (8)	1.5500 (25)
3,4-Dimethyl-3,4-hexanediol [7c]	0.021	0.12	0.24	Ether 23	43	67–68 (0.6)	—

Gerrard [1] had shown earlier that the slow addition of 0.5 mole of thionyl chloride to a mixture of pyridine (1.0 mole) and hydroxy compounds (*n*-butyl, *n*-amyl, or ethyl lactate—0.1 mole) gives pyridine hydrochloride and good yields of the sulfite (see Eq. 4). Primary and secondary alcohols with an aromatic nucleus in the α-position give chlorides in the absence of catalysts [2]. For example, diphenylmethanol gives the chloride even at $-78°C$. Sulfites derived from tertiary alcohols are not known. The further addition of thionyl chloride converted the sulfite to the chlorosulfinate (see Eq. 5). On heating, the chlorosulfinate is catalytically decomposed by pyridine hydrochloride to the corresponding alkyl chloride and sulfur dioxide (see Eq. 6). Secondary chlorosulfites give olefins even under the mildest conditions [1]. The use of excess pyridine reduces the yield of sulfite by the method described in Eq. (4).

Secondary alkyl sulfites are produced if the formed hydrogen chloride is removed by carrying out the reaction at reduced pressure (use HCl trap).

Examples of the preparation of cyclic sulfites from 1,2-diols are shown in Table I.

24-1. Preparation of Dibenzyl Sulfite [2]

$$2 \, C_6H_5CH_2OH + SOCl_2 \xrightarrow{\text{pyridine}} (C_6H_5CH_2O)_2SO \tag{8}$$

To a stirred solution of 5.4 gm (0.05 mole) of benzyl alcohol in 4.0 gm (0.05 mole) of pyridine and 30 ml of ether at $-78°C$ is added dropwise a solution of 3.0 gm (0.025 mole) of thionyl chloride in 15 ml of ether over a 20–30 min period. After 1 hr the mixture is filtered,

TABLE II[a]

PREPARATION OF SULFITES USING PREPARATION 24-1

Alcohol	Sulfite (yield %)	B.p., °C (mm Hg)	n_D (°C)
2-Phenylethanol	86	162–165 (0.5)	1.5510 (15)
3-Phenylpropanol	85	185–190 (0.3)	1.5423 (18)
1-Phenyl-2-propanol	86	158–160 (0.1)	1.5351 (21.5)

[a]Data taken from Gerrard and Shepherd [2].

concentrated under reduced pressure, and distilled to afford 5.7 gm (87%), b.p. 152°C (0.4 mm), n_D^{25} 1.5590.

Table II lists the other alcohol used, the yields of sulfites obtained, and their physical properties using a similar procedure.

REFERENCES

1. W. Gerrard, *J. Chem. Soc.* 218 (1940).
2. W. Gerrard and B. D. Shepherd, *J. Chem. Soc.* 2069 (1955).
3. J. S. Brimacombe, A. B. Foster, E. B. Hancock, W. G. Overend, and M. Stacey, *J. Chem. Soc.* 201 (1960).
3a. W. J. Myles and J. H. Prichard, U.S. Pat. 2,497,135 (1950).
3b. R. H. Garst and J. P. Henry, U.S. Pat. 3,554,986 (1971).
3c. M. E. Chiddix and R. W. Wynn, U.S. Pat. 3,169,130 (1965).
4. E. J. Grubbs and D. J. Lee, *J. Org. Chem.* **29,** 3105 (1964).
5. S. Hauptmann and K. Dietrich, *J. Prakt. Chem.* **19,** 174 (1963).
6. C. C. Price and G. Berti, *J. Amer. Chem. Soc.* **76,** 1211 (1954).
7. A. Green, *J. Chem. Soc.* 500 (1927).
7a. E. E. Hamel, U.S. Pat. 3,492,311 (1970).
7b. H. C. Vogt and P. Davis, U.S. Pat. 3,394,147 (1968).
7c. W. Reeve and S. K. Davidsen, *J. Org. Chem.* **44,** 3430 (1979).

25

ENAMINES

1. INTRODUCTION

W ittig and Blumenthal [1] in 1927 introduced, the term "en-amine" to designate the nitrogen analog of the term "enol" (structures I and II). The enamine structure had been known in the early

$$\begin{array}{ccc} >\!\!C\!\!=\!\!\overset{|}{C}\!\!-\!\!N\!\!< & & >\!\!C\!\!=\!\!\overset{|}{C}\!\!-\!\!OH \\ \text{enamine} & & \text{enol} \\ I & & II \end{array}$$

literature (pyrrole, indole, etc.), but it was not until 1954 [2, 3] that the chemical potential of this group was emphasized.

In addition to the condensation of secondary amines with aldehydes or ketones (Eq. 1), the other important methods [4] of synthesizing enamines are briefly outlined in Eq. (2).

From S. R. Sandler and W. Karo, *Organic Functional Group Preparations*, Vol. II, 2d ed. (Orlando, Florida, 1986), 95ff., by permission of Academic Press, Inc.

$$(1)$$

$$(2)$$

$$X = OH, ester, halide, NR_2, etc.$$
$$X' = halide$$

Another useful method involves the base-catalyzed isomerization of tertiary allylamines to *cis*- and *trans*-propenylamines [Eq. (3)].

$$CH_2{=}CH{-}CH_2NR_2 \xrightarrow[\text{DMSO}]{\text{K}-\text{O}-t\text{-Bu}} CH_3{-}CH{=}CH{-}NR_2 \qquad (3)$$

Enamides and thioenamides are prepared by similar methods and involve the condensation reaction of the appropriate amides with aldehydes [5] to give

$$X = O \text{ or } S$$

25-1. Preparation of Dimethylamino-1-cyclohexene [6]

$$(4)$$

To a flask equipped with a mechanical stirrer and dry ice condenser and containing 150 gm (3.4 moles) of dimethylamine in 400 ml of anhydrous ether are added 150 gm of 12-mesh anhydrous calcium chloride and 196 gm (2.0 moles) of cyclohexanone. The reaction mixture is stirred for 64 hr at room temperature under a nitrogen atmosphere, filtered, the residue washed with ether, the ether concentrated, and the residue fractionally distilled to afford 108.3 gm (52%) of the enamine; b.p. 81°C (35 mm), n_D^{25} 1.4851, and 94 gm (0.97 mole), b.p. 156°C, n_D^{19} 1.4522, of cyclohexanone. The conversion to enamine is 83% on the basis of the ketone utilized.

By a similar method, cyclopentanone reacts with dimethylamine to afford a 56% yield (or an 87% conversion based on ketone used) of dimethylamino-1-cyclopentene, b.p. 85°–86°C (104 mm), n_D^{25} 1.4801.

25-2. Preparation of 1-Morpholino-1-isobutene [7, 8]

$$CH_3-CH-CH{=}O + \underset{\underset{H}{N}}{\overset{O}{\bigcirc}} \longrightarrow O\bigcirc N-CH{=}C-CH_3 \qquad (5)$$
$$\quad\;\; CH_3 \qquad\qquad\qquad\qquad\qquad\qquad\qquad CH_3$$

To an apparatus similar to that described in Preparation 25-1 and containing 43.5 gm (0.5 mole) of morpholine is slowly added 72 gm (1.0 mole) of isobutyraldehyde. The reaction mixture is refluxed for about 3 hr until 9 ml of water separates and is then distilled to afford 66 gm (93.5%), b.p. 56°–57°C (11 mm), n_D^{20} 1.4670.

REFERENCES

1. G. Wittig and H. Blumenthal, *Chem. Ber.* **60,** 1085 (1927).
2. G. Stork, R. Terrell, and J. Szmuszkovicz, *J. Amer. Chem. Soc.* **76,** 2029 (1954).
3. G. Stork and H. K. Landesman, *J. Amer. Chem. Soc.* **78,** 5128, 5129 (1956).
4. G. Laban and R. Mayer, *Z. Chem.* **8,** 165 (1968).
5. A. Couture, R. Dubiez, and A. Lablache-Combier, *J. Org. Chem.* **49,** 714 (1984).
6. E. P. Blanchard, Jr., *J. Org. Chem.* **28,** 1397 (1963).
7. E. Benzing, *Angew. Chem.* **71,** 521 (1959).
8. E. Benzing, U.S. Pat. 3,074,940 (1963).

YNAMINES

1. INTRODUCTION
    26-1. Preparation of *N,N*-Dimethylamino-2-phenylacetylene

1. INTRODUCTION

The most important methods of synthesizing ynamines are based on the reactions shown in the scheme below.

$$R'C{\equiv}C{-}X \xrightarrow[\text{X = F, Cl, Br, OR}]{\text{HNR}_2''(\text{O}_2)} R'{-}C{\equiv}CNR_2'' \xleftarrow{R'X} Li{-}C{\equiv}CNR_2'' \qquad (1)$$

Most ynamines are water-clear liquids with a slight amine odor. They are stable and can be vacuum-distilled without decomposition.

The air oxidation of phenylacetylene and secondary amines in the presence of cupric acetate in benzene solution yields ynamines [1].

From S. R. Sandler and W. Karo, *Organic Functional Group Preparations*, Vol. II, 2d ed. (Orlando, Florida, 1986), 135 ff., by permission of Academic Press, Inc.

This reaction requires only catalytic amounts of cupric salts and gives high conversions in less than 30 min when the Cu^{+2}/phenylacetylene ratio is only 0.02. Only 1,4-diphenylbutadiyne is produced if the stoichiometric amount of cupric ion is used in the absence of oxygen. The yield of ynamine can be increased from 45 to 90% if the stoichiometric amounts of a reducing agent such as hydrazine are continuously added during the course of the reaction. The use of primary amines under similar conditions yields the acetamide derivative.

26-1. Preparation of *N,N*-Dimethylamino-2-phenylacetylene [1]

$$C_6H_5-C{\equiv}CH + HN(CH_3)_2 \xrightarrow[C_6H_6]{Cu(OAc)_2, O_2}$$

$$C_6H_5-C{\equiv}C-N(CH_3)_2 + C_6H_5-C{\equiv}C-C{\equiv}C-C_6H_5 + H_2O \quad (2)$$

To a flask containing 2.0 gm (0.01 mole) of cupric acetate in 25 ml (0.38 mole) of dimethylamine in 100 ml of benzene at 5°C are added dropwise simultaneously a solution of 5.1 gm (0.05 mole) of phenylacetylene in 100 ml of benzene over 30 min and a stream of oxygen (1.0 ft^3/hr). The oxygen stream is added for 30 min more after the phenylacetylene addition has been completed. The copper ions are precipitated by adding 100 ml of ice water. The organic layer is separated, dried, and concentrated under reduced pressure. A gas chromatograph of the crude (column 10 ft × ½ in. 410 gum rubber) showed two peaks: *N,N*-dimethylamino-2-phenylacetylene and 1,4-diphenylbutadiyne. The ynamine was obtained in 40% yield as determined by reaction of the crude mixture with dilute hydrochloric acid and isolating the resultant *N,N*-dimethyl- aminophenylacetylene. The ynamine has characteristic IR absorption bands at 2205 and 2235 cm^{-1} and NMR absorption bands at δ 7.14 (4.0 protons) and 2.73 (6.05 protons).

REFERENCE

1. L. I. Peterson, *Tetrahedron Lett.* 5357 (1968); and U.S. Pat. 3,499,928 (1970).

27

UREAS AND POLYUREAS

1. INTRODUCTION

1. INTRODUCTION

The most practical laboratory methods for the preparation of monomeric ureas involve the condensation of amines with either ureas, isocyanates, or isocyanate derivatives. The use of difunctional reactants may yield polymers (Eq. 1).

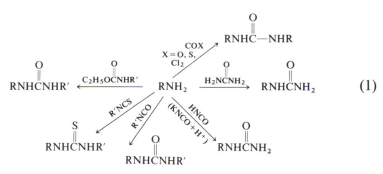

$$\text{(1)}$$

Ureas from S. R. Sandler and W. Karo, *Organic Functional Group Preparations*, Vol II, 2d ed. (Orlando, Florida, 1986), 152ff., by permission of Academic Press, Inc.
Polyureas from S. R. Sandler and W. Karo, *Polymer Syntheses*, Vol. I, 2d ed. (San Diego, California, 1992), 218ff., by permission of Academic Press, Inc.

213

If isocyanates are available, they may be reacted with water to yield symmetrical ureas (Eq. 2).

$$2\,RNCO + H_2O \longrightarrow RNH\overset{\overset{\displaystyle O}{\|}}{C}NHR + CO_2 \qquad (2)$$

The reactions which are described as giving monomeric ureas can be applied to giving polyureas by using di- and polyfunctional starting materials, as shown in Eq. (4) [1]. The reaction of diisocyanates with water can also be used since the diamine can be formed *in situ* and reacted with the diisocyanates to give the polyurea [2].

■ **CAUTION:** All reactions should be carried out in a well-ventilated hood with the use of proper personal protection. It is recommended that all Material Safety Data Sheets (MSDS) be obtained for each chemical being used and read very carefully by all those engaged with the synthesis procedures.

27-1. Preparation of *p*-Ethoxyphenylurea [3]

$$C_2H_5O-\!\!\!\!\left\langle\bigcirc\right\rangle\!\!\!\!-\overset{+}{N}H_3Cl^- + NH_2-\overset{\overset{\displaystyle O}{\|}}{C}-NH_2 \longrightarrow$$

$$C_2H_5O-\!\!\!\!\left\langle\bigcirc\right\rangle\!\!\!\!-NH\overset{\overset{\displaystyle O}{\|}}{C}-NH_2 + NH_4Cl \qquad (3)$$

To a 1 liter flask equipped with a condenser and mechanical stirrer is added a mixture of 87.0 gm (0.50 mole) of *p*-phenetidine hydrochloride and 120.0 gm (2.0 moles) of urea. To the stirred mixture is added a solution prepared from 200 ml of water, 4.0 ml of concentrated hydrochloric acid, and 4.0 ml of glacial acetic acid. The solution is heated by means of an oil bath and refluxed for $1\frac{1}{2}$ hr while the product is precipitating. At the end of the reaction period the entire contents appear to have solidified. The reaction is cooled, the solid is broken up, suspended in water, filtered, and dried to obtain 740–810 gm (82–90%) of a pale-yellow crude solid, *p*-ethoxyphenylurea [2–4]. Recrystallization from boiling water gives an 80% recovery of material, m.p. 173°–174°C. The by-product is *sym*-di(*p*-ethoxyphenyl)urea [(*p*-C$_2$H$_5$OC$_6$H$_4$NH)$_2$-CO] which is insoluble in boiling water. Prolonged boiling of *p*-ethoxyphenylurea in water will slowly convert it into *sym*-di(*p*-ethoxyphenyl)urea. Heating *sym*-di(*p*-ethoxyphenyl)urea with urea, ammonium carbamate, ammonium carbonate, or ethanol and ammonia converts it into *p*-ethoxyphenylurea again [4, 5, 6, 7].

27-2. Preparation of Poly(4-oxyheptamethyleneurea) [8]

$$\underset{\substack{\| \\ O}}{NH_2CNH_2} + H_2N-(CH_2)_3-O-(CH_2)_3-NH_2 \xrightarrow[\text{heat}]{N_2}$$

$$\underset{\substack{\| \\ O}}{+(CH_2)_3-O-(CH_2)_3-NHC-NH+_n} + 2\,NH_3 \quad (4)$$

To a test tube with a side arm are added 7.5 gm (0.125 mole) of urea and 16.5 gm (0.125 mole) of bis(γ-aminopropyl) ether. A capillary tube attached to a nitrogen gas source is placed on the bottom of the tube and the temperature is raised to 156°C and kept there for 1 hr, during which time ammonia is evolved. The temperature is raised to 231°C for 1 hr and then to 255°C for an additional hour. At the end of this time a vacuum is slowly applied to remove the last traces of ammonia.

■ **CAUTION:** Frothing may be a serious problem if the vacuum is applied too rapidly.

The polymer is cooled, the test tube broken, and the polymer isolated. The polymer melt temperature is approximately 190°C, and the inherent viscosity is approximately 0.6 in *m*-cresol (0.5% concentration, 25°C).

27-3. Preparation of Poly(decamethyleneurea) [9]

$$H_2N-(CH_2)_{10}-NH_2 + OCN-(CH_2)_{10}-NCO \longrightarrow \underset{\substack{\| \\ O}}{+(CH_2)_{10}NH-C-NH+_n} \quad (5)$$

In a three-necked, round-bottomed flask equipped with stirrer, dropping funnel, condenser, and drying tubes is placed, under a nitrogen atmosphere, 19.0 gm (0.11 mole) of freshly distilled decamethylenediamine in 39 ml of distilled *m*-cresol. While vigorously stirring,

24.8 gm (0.11 mole) of decamethylene diisocyanate is added drop-wise over a 10 min period. The dropping funnel is washed with 10 ml of *m*-cresol, and this is added to the reaction mixture. The temperature of the reaction mixture is raised to 218°C for 5 min, cooled to room temperature, and then poured into 1.5 liters of methanol while stirring vigorously. The polyurea which separates as a white solid is filtered, washed with ethanol in a blender, and dried at 60°C under vacuum to give 38–40 gm (90–95%), polymer melt temperature 210°C, inherent viscosity in *m*-cresol 8.3 (0.5% concentration at 25°C).

Ureas with heterocyclic substituents are prepared by the reaction of heterocyclic amines with aryl or alkyl isocyanates. The pyridyl ureas possess excellent anti-inflammatory properties and the 3-ureido-pyrrolidines have novel pharmacological properties (analgesic, central nervous system, and psychopharmacologic activities).

Polyureas can be prepared by the interfacial polycondensation of phosgene in an organic solvent with an aqueous solution of a diamine and alkali as described in Preparation 27-4 [10, 11].

27-4. Preparation of Poly(hexamethyleneurea) [10]

$$H_2N(CH_2)_6NH_2 + COCl_2 \longrightarrow \left[-(CH_2)_6NH\overset{\overset{\displaystyle O}{\|}}{C}NH-\right]_n + 2HCl \qquad (6)$$

■ **CAUTION:** This reaction should be run in a well-ventilated hood. (We recommend the presence of small amounts of ammonia vapors in the air to reduce the danger of escaping phosgene.) Phosgene is extremely toxic.

To a three-necked, round-bottomed flask equipped with a mechanical stirrer, dropping funnel, and condenser is added a solution of 4.95 gm (0.05 mole) of phosgene in 200 ml of dry carbon tetrachloride. The solution is vigorously stirred while the rapid addition of 5.8 gm (0.05 mole) of hexamethylenediamine and 4.0 gm (0.10 mole) of sodium hydroxide in 70 ml of water takes place. The reaction is exothermic while the polyurea forms. After 10 min, the carbon tetrachloride is evaporated off on a steam bath or with the aid of a water aspirator. The polyurea is washed several times in a blender and air-dried overnight to obtain 5.0 gm (70%), inherent viscosity 0.90 (in *m*-cresol, 0.5% conc. at 30°C), polymer melt temperature approximately 295°C.

REFERENCES

1. C. I. Chiriac, *Encycl. Polym. Sci. Eng.*, **13,** 212 (1988).
2. H. G. J. Overmars, *Encycl. Polym. Sci Technol.* **11,** 464 (1969); O. E. Snider and R. J. Richardson, *ibid.*, p. 495; M. A. Dietrich and H. W. Jacobson, U.S. Patent 2,709,694 (1955).
3. F. Kurzer, *Org. Syn. Coll. Vol.* **4,** 52 (1963).
4. J. D. Riedel, German Pat. 73,083 [*Frdl.* **3,** 907 (1890–1894)].
5. J. D. Riedel, German Pat. 77,310 [*Frdl.* **4,** 1271 (1894–1897)].
6. K. Findelsen, R. Freimuth, and K. Wagner, U.S. Pat 4,310,692 (Jan. 12, 1982).
7. P. E. Throckmorton, S. Freyard, and D. Grote, U.S. Pat. 3,937,727 (Feb. 10, 1976).
8. E. I. du Pont de Nemours, Brit. Pat. 530,267 (Dec. 9, 1940).
9. E. I. du Pont de Nemours, Brit. Pat. 535,139 (March 31, 1941).
10. E. L. Wittbecker, U.S. Patent 2,816,879 (1957).
11. L. Alexandra and L. Dascalu, *J. Polym. Sci.* **52,** 331 (1961).

28

CARBODIIMIDES

1. INTRODUCTION
    28-1. Preparation of Diphenylcarbodiimide

1. INTRODUCTION

T he most practical methods for the preparation of carbodiimides are summarized in the scheme below. The carbodiimide structure may be viewed as a dehydrated urea.

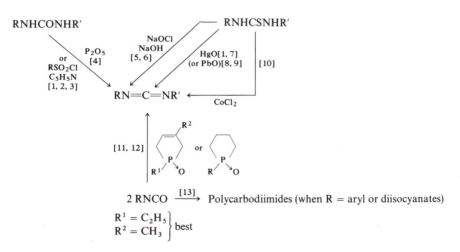

$$R^1 = C_2H_5$$
$$R^2 = CH_3$$ } best

From S. R. Sandler and W. Karo, *Organic Functional Group Preparations*, Vol. II, 2d ed. (Orlando, Florida, 1986), 233ff., by permission of Academic Press, Inc.

28-1. Preparation of Diphenylcarbodiimide [11]

$$2 \ C_6H_5NCO \xrightarrow{\hspace{2cm}} C_6H_5N{=}C{=}NC_6H_5 + CO_2 \qquad (1)$$

(a) Catalyst Formation. Preparation of 1-ethyl-3-methyl-3-phospholine-1-oxide.

$$C_2H_5PCl_2 + CH_2{=}C{-}CH{=}CH_2 \longrightarrow \qquad \xrightarrow{H_2O_2} \qquad (2)$$

To a 2 liter, four-necked flask equipped with a spiral condenser topped with a dry ice condenser, 1 liter dropping funnel with pressure-equalizing arm, and a magnetic stirrer are added 1 gm of copper stearate, 780 gm (5.96 moles) of dichloroethylphosphine, and, from the dropping funnel, 447 gm (6.56 moles) of freshly distilled isoprene. The stirred solution is refluxed under a nitrogen atmosphere for 42 hr, cooled, allowed to stand for 2 days, and then refluxed an additional 5 days. The excess isoprene is removed by distillation and 850 ml of water is added dropwise while cooling the stirred solution in an ice bath. The dark-brown aqueous solution is transferred to a 5 liter flask and then 1250 ml of 30% sodium hydroxide solution is gradually added so that the solution obtains a pH of 8. The mixture is filtered and the aqueous solution is continuously extracted with chloroform for 12 days. The chloroform is dried and then distilled to yield a residue which on vacuum distillation affords 435 gm (51%), b.p. 115°–120°C (1.2 mm). The product is further purified by oxidation at 50°C with an excess of 3% hydrogen peroxide for 6 hr. The aqueous layer is continuously extracted with benzene, dried, and then concentrated to give a residue which distills as follows: b.p. 115°–119°C (1.2–1.3 mm), n_D^{25} 1.5050.

(b) Carbodiimide Formation. A solution consisting of 50 ml of anhydrous benzene, 54 gm (0.45 mole) of phenyl isocyanate, and 0.20 gm of 1-ethyl-3-methyl-3-phospholine-1-oxide catalyst is refluxed for approximately 2.5 hr or until carbon dioxide evolution ceases. The

benzene is removed under reduced pressure and the residue on distillation affords 41.1 gm (94%) of an oil, b.p. 119°–121°C (0.40–0.45 mm), n_D^{25} 1.6372. The product is stable for several weeks at 0°C. At room temperature it gradually solidifies to a mixture of trimer and polymer.

REFERENCES

1. J. C. Sheehan, P. A. Cruickshank, and G. L. Boshart, *J. Org. Chem.* **26,** 2525 (1961).
2. G. Amiard and R. Heymes, *Bull. Soc. Chim. Fr.* 1360 (1956).
3. J. C. Sheehand and P. A. Cruickshank, *Org. Syn.* **48,** 83 (1968).
4. C. L. Stevens, G. H. Spinghal, and A. B. Ash, *J. Org. Chem.* **32,** 2895 (1967).
5. E. Schmidt and R. Schnegg, U.S. Pat. 2,656,383 (1953).
6. E. Schmidt and R. Schnegg, Ger. Pat. 823,445 (1951).
7. H. G. Khorana, *J. Chem. Soc.* 2081 (1952).
8. W. Weith, *Chem. Ber.* **8,** 1530 (1875).
9. E. Schmidt, W. Striewsky, and F. Hitzler, *Ann. Chem.* **560,** 222 (1948).
10. A. A. R. Sayigh and H. Ulrich, U.S. Pat. 3,301,895 (1967).
11. T. W. Campbell, J. J. Monagle, and V. S. Foldi, *J. Amer. Chem. Soc.* **84,** 3673 (1962).
12. W. B. McCormack, *Org. Syn* **43,** 73 (1963).
13. R. Sundermann and W. E. Slack, U.S. Pat. 4,154,752 (1979).

29

N-CARBAMATES
(URETHANES)

1. INTRODUCTION

T he electronic structure of the isocyanate group indicates the following possible resonance structure (Eq. 1):

$$R\ddot{\overset{-}{N}}-\overset{+}{C}=\ddot{\overset{..}{O}} \leftrightarrow R-\ddot{N}=C=\ddot{\overset{..}{O}} \leftrightarrow R-\ddot{N}=\overset{+}{C}-\ddot{\overset{-}{O}}: \qquad (1)$$

I

The usual reactions with isocyanates involve the reaction of active hydrogen with the nitrogen of the isocyanate group as is shown in Eq. 2.

$$R-N=C=O + HA \longrightarrow RNH-\overset{\overset{\displaystyle O}{\|}}{C}-A \qquad (2)$$

From S. R. Sandler and W. Karo, *Organic Functional Group Preparations*, Vol. II, 2d ed. (Orlando, Florida, 1986), 260–273, by permission of Academic Press, Inc.

Some of these products are stable, while others can be decomposed easily to the starting material or other products. The reaction of isocyanate groups with alcohols is discussed in detail in this chapter (Eq. 3). Reference is also made to the preparation of polymers using difunctional starting materials (see pp. 223–224).

$$R\text{—}NCO + R'OH \longrightarrow RNH\text{—}\overset{\overset{\displaystyle O}{\|}}{C}OR' \qquad (3)$$

The synthesis of O-alkyl carbamates $NH_2\overset{}{C}\text{—}OR$, is discussed

$$\overset{\|}{O}$$

in Chapter 30. This chapter mainly describes N-alkyl carbamates, $RNH\overset{}{C}OR$, as derived from the condensation reactions of isocya-

$$\overset{\|}{O}$$

nates. The conversion of O-alkyl carbamates to N-alkyl carbamates is briefly described (see Eq. (4) and (5)).

$$NH_2\underset{\underset{O}{\|}}{C}OR + RNH_2 \xrightarrow{R_2N} RNH\underset{\underset{O}{\|}}{C}OR + NH_3 \qquad (4)$$

$$NH_2\underset{\underset{O}{\|}}{C}OR + RCH{=}CH_2 \xrightarrow{H^+} RCH_2CH_2NHC\underset{\underset{O}{\|}}{}OR \qquad (5)$$

■ **CAUTION:** The handling of isocyanates should be done with great care because of their known toxicity hazard. The hazards and chemistry of methyl isocyanate have recently been reported. Toluene diisocyanate (TDI), methylene diisocyanate (MDI), and methyl isocyanate (MIC) are flammable, very reactive, and have in common low maximum allowable concentrations for employee exposure by the Occupational Safety and Health Administration (OSHA).

2. CONDENSATION REACTIONS

A. REACTION OF ISOCYANATES WITH HYDROXY COMPOUNDS

For the most part, the preparation of monomeric and polymeric carbamates (urethanes), semicarbazides, and ureas consists of condensation reactions of isocyanates with alcohols, hydrazines, or amines.

Isocyanates readily react with primary alcohols at 25°–50°C, whereas secondary alcohols react 0.3 as fast, and tertiary alcohols approximately 0.005 as fast as the primary. Thus, the effect of steric hindrance shows a pronounced effect on these reactions. For example, triphenylcarbinol has been reported to be completely unreactive. Other tertiary alcohols such as *t*-butanol may react under uncatalyzed conditions with isocyanates to give olefin formation, as shown in Eq. (6), using phenyl isocyanate.

$$2 \text{ C}_6\text{H}_5\text{NCO} + (\text{CH}_3)_3\text{C—OH} \longrightarrow$$
$$\text{C}_6\text{H}_5\text{NHCONHC}_6\text{H}_5 + \text{CO}_2 + (\text{CH}_3)_2\text{C}{=}\text{CH}_2 \quad (6)$$

29-1. Preparation of the Phenylurethane of Ethanol [2]

$$(7)$$

To a dry Erlenmeyer flask are added 29.6 gm (0.25 mole) of phenyl isocyanate, 1 drop of pyridine, and 11.5 gm (0.25 mole) of absolute ethanol. The reaction mixture, after standing for 5–10 min, becomes hot and is cooled and stirred. After the reaction appears to be complete, the contents are heated to 60°–70°C for ½ hr. The contents are cooled and a few drops are placed on a watchglass to get a seed crystal. The seed crystal is added to the rest of the flask and, on stirring the contents, the urethane precipitates to yield 40.0 gm (97%), m.p. 55°C.

Other examples of the catalytic activity of amines and other catalysts are given in Table I, which shows that their activity is not dependent on base strength.

In the reaction of 2,4-toluene diisocyanate the 4-position isocyanate group reacts first and then the one at the 2-position.

Polyurethane resins are usually prepared by the reaction of a long-chain diol with an excess of the diisocyanate to obtain a "prepolymer" with terminal isocyanate groups (Eq. 8). The prepolymer can react

$$\text{R(NCO)}_2 + \text{HO}\text{—}\text{\small wwwww}\text{—OH} \longrightarrow$$
$$\text{OCN—R—NHCOO}\text{\small wwwww}\text{OCONH—R—NCO} \quad (8)$$

TABLE I[c]

COMPARISON OF THE CATALYTIC ACTIVITY OF SOME ORGANOTIN COMPOUNDS
WITH THAT OF OTHER CATALYSTS[a] [3]

Catalyst	Catalyst (mole %)	Relative activity at 1.0 (mole %)
None		1.0^b
N-Methylmorpholine	1.0	4
Triethylamine	1.0	8
N,N,N',N'-Tetramethyl-1,3-butanediamine	1.0	27
1,4-Diazabicyclo[2,2,2]octane	1.0	120
Stannous chloride	0.10	2,200
Stannic chloride	0.10	2,600
Ferric acetylacetonate	0.01	3,100
Tetra-n-butyltin	1.0	160
Tetraphenyltin	1.0	9
Tri-n-butyltin acetate	0.001	31,000
Dimethyltin dichloride	0.001	78,000
Di-n-butyltin diacetate	0.001	56,000
Di-n-butyltin dichloride	0.001	57,000
Di-n-butyltin dilaurate	0.001	56,000
Di-n-butyltin dilaurylmercaptide	0.001	71,000
Bis(2-ethylhexyl)tin oxide	0.001	35,000
Di-n-butyltin sulfide	0.001	20,000
2-Ethylhexylstannonic acid	0.001	30,000

[a]Phenyl isocyanate–butanol reaction, both 0.25 M in dioxane at 70°C.

[b]$K_0 = 0.37 \times 10^{-4}$ 1 mole^{-1} sec^{-1} at 70°C.

[c]Reprinted from F. Hostettler and E. F. Cox, Ind. Eng. Chem. **52,** 609 (1960). (Copyright 1960 by the American Chemical Society. Reprinted by permission of the copyright owner.)

separately with diols or diamines of low molecular weight to cause further chain extension or polymerization (curing of the prepolymer). On reaction with water, the prepolymer can also be used to give foams, amines or tin compounds being used as catalysts in the foaming process. The properties of the polyurethane resin are controlled by the choice of the diisocyanate and polyol.

The reaction of α-naphthyl isocyanate with alcohols has been reported to be a convenient analytical method for the preparation of solid derivatives. In addition, the by-product dinaphthylurea is very insoluble in hot ligroin (b.p. 100°–120°C). The urethanes are readily soluble in hot ligroin, and on cooling the solution they recrystallize to sharp-melting solids. It is recommended that two recrystallizations be performed to obtain substances for analysis. Primary alcohols react well without the need for heating the reaction mixture. Secondary alcohols require additional heat, and the yields of urethane oft are smaller than when primary alcohols are used. Tertiary alcohols other than *t*-butyl or *t*-amyl were not able to react under the conditions used. Table II lists some representative alcohols and their α-naphthylurethane derivatives.

TABLE II[a]

α-NAPHTHYLURETHANE DERIVATIVES OF ALCOHOLS [4]

Alcohol	α-Naphthylurethane, m.p. (°C)
Methyl	124
Benzyl	134.5
Cinnamyl	114
Phenylethyl	119
Lauryl	80
n-Amyl	68
Furfuryl	129–130
m-Xylyl	116
o-Methoxybenzyl	135–136
Ethylene glycol	176
Trimethylene glycol	164
Glycerol	191–192
Ethylene bromohydrin	86–87
Trimethylene bromohydrin	73–74
Ethylene chlorohydrin	101
Trimethylene chlorohydrin	76
Phenylmethyl carbinol	106
Phenylethyl carbinol	102
Menthol	119
Borneol	127
Isoborneol	130
Cholesterol	160
Benzoin	140
Diphenyl carbinol	135–136
Cyclohexanol	128–129
2-Methylcyclohexanol	154–155
3-Methylcyclohexanol	122
4-Methylcyclohexanol	159–160
Methylhexyl carbinol	63–64
Diethyl carbinol	71–73
Triphenyl carbinol	No reaction
Diethylmethyl carbinol	No reaction
Citronellol	No reaction

[a]Reprinted in part from V. T. Bickel and H. E. French, *J. Amer. Chem. Soc.* **48**, 747 (1926). (Copyright 1926 by the American Chemical Society. Reprinted by permission of the copyright owner.)

TABLE III[c]

α-NAPHTHYLURETHANE DERIVATIVES OF SOME
REPRESENTATIVE PHENOLS [5]

Phenolic compound	Urethane, m.p. (°C)	Catalyst $(C_2H_5)_3N$ added
m-Cresol	127–128	No
p-Cresol	146	No
Thymol	160	No
Carvacrol	116	No
o-Nitrophenol	112–113	Yes
m-Nitrophenol	167	No
p-Nitrophenol	150–151	No
o-Chlorophenol	120	Yes
m-Chlorophenol	157–158	Yes
p-Chlorophenol	165–166	No
o-Bromophenol	128–129	Yes
p-Bromophenol	168–169	Yes
2,4,6-Tribromophenol	153	Yes
2-Chloro-5-hydroxytoluene	153–154	Yes
4-Hydroxy-1,2-dimethylbenzene	141–142	Yes
4-Hydroxy-1,3-dimethylbenzene	134–135	Yes
2-Hydroxy-1,4-dimethylbenzene	172–173	Yes
Resorcinol–monomethyl ether	128–129	Yes
Guiacol	118	No
Eugenol	122	No
Isoeugenol	149–150	Yes
Orcinol[a]	160	Yes
Resorcinol	No reaction	—
Hydroquinol	No reaction	—
Catechol	No reaction	—
Pyrogallol	No reaction	—
α-Naphthol	152	Yes
β-Naphthol	156–157	Yes
1-Nitro-2-naphthol[a]	128–129	Yes
o-Aminophenol[b]	201	No

[a]Reacts with great difficulty and gives low yields of the urethanes.
[b]Reacts readily at room temperature.
[c]Reprinted in part from H. E. French and A. F. Wirtel *J. Amer. Chem. Soc.* **48,** 1736 (1926). (Copyright 1926 by the American Chemical Society. Reprinted by permission of the copyright owner.)

Phenols but not polyhydroxybenzenes also react with α-naphthyl isocyanate to give α-naphthylurethanes. In some cases in which solid derivatives were difficult to obtain, the addition of 1 or 2 drops of triethylamine solution in ether caused a solid derivative to form rapidly.

Table III gives several phenols and the respective melting points for their α-naphthylurethane derivatives.

REFERENCES

1. W. Worthy, *Chem. Eng. News*, pp.27–33 (Feb. 11, 1985).
2. S. R. Sandler, unpublished data (1970).
3. F. Hostettler and E. F. Cox, *Ind. Eng. Chem.* **52,** 609 (1960).
4. V. T. Bickel and H. E. French, *J. Amer. Chem. Soc.* **48,** 747 (1926).
5. H. E. French and A. F. Wirtel, *J. Amer. Chem. Soc.* **48,** 1736 (1926).

30

O-CARBAMATES

1. INTRODUCTION

O-Carbamates (structure I) represent a class of compounds distinct from *N*-carbamates (substituents on the nitrogen atom as in structure II). In this chapter only carbamates with varying substituents on the oxygen atom are discussed.

$$
\begin{array}{cc}
\overset{\text{O}}{\underset{\|}{}} & \overset{\text{O}}{\underset{\|}{}} \\
\text{ROC—NH}_2 & \text{ROC—NHR} \\
\textit{O}\text{-carbamate} & \textit{N}\text{-carbamate} \\
\text{I} & \text{II}
\end{array}
$$

The recent industrial applications of carbamates such as those reported as a tranquilizer (structure III), a crease-resistant agent (when

From S. R. Sandler and W. Karo, *Organic Functional Group Preparations*, Vol. II, 2d ed. (Orlando, Florida, 1986), 274–290, by permission of Academic Press, Inc.

TABLE I
MELTING POINTS OF
O-CARBAMATES

Carbamate	M.p. (°C)
Methyl	54.2
Ethyl	48.2
n-Propyl	60
Isopropyl	95
n-Butyl	53
Lauryl	81–82
n-Octyl	67
n-Stearyl	94–95

reacting with formaldehyde) in the textile industry, a solvent, hair conditioners, a plasticizer, and a fuel additive have stimulated interest in the synthesis of various *O*-carbamates.

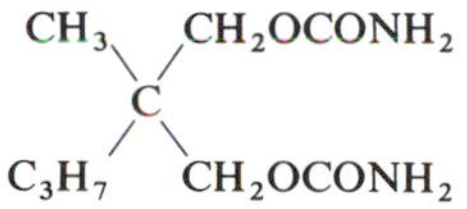

$$CH_3 \diagdown \quad \diagup CH_2OCONH_2$$
$$C$$
$$C_3H_7 \diagup \quad \diagdown CH_2OCONH_2$$

Meprobamate, a tranquilizer

III

The *O*-carbamates are usually solids; some representative examples are shown in Table I.

The most important methods of preparing *O*-carbamates are shown in Eq. (1).

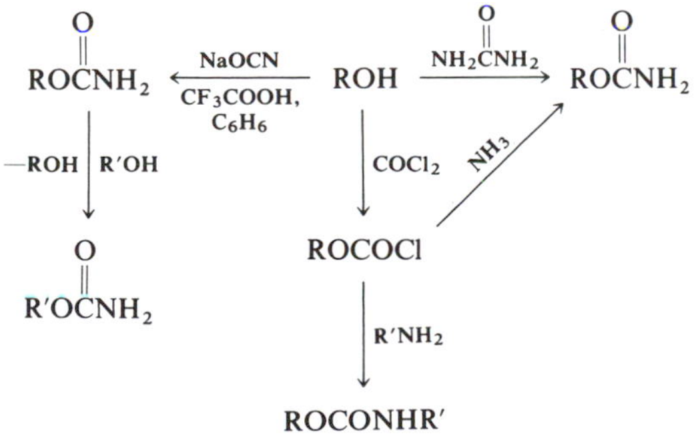

$$
\overset{O}{\overset{\|}{ROCNH_2}} \xleftarrow[\substack{CF_3COOH, \\ C_6H_6}]{NaOCN} ROH \xrightarrow{\overset{O}{\overset{\|}{NH_2CNH_2}}} \overset{O}{\overset{\|}{ROCNH_2}}
$$

$$
-ROH \Big\downarrow R'OH \qquad\qquad \Big\downarrow COCl_2 \quad \nearrow^{NH_3}
$$

$$
\overset{O}{\overset{\|}{R'OCNH_2}} \qquad\qquad ROCOCl \tag{1}
$$

$$
\Big\downarrow R'NH_2
$$

$$
ROCONHR'
$$

2. CONDENSATION REACTIONS

A. REACTIONS OF ALCOHOLS WITH UREA

The reaction of primary alcohols with urea gives carbamates when the reaction is carried out at 115°–150°C (Eqs. 2, 3). Since 150°C is the temperature for the optimum dissociation of urea to cyanic acid and ammonia, lower-boiling alcohols (methyl, ethyl, and propyl) must be heated under pressure. Refluxing urea and n-butanol at 115°–120°C requires a 40-hr reaction time to give a 75% yield of butyl carbamate.

$$NH_2\overset{\overset{\text{O}}{\|}}{C}NH_2 \underset{}{\overset{\text{heat}}{\rightleftharpoons}} HNCO + NH_3 \qquad (2)$$

$$HNCO + ROH \longrightarrow RO-\underset{\underset{\text{O}}{\|}}{C}NH_2 \qquad (3)$$

In order to shorten the reaction time, various heavy metal salts (zinc, lead, and manganese acetates) of weak organic acids, zinc or cobalt and tin chlorides are added to the reaction mixture. For example, refluxing an uncatalyzed mixture of 3 moles of isobutyl alcohol and urea for 150 hr at 108°–126°C gives a 49% yield of the carbamate. Adding lead acetate or cobalt chloride to the same reaction lowers the reaction time to 75 hr, at which point an 88–92% yield is obtained. In another example, ethylene glycol (1 mole) and urea (2 moles) are heated for 3 hr at 135°–155°C with $Mn(OAc)_2$ to give a 78% yield of the diurethane. The commercial production of butyl carbamate uses catalytic quantities of cupric acetate [2].

30-1. Preparation of Ethyl Carbamate [1]

$$C_2H_5OH + NH_2\overset{\overset{\text{O}}{\|}}{C}NH_2 \overset{BF_3}{\longrightarrow} C_2H_5O\overset{\overset{\text{O}}{\|}}{C}NH_2 + NH_3 \qquad (4)$$

To a tared flask containing 600.6 gm (10 moles) of urea dissolved in 2073.2 gm of ethanol at 70°C is added, via a gas addition tube, boron trifluoride until 678.2 gm (4.3 moles) is absorbed. The ethanol is removed by distillation until the pot temperature reaches 100°C, then the hot reaction mixture is filtered. The solids, which consist of the BF_3-NH_3 complex, are washed with ethanol and then the ethanol is added to the original filtrate. The ethanol is then distilled off again as above and filtered again to give more BF_3-NH_3 solids. The total BF_3-NH_3 solids weigh 535 gm. This solid is washed with hot ethanol and the total alcohol washing and remaining ethanol filtrates cooled to give 255 gm of $BF_3-4(NH_2CONH_2)$ complex, m.p. $96°-98°C$. The combined filtrates are freed of alcohol by distillation, extracted with benzene, and the extract on distillation under reduced pressure affords 380 gm (99%) of $C_2H_5OCONH_2$, b.p. $115°C$ (100 mm), m.p. $48°-50°C$. The tacky residue weighs 150 gm.

30-2. Preparation of *n*-Butyl Carbamate [2]

$$\underset{\text{H}_2\text{N}\overset{\displaystyle\text{O}}{\overset{\|}{\text{C}}}\text{NH}_2}{} + n\text{-C}_4\text{H}_9\text{OH} \xrightarrow{\text{Cu(OAc)}_2} \underset{n\text{-C}_4\text{H}_9\text{O}\overset{\displaystyle\text{O}}{\overset{\|}{\text{C}}}\text{NH}_2}{} + \text{NH}_3 \tag{5}$$

To a flask is added 60 gm (1 mole) of urea, 100 ml of *n*-butanol (1.1 mole), and 2.0 gm (1.01 mmole) of cupric acetate. The contents are heated to reflux ($118°-160°C$) for 4 hr until 99.5% of the theoretical amount of ammonia evolves. After the reaction, 83.2 gm (71%) of *n*-butyl carbamate is obtained having a melting point of $52°C$.

B. TRANSESTERIFICATION OF CARBAMATES

Carbamates have been prepared by heating ethyl carbamate with a higher-boiling alcohol in the presence or absence of catalysts [3, 4, 5]. Aluminum isopropoxide has been reported [6] to be an excellent catalyst for the interchange reaction between ethyl carbamate and benzyl alcohol. The interchange reaction is also effective for *N*-alkyl carbamates as well as unsubstituted carbamates. This catalyst is effective in preparing mono- and dicarbamates in excellent yields from primary and secondary alcohols and diols. Other effective catalysts are: dibutyltin dilaurate, dibutyltin oxide, sulfuric acid or *p*-toluenesulfonic acid, and sodium metal (reacts with alcohols to give the alkoxide catalyst) [5].

Examples of the transesterification reaction of ethyl carbamates and alcohols are given in Table II.

TABLE II[a]

TRANSESTERIFICATION REACTION OF ETHYL CARBAMATES AND ALCOHOLS

Alcohol	Mole ratio of ethyl carbamate/ alcohol	Catalyst	Temp. (°C)	Time (hr)	Yield (%)	M.p. (°C)
Isobutyl	1/6	—	110–120	103	—	—
	1/6	H_2SO_4[e]	110–120	19	87 (56)[b]	63–64
	1/6	(*n*-Bu)₃N	110–120	8	—	—
sec-Butyl	1/6	H_2SO_4	105–110	16	37	92–93
t-Butyl	1/6	H_2SO_4	85–90	—	—	—
Benzyl	1/1.5	—	190–230	19	70 (53)[b]	85–86
	1/1.5	H_2SO_4	145–240	5	9	82–84
	1/1[c]	Al (iso-PrO)₃	130–140[d]	5–10	86	86–87

[a]Data taken from Gaylord and Sroog [3] except where noted.
[b]Recrystallized product.
[c]Data from Kraft [6].
[d]Bath temperature.
[e]Approximately 2 ml concentrated H_2SO_4/0.5 mole ethyl carbamate used.

Tertiary alcohols and phenols do not undergo the transesterification reaction of carbamates with acidic or basic catalysts.

30-3. Preparation of Benzyl Carbamate [6]

$$C_2H_5O\overset{\overset{\text{O}}{\|}}{C}NH_2 + C_6H_5CH_2OH \xrightarrow{\text{Al (iso-PrO)}_3} C_6H_5CH_2O\overset{\overset{\text{O}}{\|}}{C}NH_2 + C_2H_5OH \quad (6)$$

To a 250 ml three-necked flask equipped with a thermometer and a 20 cm distillation column filled with glass beads are added 44.5 gm (0.5 mole) of ethyl carbamate, 54.1 gm (0.5 mole) of benzyl alcohol, and 60 ml of toluene. The reaction flask is heated with an oil bath at 110°–125°C in order to remove any water in the reagents. The bath is cooled to 100°C, and 2.0 gm (0.01 mole) of aluminum isopropoxide is added all at once. The reaction is heated with the oil bath set at 130°–140°C in order to remove about 50 ml of the ethanol–toluene azeotrope at 77°C. The residue is recrystallized from toluene to yield 64.5 gm (85.5%), m.p. 86°–87°C.

NOTE: Substituting sodium methoxide for aluminum isopropoxide gave poor yields.

REFERENCES

1. F. J. Sowa, U.S. Pat. 2,834,799 (1958).
2. S. Beinfest, P. Adams, and J. Halpern, U.S. Pat. 2,837,561 (1958).
3. N. G. Gaylord and C. C. Sroog, *J. Org. Chem.* **18,** 1632 (1953).
4. G. Heilner, Ger. Pat. 551,777 (1932).
5. M. Metayer, *Bull. Soc. Chim. Fr.* 802 (1951).
6. W. M. Kraft, *J. Amer. Chem. Soc.* **70,** 3569 (1948).

31

IMINES

1. INTRODUCTION
    31-1. Preparation of Benzophenone *N*-methylimine
    31-2. Preparation of Benzalaniline (Benzylideneaniline) (Schiff Base Formation)
    31-3. Preparation of 3-Iminobutyronitrile (Thorpe Reaction)

1. INTRODUCTION

T he nomenclature of imines is highly variable. For example, one recent article identified a compound as (3-methoxy-3-methyl-2-butylidene)isopropylamine, i.e. an–ylideneamine system [1]. Another recent article in the same journal talks of compounds such as cinnamaldehyde *N*-isopropylimine—an aldehyde–imine system [2]. The "ylideneamine" system is probably the most systematic one, although the synthetic chemist would probably opt for the latter method, which shows the carbonyl compound and the amine from which the Schiff base may be prepared. Table I lists various nomenclature systems found in the literature.

31-1. Preparation of Benzophenone *N*-methylimine [3, 4]

$$\text{(C}_6\text{H}_5)_2\text{C=O} + \text{CH}_3\text{NH}_2 \xrightarrow{180^\circ - 185^\circ\text{C}} (\text{C}_6\text{H}_5)_2\text{C=N-CH}_3 + \text{H}_2\text{O} \quad (1)$$

From S. R. Sandler and W. Karo, *Organic Functional Group Preparations*, Vol. II, 2d ed. (Orlando, Florida, 1986), 291ff., by permission of Academic Press, Inc.

TABLE I

NOMENCLATURE USED TO DESCRIBE THE VARIOUS IMINES OF STRUCTURE RR'C=NR"

Name	Source	Substituents
Imine	*Chemical Abstracts*	R" = hydrogen (H)
Amine[a]	*Chemical Abstracts*	R" = alkyl (R) or aryl (Ar) group
Aldimine	Common usage	R = R or Ar; R' = H
Ketimine	Common usage	R, R' = R or Ar
Schiff base	Common usage	R = Ar; R' = H; R" = R or Ar
Anils	Common usage	R, R'=R, Ar, H; R"=phenyl or substituted phenyl

[a]That is, "ylideneamine" systems.

In a well-ventilated hood, in a suitable reaction vessel 66.5 gm (0.35 mole) of benzophenone is melted and heated to 180°–185°C. A stream of methylamine is passed through the melt for about 10 hr. The reaction is considered over when the evolution of water has stopped. The oily flask residue is allowed to cool and dissolved in ether. The ether solution is then extracted with several portions of ice-cold 2 N hydrochloric acid. Each portion of the acid extract is made alkaline as quickly as possible with 40% aqueous sodium hydroxide. The product oil is extracted with ether. The ether solution is dried over anhydrous pentoxide, filtered, and freed of solvent on a rotary evaporator. The pure imine is finally distilled under reduced pressure at 126°–128°C/2.5 mm. Yield: 55.3 gm (82%).

By a similar procedure 4-methylbenzophenone was reacted with methylamine. Yield: 68%; b.p. 140°–142°C/2.4 mm.

The preparation of benzalaniline is an example of an aldimine preparation without any catalyst and without extraneous solvents.

31-2. Preparation of Benzalaniline (Benzylideneaniline) [5]
Schiff Base Formation

$$C_6H_5CH=O + C_6H_5NH_2 \longrightarrow C_6H_5CH=N-C_6H_5 + H_2O \qquad (2)$$

To a 500 ml, three-necked flask containing 106 gm (1.0 mole) of benzaldehyde is added 93 gm (1.0 mole) of aniline while stirring rapidly. The exothermic reaction starts immediately as water begins to separate. After 15 min the mixture is poured into a beaker containing 165 ml of vigorously stirred 95% ethanol. Crystallization begins in about 5 min, and after 10 min the mixture is placed in an ice bath for $\frac{1}{2}$ hr. The semisolid is filtered by suction and air-dried to yield 152–158 gm (84–87%) of benzalaniline, m.p. 52°C. From concentrating the filtrate under reduced pressure an additional 10 gm, m.p. 51°C is obtained. Further recrystallization may be made from 95% ethanol.

31-3. Preparation of 3-Iminobutyronitrile (Thorpe Reaction) [6, 7, 8]

$$CH_3CN + Na \xrightarrow[-CH_4, \ -NaCN]{\substack{C_6H_6 \\ \text{reflux temp.}}} CH_3\overset{\overset{\displaystyle NNa}{\|}}{C}-CH_2CN \xrightarrow[-NaOH]{H_2O} CH_3\overset{\overset{\displaystyle NH}{\|}}{C}-CH_2CN \qquad (3)$$

In a three-necked flask equipped with a long reflux condenser, stirrer, and dropping funnel is placed 78 gm (3.9 gm-atom) of powdered sodium in 800 ml of dry benzene. Acetonitrile (246 gm, 6.0 moles) is slowly added over a period of $2\frac{1}{2}$ hr at a rate to maintain the reaction mixture boiling. Once the reaction starts, it is very exothermic and some cooling may be necessary. Methane is evolved, and care should be taken to prevent loss of benzene.

After the addition of acetonitrile the mixture is refluxed for 3 hr or longer until, in cooling, the sodium salt or 3-iminobutyronitrile and sodium cyanide crystallize from the reaction mixture. The product salts are filtered, suspended in 1 liter of ether, and water is slowly added until the salts dissolve. The ether layer is separated, dried, and the ether removed by atmospheric distillation. After a few hours the residue crystallizes to yield 121 gm (49.7%) of crude product. Recrystallization from benzene yields 91 gm (37.4%), m.p. 63°–71°C. Von Meyer [9] has reported that this compound exists in a stable form with m.p. 50–54°C and in a labile form, m.p. 79°–84°C.

REFERENCES

1. N. DeKimpe, P. Sulmon, R. Verhé, L. DeBuyck, and N. Schamp, *J. Org. Chem.* **48**, 4320 (1983).
2. W. T. Brady and C. H. Shieh, *J. Org. Chem.* **48**, 2499 (1983).
3. C. R. Hauser and D. Lednicer, *J. Org. Chem.* **24**, 46 (1959).
4. R. L. Ehrhardt, G. Gopalakrishnan, and J. L. Hogg, *J. Org. Chem.* **48**, 1586 (1983).
5. L. A. Bigelow and H. Eatough, *Org. Syn. Coll. Vol.* **1**, 80 (1941).
6. H. Adkins and G. M. Whitman, *J. Amer. Chem. Soc.* **64**, 150 (1942).
7. E. F. G. Atkinson and J. F. Thorpe, *J. Chem. Soc.* **89**, 1906 (1906).
8. N. Lees and J. F. Thorpe, *J. Chem. Soc.* **91**, 1282 (1907).
9. E. von Meyer, *J. Prakt. Chem.* (2) **52**, 84 (1895).

32

AZIDES

1. INTRODUCTION

The most generally useful method of preparing azide and acyl azides makes use of the displacement of other functional groups by azide ions (Eq. 1).

$$RX + NaN_3 \longrightarrow RN_3 + NaX \tag{1}$$

X = halide, acyl halide, sulfonyl halide, bridgehead hydroxyl, etc.

Other useful preparations involve diazo transfer reactions. In the aromatic series diazonium salts may be reacted with sodium azides (Eq. 2).

$$Ar\overset{+}{N}_2X^- \xrightarrow{\ NaN_3\ } ArN_3 \tag{2}$$

From S. R. Sandler and W. Karo, *Organic Functional Group Preparations*, Vol. II, 2d ed. (Orlando, Florida, 1986), 323ff., by permission of Academic Press, Inc.

237

Hydrazine derivatives may be treated with nitrous acid to form azides, a reaction which is of particular value in preparing polymeric acrylazide.

Hydrazoic acid may be added to activated olefinic bonds to yield azides. Some epoxy compounds have also been reacted with sodium azide to form hydroxy azides.

A. HAZARDS AND SAFE HANDLING PRACTICES

■ **CAUTION:** Although organic azides are reputed to be explosive materials, detailed information on hazards and safety precautions is sparse. Smith called acetyl azide "treacherous" [1].

The explosive hazard may be a function of the size of the molecule. For example, methyl azide is reported to be handled in a routine manner (but not in the presence of mercury) [1]. Yet, while this chapter was in preparation, Burns and Smith reported an explosion during the preparation of this very compound from dimethyl sulfate and sodium azide while sodium hydroxide was being added [2]. They attributed the explosion to the formation of hydrazoic acid during the preparation, when the pH of the reaction mixture may have dropped below 7. They therefore recommend adding the indicator bromthymol blue to the reaction mixture. This indicator changes color from deep blue at pH 6.5 to yellow at pH 8. The rate of base addition can then be monitored readily by observing the color of the reaction mixture and maintaining the pH at 8 or higher, throughout the reaction. Since the indicator may fade during the process, additional quantities will have to be added from time to time (approximately once each hour).

Burns and Smith more recently [3] reported an explosion during their preparation of 20 gm ethyl azide following their precautions cited above for methyl azide [2]. The explosion took place just after approximately 0.5 ml of additional indicator solution was added. They postulated that either the acidic indicator or exposure to the ground glass joint initiated the detonation. The force of the detonation of this 20 gm ethyl azide left a 1-cm deep depression in their 16-gauge stainless steel hood floor and shattered the safety glass around the hood's fluorescent lamp.

The present authors are not in a position to comment on either the validity of the hypothesis of Burns and Smith or their recommendation. Obviously, all reactions involving the preparation, use, and disposal of solvents and by-products of azides and related compounds must be carried out on a very small scale with suitable protection of personnel even if a particular reaction has been repeatedly carried out without incident. The avoidance of ground glass joints, the protection from strong light, and the use of dilute solutions (as with diazo methane preparations) are additional precautions suggested by Burns and Smith [3].

In the case of "triflyl" azide, the recommendation has been made that the compound not be allowed to be completely free of solvent and that it not be stored for any length of time [4]. Ethyl azidoformate could be distilled at about 100°C. It did not detonate until 160°C. On the other hand, the vapors of this azide are toxic, leading to vertigo, severe headaches, and sometimes vomiting [5].

Of the 1,2-diazobenzenes, the parent compound, 1,2-diazobenzene, could be detonated on an anvil with a hammer. However, the 3-methyl-4-methyl, 4-methoxy, and 4-chloro derivatives were not said to be shock-sensitive [6].

These observations show that extreme care must be exercised in the handling and preparation of azides and their derivatives.

In addition, sodium azide and hydrazoic acid must be handled safely. In working with sodium azide, the salt must not come in contact with copper, lead, mercury, silver, gold, their alloys, and their compounds. All of these form sensitive explosive azides. Azide salts must not be thrown into sinks or sewers since all azide salts are highly toxic, react with acids to form explosive, toxic, and gaseous hydrazoic acid, and react with copper and lead pipes. Azides and hydrazoic acid are thought to be more toxic than cyanides and hydrogen cyanide [7a].

Decontamination of rags, filter paper, solutions containing sodium azide, and apparatus which has been contaminated with sodium azide should be done by soaking, in a fume hood, with acidified sodium nitrite until the azides have been destroyed, followed by washing or other disposal [7b]. In general, the handling and disposition of azides must be in conformity with all appropriate laws and regulations.

While the use of silver azide has been recommended in some syntheses, in most cases this does not appear to be necessary. In view of the explosive hazards associated with heavy metal azides, use of this azide is best avoided.

In view of the general hazards of handling azides, the scale of reaction should probably be reduced considerably. The removal of unreacted alkyl halide with silver nitrate may lead to silver azide formation and should, therefore, be replaced by another procedure.

B. METHODS

32-1. Preparation of *n*-Butyl Azide [8]

$$n\text{-}C_4H_9Br + NaN_3 \xrightarrow[\text{CH}_3\text{OH}]{\text{H}_2\text{O}} n\text{-}C_4H_9N_3 + NaBr \qquad (3)$$

With suitable safety precautions, to a flask containing 34.5 gm (0.53 mole) of sodium azide in 70 ml of water and 25 ml of methanol is added 68.5 gm (0.50 mole) of *n*-butyl bromide while stirring at room temperature. The resulting mixture is heated and stirred on a steam bath for 24 hr. The bottom layer of *n*-butyl bromide disappears after this time and a top layer of crude *n*-butyl azide forms. The crude azide is separated and then treated overnight with alcoholic silver nitrate to remove traces of butyl bromide. The mixture is then filtered, washed with water, and distilled behind a safety barricade to yield 40.0 gm (90%) of *n*-butyl azide, b.p. 106.5°C (760 mm), $n_D^{29.5}$ 1.4152, $d^{29.5}$ 0.8649. [NOTE: *n*-Butyl azide and methanol form an azeotrope (b.p. 60°C) from which the azide is liberated by the addition of a saturated solution of calcium chloride.]

32-2. Preparation of Phenyl Azide [9]

$$\langle\rangle\text{—NHNH}_2 + HNO_2 \longrightarrow \langle\rangle\text{—N}_3 + 2H_2O \qquad (4)$$

To an ice-cooled (0°C to −10°C) stirred flask containing 300 ml of water and 55.5 ml of concentrated hydrochloric acid is added dropwise over a 10 min period 33.5 gm (0.31 mole) of freshly distilled phenylhydrazine. Phenylhydrazine hydrochloride crystals precipitate as they are formed. Addition of 100 ml of ether at 0°C is followed by the dropwise addition (25 min) of a solution of 25 gm (0.36 mole) of sodium nitrite in 30 ml of water. At all times the reaction temperature is kept below 5°C.

The product is isolated by carrying out a steam distillation of the reaction mixture to yield 400 gm of distillate. The ether layer is separated from the distillate, and the water layer is extracted with 25 ml of ether. The combined ether layers are concentrated at 25°–30°C under reduced pressure. The residue is distilled under reduced pressure to afford 24–25 gm (65–68%) of phenyl azide, b.p. 49°–50°C (5 mm).

■ **CAUTION:** Phenyl azide decomposes violently when heated at 80°C or above. Care must be taken that the bath temperature never exceeds 60°–70°C. The product should be stored in a cool place in brown bottles.

Preparation 32-3 illustrates the preparation of a polyacrylazide from cross-linked polyacrylamide. A similar procedure for the preparation of a polymeric acyl azide from a monodispersed poly(styrene-coacrylamide) latex is described in the patent cited in [11].

32-3. Preparation of Cross-Linked Poly(acrylazide) [10]

$$\left(\begin{array}{c} CH_2-CH \\ | \\ CONH_2 \end{array} \right)_n + NH_2NH_2 \longrightarrow \left(\begin{array}{c} CH_2-CH \\ | \\ CONHNH_2 \end{array} \right)_n \tag{5}$$

$$\left(\begin{array}{c} CH_2-CH \\ | \\ CONHNH_2 \end{array} \right)_n + HNO_2 \longrightarrow \left(\begin{array}{c} CH_2-CH \\ | \\ CON_3 \end{array} \right)_n \tag{6}$$

(a) Preparation of Poly(Acrylhydrazide). In a siliconized glass-stoppered Erlenmeyer flask, 1 gm of cross-linked polyacrylamide beads are allowed to swell overnight in an excess of distilled water equal to approximately 1.3 times the bed volume of the gel. The flask is then suspended in a constant-temperature bath maintained at 47°C. At the same time, a glass-stoppered cylinder containing six times the number of equivalents of the acrylamide in the resin of hydrazine hydrate is immersed in the constant-temperature bath. After about 45 min, the hydrazine is added to the swollen polyacrylamide, a magnetic stirrer is inserted in the flask, the flask is stoppered, and the mixture is stirred at 47°C for 7 hr.

In a fume hood, the gel is washed with 0.1 M aqueous sodium chloride on a Büchner funnel and finally by sedimentation. This washing operation is repeated until the aqueous supernatant solution is free of hydrazine. The gel is finally washed and suspended in a storage buffer at pH 7.3, which is 0.20 M sodium chloride, 0.002 M Na$_2$EDTA, 0.10 M boric acid, 0.005 M sodium hydroxide, and 5 × 10^{-6} M pentachlorophenol. The resin is reported to contain approximately 4 milliequivalents of hydrazide per gram of dry resin.

(b) Preparation of Poly(Acrylazide). About 50 ml of the polymer from step (*a*) is washed with 0.1 M aqueous sodium chloride and 0.25 N hydrochloric acid and resuspended to a 32-ml volume with

0.25 N hydrochloric acid. The suspension is cooled to 0°C. Then 8 gm of crushed ice is added. The container is placed in an ice bath and, with efficient magnetic stirring, 4.0 ml of 1.0 M aqueous sodium nitrite is added.

If a biochemical reagent such as a protein with a free amino group is to be coupled to the resin, about 90 seconds later, this reagent is added rapidly to the reacting system while maintaining a temperature of 0°C. Stirring is continued for the reaction period required for the specific protein involved (see [10] for typical examples).

The excess unreacted azide is reconverted into the hydrazide and then to the stable acetyl hydrazide by adding 1.5 ml of hydrazine hydrate and stirring for 0.5 to 1 hr, followed by washing on a Büchner funnel in turn with 100 ml of 0.1 M aqueous sodium chloride, 100 ml of 0.2 M aqueous sodium acetate in which 4 ml of acetic anhydride was dissolved immediately before the washing, 100 ml of 0.1 M aqueous sodium chloride, 50 to 100 ml of 2 M aqueous sodium chloride, and a storage buffer.

REFERENCES

1. P. A. S. Smith, "Open-Chain Nitrogen Compounds," Vol. 2, p. 214, Benjamin, New York, 1966.
2. M. E. Burns and R. H. Smith, Jr., *Chem. Eng. News* p. 2 (Jan. 9, 1984).
3. M. E. Burns and R. H. Smith, Jr., *Chem. Eng. News* p. 2 (Dec. 16, 1985).
4. J. Zaloom and D. C. Roberts, *J. Org. Chem.* **46,** 5173 (1981).
5. W. Lwowski and T. W. Mattingly, Jr., *J. Amer. Chem. Soc.* **87,** 1947 (1965).
6. J. H. Hall and E. Patterson, *J. Amer. Chem. Soc.* **89,** 5856 (1967).
7a. Lonza, Inc., Fairlawn, New Jersey, "Sodium Azide, Determination of Hydrazoic Acid and Azide in the Atmosphere of Azide Plants" (October, 1983).
7b. Military Specification MIL-S-20552A. "Sodium Azide, Technical" (July 24, 1952).
8. J. H. Boyer and J. Hamer, *J. Amer. Chem. Soc.* **77,** 951 (1955).
9. R. O. Lindsay and C. F. H. Allen, *Org. Syn. Coll. Vol.* **3,** 710 (1955).
10. J. K. Inman and H. M. Dintzis, *Biochemistry* **8,** 4074 (1969).
11. L. C. Dorman, U.S. Patent 4,046,723 (Sept. 6, 1977).

33

AZO COMPOUNDS
AND AZOXY
COMPOUNDS

1. INTRODUCTION

A. HAZARDS AND SAFE HANDLING PRACTICES

■ **CAUTION:** While azo dyes have been in use for many years, only recently questions of health hazards associated with dyes generally have been raised. Derivatives of *p*-dimethylaminoazobenzene (DMAB, butter yellow) are known carcinogens [1], [2]. Consequently, we suggest that other azo compounds, especially aromatic azo compounds, be handled with great caution.

B. COUPLING REACTIONS

Perhaps the best-known method of preparing aromatic azo compounds involves the coupling of diazonium salts with sufficiently reactive aromatic compounds such as phenols, aromatic amines, phenyl

From S. R. Sandler and W. Karo, *Organic Functional Group Preparations*, Vol. II, 2d ed. (Orlando, Florida, 1986), Azo Compounds: 353ff.; Azoxy Compounds: 414ff., by permission of Academic Press, Inc.

ethers, the related naphthalene compounds, and even sufficiently reactive aromatic hydrocarbons. Generally, the coupling must be carried out in media which are neutral or slightly basic or which are buffered in the appropriate pH range. The reaction may also be carried out in nonaqueous media. While some primary and secondary aromatic amines initially form an *N*-azoamine, which may rearrange to the more usual amino-*C*-azo compound, tertiary amines couple in a normal manner.

Under some conditions, phenolic ethers are dealkylated during coupling. However, the dealkylation follows the coupling step and is acid catalyzed. Consequently, use of an excess of sodium acetate as a buffer or use of a nonaqueous medium obviates the dealkylation.

Two molecules of a diazonium salt may couple with loss of some nitrogen. In the Bogoslovskii reaction, this reaction has been developed as a means of preparing *o,o'*-dihydroxyazo compounds, which are difficult to obtain by other means. This reaction involves the use of a cuprous complex as the reaction catalyst. Self-coupling of diazonium salts also takes place in the presence of sodium sulfite.

Several intramolecular couplings of diazonium salts with ortho substituents bearing an active methylene group give rise to cinnolines, a class of cyclic azo compounds (the Borsche, von Richter, and Widman–Stoermer syntheses).

33-1. Preparation of 2,2′-Dihydroxyazobenzene (Bogoslovskii Reaction) [3, 4]

$$\text{(1)}$$

$$\text{(2)}$$

(a) Preparation of Catalyst Stock Solution. Copper(II) sulfate pentahydrate, 28.5 gm, is dissolved in 100 ml of hot water, cooled to room temperature, and then enough concentrated aqueous ammonia is added dropwise until the soluble complex has been formed completely. To this solution is added a solution of 7 gm of hydroxylamine hydrochloride in 20 ml of water to reduce all of the copper(II) complex to the colorless copper(I) complex.

(b) Diazotization and Coupling. To a solution of 11.0 gm (0.1 mole) of *o*-aminophenol and 7.0 gm (0.1 mole) of sodium nitrite in 250 ml of 5% aqueous sodium hydroxide is dropwise added concentrated hydrochloric acid until a positive reaction with starch sodium iodide test paper is observed. The solution of the diazo compound is added rapidly with stirring to the catalyst solution contained in a large vessel. The excess foaming may be controlled by the addition of a small quantity of ether. The reaction mixture is allowed to stand at room temperature with occasional stirring for 1 hr. Then the brown solid is removed by filtration.

The resultant chelated complex is mixed with 500 ml of concentrated hydrochloric acid and warmed gently. The mixture is cooled, diluted with ice water, and filtered. The product is recrystallized three times from benzene to give 5.7 gm (53%) of yellow-orange needles, m.p. 172°–172.7°C.

The yield may be increased to 78% by continuous liquid–liquid extraction of the dilute hydrochloric acid filtrate solution with ether.

It is self-evident that one of the simpler methods of preparing unsymmetrically substituted azoxy compounds must involve the condensation of two distinctly different starting materials. In principle, the reaction of *C*-nitroso compounds with hydroxylamines meets this requirement (Eq. 3).

$$R-NO + R'NHOH \longrightarrow R-NON-R' + H_2O \tag{3}$$

Historically this reaction developed from the assumption that the formation of azoxy compounds by the reduction of aromatic nitro compounds probably involved the intermediate formation of *C*-nitroso compounds and hydroxylamines. In the all-aliphatic series, this reaction appears to be quite general. Symmetrically and unsymmetrically substituted azoxy compounds have been prepared by it, the only major problems being the usual ones of developing procedures that afford good yields and of determining the exact position of the azoxy oxygen in unsymmetrically substituted products.

In this reaction the source of the azoxy oxygen appears to be the nitroso group [5]. The preparation of *t*-butyl-*ONN*-azoxymethane (*N*-methyl-*N'*-*t*-butyldiazene-*N'*-oxide) is an example of a preparation of an unsymmetrical azoxy compound which is quite generally applicable. The structure assignment is based on NMR data.

33-2. Preparation of *t*-Butyl-*ONN*-azoxymethane
(*N*-Methyl-*N*'-*t*-butyldiazene-*N*'-oxide) [5, 6]

$$
\begin{array}{c}
\text{CH}_3 \\
| \\
\text{CH}_3-\text{C}-\text{NO} + \text{CH}_3\text{NHOH}\cdot\text{HCl} \\
| \\
\text{CH}_3
\end{array}
\xrightarrow{\text{KOH}}
\begin{array}{c}
\text{CH}_3 \quad \text{O} \\
| \quad \nearrow \\
\text{CH}_3\text{C}-\text{N}=\text{N}-\text{CH}_3 + \text{HCl} + \text{H}_2\text{O} \\
| \\
\text{CH}_3
\end{array}
\qquad (4)
$$

■ **CAUTION:** *C*-Nitroso compounds and hydroxylamine derivatives must be handled with due caution. The final product may also have adverse physiological properties [7].

To a stirred suspension of 1.9 gm (0.033 mole) of powdered potassium hydroxide in 25 ml of anhydrous ether in a reflux setup is added gradually 2.6 gm (0.03 mole) of *N*-methylhydroxylamine hydrochloride. The ether is evaporated at reduced pressure, leaving a slightly yellow, curdy solid. To this residue is rapidly added 2.4 gm (0.03 mole) of *t*-nitrosobutane. The reaction mixture is cautiously warmed. The exothermic reaction which may develop is moderated by cooling the flask as required. After the reaction has been brought under control, the reaction system is heated first for 1 hr at 85°C, followed by heating at 110°C for 2 hr. After the reaction mixture has been cooled, it is diluted with approximately 10 ml of water. The product is separated from the aqueous layer by repeated extraction with ether. The ether extracts are combined and dried over anhydrous sodium sulfate. After filtration and removal of the ether by evaporation, the residue is distilled under reduced pressure. The colorless product is isolated at 60°–62°C (110 mm Hg): yield 2.2 gm (63%), n_D^{20} 1.4265.

By a similar procedure the symmetrical 2-azoxy-2,3-dimethylhexane is also produced [8].

A convenient reaction involves preparing the toluene sulfonyl derivatives of the nitrosohydroxylamines and treating these "tosylates" with Grignard reagents. This procedure permits the preparation of a variety of unsymmetrical azoxy compounds [9]. The cited reference shows only the preparation of aromatic azoxy compounds and arylazoxyalkanes by the reaction of Grignard reagents with tosylates derived from aromatic nitrosohydroxylamines. Aliphatic nitrosohydroxylamine tosylates did not undergo this reaction, thus precluding the possibility of using this approach for the preparation of totally aliphatic azoxy compounds.

The tosylates may be prepared by treating the nitrosohydroxyl-amine with p-toluenesulfonyl chloride in aqueous sodium bicarbonate solution, in acetone–aqueous sodium hydroxide mixtures, or in benzene solution with the preformed salts of the nitrosohydroxylamines.

The Grignard reagents used in the reaction may be either those derived from aryl halides or those formed from alkyl halides. Phenyllithium reacted with the tosylate to give sulfones rather than azoxy compounds.

Traces of azo compounds were detected in the reaction mixture. They probably were formed by the reduction of the azoxy compound by the Grignard reagent.

33-3. Preparation of 4′-Methylazoxybenzene (p-Tolyl-NNO-azoxybenzene) [9]

$$\left[\ce{C6H5-N(N=O)(N=O)} \right]^- NH_4^+ + CH_3-C_6H_4-SO_2Cl \xrightarrow{OH^+}$$

$$C_6H_5-\overset{O}{\underset{\uparrow}{N}}=NOSO_2-C_6H_4-CH_3 \quad (5)$$

$$C_6H_5-\overset{O}{\underset{\uparrow}{N}}=NOSO_2-C_6H_4-CH_3 + CH_3-C_6H_4-MgBr \longrightarrow$$

$$C_6H_5-\overset{O}{\underset{\uparrow}{N}}=N-C_6H_4-CH_3 \quad (6)$$

■ **CAUTION:** The physiological effects of the reagents and products involved in this synthesis are not well understood. We suggest that full safety precautions be exercised.

(a) Preparation of Phenylnitrosohydroxylamine Tosylate. To a stirred solution of 16 gm (0.10 mole) of Cupferron (ammonium salt of N-phenyl-N-nitrosohydroxylamine) in 200 ml of a 10% sodium bicarbonate solution in water at room temperature is added 22 gm (0.11 mole) of p-toluenesulfonyl chloride. The reaction mixture is vigorously stirred overnight and extracted with methylene chloride.

After removal of the methylene chloride from the extract by evaporation, the dark residue is stirred with 30 ml of methanol. The solid which forms is separated by filtration. Recrystallization of the solid affords 12 gm of *N*-phenyl-*N*'-tosyloxydiimide *N*-oxide (65%), m.p. 130°–137°C dec.

(b) Preparation of 4'-Methylazoxybenzene. To a stirred solution of 2.20 gm (7.5 mmoles) of *N*-phenyl-*N*'-tosyldiimide *N*-oxide in 40 ml of tetrahydrofuran at room temperature is added dropwise 9 ml of a 1.2 *M* solution of *p*-tolyl-magnesium bromide in tetrahydrofuran. After the addition has been completed, the reaction mixture is stirred at 50°–60°C for 2 hr, cooled, and then poured into a mixture of ice and dilute hydrochloric acid. The crude product is extracted with methylene chloride. After concentration of the organic layer by evaporation at the aspirator, the residue is percolated through a silica-gel column. Elution of the column with pentane–methylene chloride (3:1) affords first 0.06 gm of 4-methylazobenzene (m.p. 69°–70°C). Continued elution first with pentane–methylene chloride (2:1) and then at a ratio of 1:1 finally gives 1.16 gm of 4'-methylazoxybenzene: yield 73%. After recrystallization from hexane, 1.05 gm of product is isolated, m.p. 50°–51°C.

Alkyllithium and Grignard reagents have been reacted with *N*-nitroso-*O*,*N*-dialkylhydroxylamines to form regiospecific azoxy alkanes. Usually the (Z)-stereoisomers were isolated. However, in one case the (E)-form was the major product [10].

33-4. Preparation of Azoxycyclohexane [11]

$$2 \bigcirc\!\!-NHOH \xrightarrow{\text{O}_2} \bigcirc\!\!-\overset{\overset{\text{O}}{\uparrow}}{N}=N-\bigcirc \qquad (7)$$

Through a stirred solution of 25 gm (0.22 mole) of cyclohexylhydroxylamine, 100 ml of methanol, and 2 gm of lead acetate maintained between 0°C and 10°C with an ice bath for 45 hr is bubbled a steady stream of air. During this period the evaporated solvent is replaced from time to time.

On vacuum fractional distillation of the product mixture, 1 gm of cyclohexanone oxime and 19.5 gm (85%) of crude azoxycyclohexane are isolated. After two fractional distillations of the product fraction, the following physical properties are observed for azoxycyclohexane: b.p. 160°–161°C (14 mm Hg), m.p. 22°–23°C, n_D^{20} 1.497, d_D^{20} 1.007.

REFERENCES

1. J. A. Miller and E. C. Miller, *Advan. Cancer Res.* **1,** 339 (1953).
2. J. A. Miller and G. C. Finger, *Cancer Res.* **17,** 387 (1957).
3. D. C. Freeman, Jr., and C. E. White, *J. Org. Chem.* **21,** 379 (1956).
4. B. M. Bogoslovskii, *J. Gen. Chem. USSR* **16,** 193 (1946).
5. J. P. Freeman, *J. Org. Chem.* **28,** 2508 (1963).
6. J. G. Aston and D. M. Jenkins, *Nature (London)* **167,** 863 (1951).
7. J. P. Snyder, V. T. Bandurco, F. Darack, and H. Olsen, *J. Amer. Chem. Soc.* **96,** 5158 (1974).
8. J. G. Aston and D. E. Ailman, *J. Amer. Chem. Soc.* **60,** 1930 (1938).
9. T. E. Stevens, *J. Org. Chem.* **29,** 311 (1964).
10. A. C. M. Meesters, H. Rueger, K. Rajeswari, and M. H. Benn, *Can. J. Chem.* **59,** 264 (1981).
11. H. Meister, *Ann. Chem.* **679,** 83 (1964).

34

ACETALS
AND KETALS

1. INTRODUCTION

Acetals possess a terminal 1,1-diether group, and ketals possess the same type of group in an internal position. These groups may be thought to be diether derivatives of the parent geminal dihydroxy compounds (hydrated aldehydes or ketones).

A wide variety of methods are available for the preparation of acetals and ketals, and the most important synthetic routes are summarized in Schemes 1 and 2.

Acetals and ketals are important functional groups that find use in the preparation of novel heterocyclic compounds, in polymers, and in the protection of carbonyl compounds or alcohols. Acetals and ketals are stable under basic conditions but hydrolyze easily under acidic conditions to the starting carbonyl compound and alcohol.

A brief review of the chemistry of acetals appears in the literature [1]. The chemistry of the condensation products of glycerol with aldehyde and ketones to give cyclic aldehydes has been reviewed [2].

From S. R. Sandler and W. Karo, *Organic Functional Group Preparations*, Vol. III, 2d ed. (San Diego, California, 1989), 1–85, by permission of Academic Press, Inc.

SCHEME 1
PREPARATION OF ACETALS

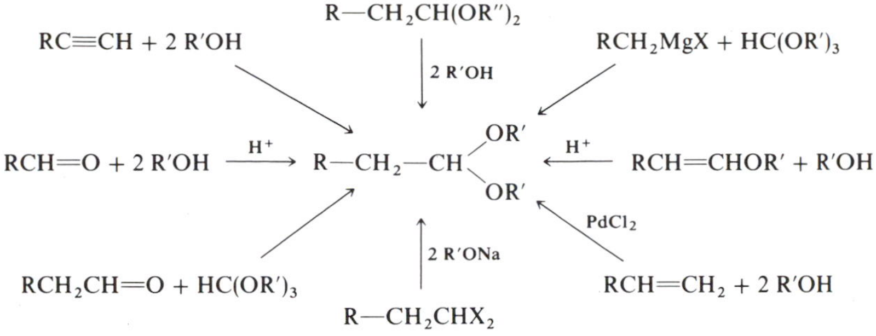

$$R{-}CH_2CH(OR'')_2$$

$$RC{\equiv}CH + 2\ R'OH \qquad\qquad\qquad RCH_2MgX + HC(OR')_3$$

↓ 2 R'OH

$$RCH{=}O + 2\ R'OH \xrightarrow{H^+} R{-}CH_2{-}CH{\overset{OR'}{\underset{OR'}{\Big\langle}}} \xleftarrow{H^+} RCH{=}CHOR' + R'OH$$

↑ 2 R'ONa   ↘ PdCl₂

$$RCH_2CH{=}O + HC(OR')_3 \qquad\qquad RCH{=}CH_2 + 2\ R'OH$$

$$R{-}CH_2CHX_2$$

SCHEME 2
PREPARATION OF KETALS

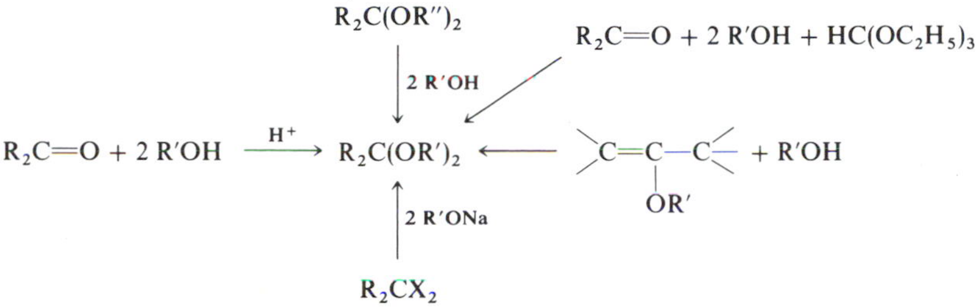

$$R_2C(OR'')_2$$

$$R_2C{=}O + 2\ R'OH + HC(OC_2H_5)_3$$

↓ 2 R'OH

$$R_2C{=}O + 2\ R'OH \xrightarrow{H^+} R_2C(OR')_2 \longleftarrow {\underset{OR'}{\overset{}{\Big\rangle}}}C{=}C{-}C{\Big\langle} + R'OH$$

↑ 2 R'ONa

$$R_2CX_2$$

The *Chemical Abstracts* nomenclature is used for most of the acetals described in this chapter. The compounds are named either as dialkoxy derivatives or as derivatives of acetals or ketals. Confusion exists in the earlier literature on naming cyclic acetals. For example, the acetals prepared from glycerol and an aldehyde were at one time referred to as 1,2- or 1,3-alkylidene (or arylidene) glycerol; however, today they are named as shown below:

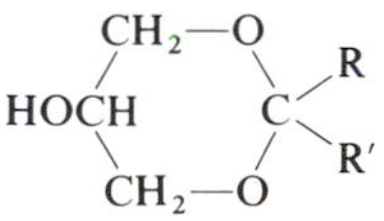

$$\begin{array}{c} CH_2{-}O \\ HOCH \qquad C{\big\langle}{\overset{R}{R'}} \\ CH_2{-}O \end{array}$$

(old) 1,3-Alkylidene glycerol
(new) 2-Alkyl-*m*-dioxan-5-ol

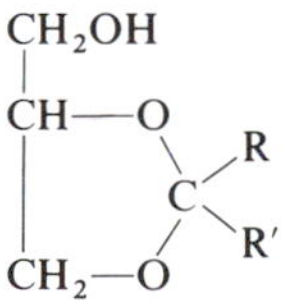

$$\begin{array}{c} CH_2OH \\ CH{-}O \\ | \qquad\quad C{\big\langle}{\overset{R}{R'}} \\ CH_2{-}O \end{array}$$

(old) 1,2-Alkylidene glycerol
(new) 2-Alkyl-4-hydroxymethyl-
1,3-dioxolane

Polyacetals prepared from formaldehyde are engineering thermo-plastics, which have found use in traditional metal applications. Some trade names of these polymers are Delrin acetal homopolymer (DuPont); Celcon acetal copolymer (Celanese/Hoechst); Duracon acetal copolymer (Celanese and Diacel—joint venture); Tenac acetal homopolymer by Asahi Chemical in Japan; and Ultraform acetal copolymer jointly by BASF and Degussa, in Germany. The polymers have the basic structure shown below:

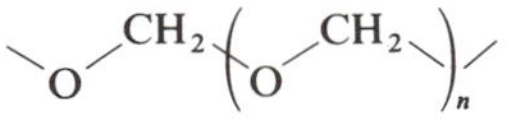

$$\sim_O\diagup^{CH_2}\diagdown_O\diagup^{CH_2}\diagdown\Big)_n\diagup$$

The number average molecular weight of the commerical polymers are in the range of 25,000–75,000. The chemistry of these polyacetal resins has been reviewed [3].

Where possible, this chapter also includes preparations of thioace-tals and thioketals.

Asahi Chemical has also reported a process for the production of thioacetal polymers prepared by the polymerization of trithiane.

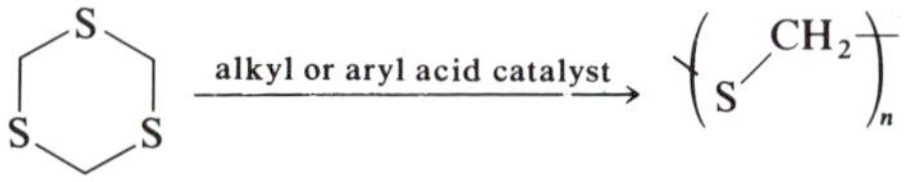

$$\underset{S}{\overset{S}{\bigcirc}}\underset{S}{}\quad\xrightarrow{\text{alkyl or aryl acid catalyst}}\quad \Big(S\diagup^{CH_2}\Big)_n \tag{1}$$

Trithiane can be used as a means to introduce the aldehyde functionality by reacting with halides, as shown below.

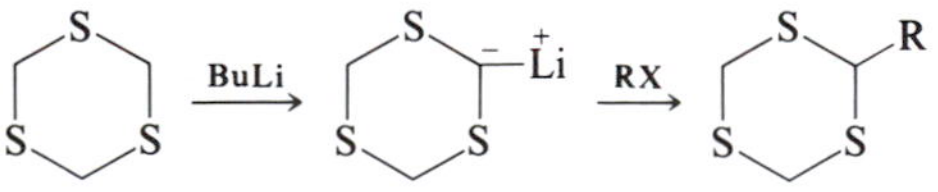

$$\underset{S}{\overset{S}{\bigcirc}}\underset{S}{}\quad\xrightarrow{\text{BuLi}}\quad \underset{S}{\overset{S}{\bigcirc}}\overset{-}{\underset{S}{}}{\overset{+}{-}\text{Li}}\quad\xrightarrow{\text{RX}}\quad \underset{S}{\overset{S}{\bigcirc}}\overset{R}{\underset{S}{}} \tag{2}$$

$$\xrightarrow[\text{CH}_3\text{OH}]{\text{HgCl}_2\cdot\text{HgO}}\quad (CH_3O)_2CHR \xrightarrow{H_3O^-} RCH{=}O \tag{3}$$

$$\text{50–55\% overall yield}$$

Thioacetals and thioketals have also been reported to have organoleptic properties.

Thioketals have been widely used as protective groups for ketones.

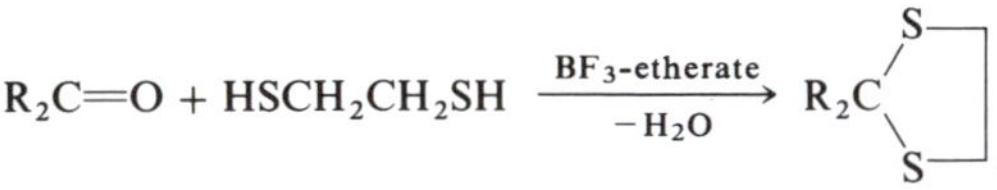

$$R_2C{=}O + HSCH_2CH_2SH \xrightarrow[-H_2O]{BF_3\text{-etherate}} R_2C\overset{S-}{\underset{S-}{\diagup}}\Big] \tag{4}$$

The protective thioketal group can be removed to regenerate the ketone. In addition, the thioketals can be desulfurized with Raney nickel to give the overall conversion $R_2C{=}O \rightarrow R_2CH_2$.

The pyrolysis of acetals has been used to give vinyl ethers. For example,

$$CH_3CH_2CH(OCH_3)_2 \longrightarrow CH_3CH{=}CH{-}OCH_3 \qquad (5)$$

2. CONDENSATION REACTIONS

Acetals or ketals can be prepared by several types of condensation reactions involving either the condensation of alcohols, ortho esters, or dihalomethylene compounds with a variety of starting materials, as described below. The most common method of preparing acetals or ketals involves the condensation of alcohols with either aldehydes or ketones, respectively. This reaction is highly recommended since the yields are usually high. However, the other methods are also widely used, and their use often depends on the starting material at hand or the structure of the acetal or ketal desired.

Acetals and ketals with side chain groups or unsaturated groups can be made to undergo a variety of condensation reactions, and some of the important ones have been described.

A. CONDENSATION OF ALCOHOLS WITH ALDEHYDES AND KETONES

Aldehydes and ketones react with alcohols to give an equilibrium mixture of the acetal and water (Eq. 6). The reaction is shifted toward completion by azeotropic removal of water or by using special drying agents. The reaction usually proceeds without the aid of catalysts in the case of aldehydes, but ketones require acids in order to obtain the ketals.

$$RR'C{=}O + 2R''OH \rightleftharpoons RR'C(OR'')_2 + H_2O \qquad (6)$$

For aldehydes: $R = $ alkyl or aryl, $R' = H$
For ketones: $R = R'$ or $R \neq R' = $ alkyl or aryl
$R'' = $ alkyl as in diols, glycols, polyols,
and hemiacetals as in β-D($+$)-glucose

In some cases, evidence has been reported to indicate that hemi-acetals are first formed by the reaction of alcohols in the presence of aldehydes or ketones in neutral, basic, or acidic solutions (Eq. 7).

$$RCH{=}O + ROH \;\rightleftharpoons\; RCH\!\!\begin{smallmatrix}OH\\OR\end{smallmatrix} \;\overset{ROH}{\rightleftharpoons}\; RCH\!\!\begin{smallmatrix}OR\\OR\end{smallmatrix} + H_2O \qquad (7)$$

Acetals and ketals are converted to aldehydes or ketones by hydrolysis in acid solution (Eq. 8–10)

$$R_2C(OR')_2 + H_3O^+ \;\overset{fast}{\rightleftharpoons}\; \left[R_2C\!\!\begin{smallmatrix}H\\OR'\\OR'\end{smallmatrix}\right]^+ + H_2O \qquad (8)$$

$$\left[R_2C\!\!\begin{smallmatrix}H\\OR'\\OR'\end{smallmatrix}\right]^+ \;\overset{slow}{\longrightarrow}\; R_2C^+{-}OR' + R'OH \qquad (9)$$

$$R_2C^+{-}OR' + H_2O \;\overset{fast}{\longrightarrow}\; R_2C{=}O + R'OH + H_3O^+ \qquad (10)$$

Acids such as hydrogen chloride, sulfuric acid, acid ion-exchange resins, phosphoric acid, or p-toluenesulfonic acid catalyze the acetal or ketal formation reaction and aid in the water removal. This is especially important in the case of ketones. A Dean–Stark trap is useful when azeotropic removal of water is attempted. After the reaction, the acid catalyst is neutralized, and the acetal or ketal product is distilled. Ammonium chloride and ammonium nitrate also have been reported to act as catalysts.

Primary alcohols give better yields of acetals than secondary or tertiary alcohols. Highly branched aldehydes or ketones also give poor yields of acetals or ketals, respectively.

The preparation of acetal (1,1-diethoxyethane) is described in Preperation 34-1 and the preparation of n-butylal (1,1-dibutoxymethane) in 34-2.

34-1. Preparation of Acetal (1,1-Diethoxyethane) [4, 5]

$$CH_3CH{=}O + 2C_2H_5OH \;\longrightarrow\; CH_3CH(OC_2H_5)_2 + H_2O \qquad (11)$$

To a pressure bottle containing 20 gm (0.18 mole) of anhydrous calcium chloride is added 105 gm of 95% (2.17 moles) ethanol and the mixture cooled to 8°C. Then 50 gm (1.14 moles) of cold acetaldehyde is slowly poured down the wall of the bottle. The bottle is closed and shaken vigorously for 5–10 min, with cooling if necessary. The mixture is allowed to stand at room temperature with intermittent shaking for 24 hr. The upper layer, which has separated, weighs 128–129 gm. It is washed three times with 30–40 ml of water. The organic layer is dried over 3 gm of anhydrous potassium carbonate and distilled through a 1 ft column, to afford 70–72 gm (59–60%), b.p. 101–103.5°C. The low-boiling fractions are washed again with water, dried and again fractionally distilled to give another 9.0–9.5 gm (7.9–8.1%), b.p. 101–103.5°C. Therefore, the total yield amounts to 79–81.5 gm (67–69%), n_D^{25} 1.3819, d_4^{20} 0.8314.

A procedure identical to 34-1, but using 132 gm (2.28 moles) of allyl alcohol in place of ethanol, affords allyl acetal (110.5 gm) in 68.2% yield, b.p. 146–150°C [13].

34-2. Preparation of *n*-Butylal (Dibutoxymethane) [6]

$$2\ C_4H_9OH + CH_2{=}O \longrightarrow CH_2 \underset{OC_4H_9}{\overset{OC_4H_9}{}} + H_2O \tag{12}$$

A flask containing 15 gm (0.5 mole) of paraformaldehyde, 74 gm (1.0 mole) of *n*-butyl alcohol, and 2.0 gm of anhydrous ferric chloride is refluxed for 10 hr. The lower layer of 3–4 ml of material is discarded and then 50 ml of 10% aqueous sodium carbonate solution is added to remove the ferric chloride as ferric hydroxide. The product is shaken with a mixture of 40 ml of 20% hydrogen peroxide and 5 ml of 10% sodium carbonate solution at 45°C in order to remove any remaining aldehyde. The product is also washed with water, dried, and distilled from excess sodium metal to afford 62 gm (78%), b.p. 180.5°C (760 mm Hg).

B. CONDENSATION OF ALCOHOLS WITH ACETYLENES

Acetylene reacts with alcohols in the presence of boron trifluoride and mercuric oxide to afford acetals. Substituted acetylenes react with alcohols to give ketals. The reaction probably proceeds via the intermediate vinyl ether as shown in Eq. (13).

TABLE I[c]

PREPARATIVE DATA AND PHYSICAL CONSTANTS OF CYCLIC ACETALS[a] FROM THE REACTION
OF ACETYLENE WITH ALCOHOLS (STOICHIOMETRIC QUANTITIES)

Starting alcohol	Taken (gm)	Yield (gm)	Yield (%)	B.p., °C (mm)	d_4	n_D
Ethylene glycol	102	88	62	82.3	0.9770^{24}	1.3945^{24}
Trimethylene glycol	93	55	45	108–111	0.9675^{23}	1.4160^{23}
Pinacol	—	—	—	133–134	—	—
Methylethyl pinacol	—	—	—	150–180	—	—
Glycol methyl ether	150	77	44	87–91 (15)	0.9691^{25}	1.4181^{25}
Glycol ethyl ether	300	255	74	110–114 (14)	0.9328^{25}	1.4163^{25}
Glycol butyl ether	200	115	52	142–146 (14)	0.9072^{26}	1.4263^{26}
Diethylene glycol	—	—	—	250 (14)	—	—
Diethylene glycol ethyl ether	—	—	—	140–145 (14)	—	—
Glycerol	—	—	(65–75)	189–191[b]	1.1193^{24}	1.4395^{24}
tert-Ethylidenebisglycerol	542	322	41.7	160–162 (14)	1.1067^{24}	1.4482^{20}
Glycerol methyl ether	141	105	60	145–147	1.0098^{24}	1.4145^{24}
Glycerol ethyl ether	—	—	—	170–171	—	—
Glycerol phenyl ether	150	119	68	142–144 (14) (m.p. 29°)	—	—
Bisethylidenepentaerythritol	69	86	90	113 (14) (m.p. 40°)	—	—
tert-Ethylidenemannite	100	67	47	165–168 (17) sublimed at 90°	—	—
Ethylene chlorohydrin	304	250	71	106–107 (14)	—	—
Trimethylene chlorohydrin	157	70	38	127–129 (14)	—	—
Glycerol chlorohydrin	611	543	72	147–149	1.1720^{24}	1.4410^{24}
Glycerol monoacetin	260	153	49	91.5–92 (14) 200–201 (760)	1.1110^{26}	1.4323^{26}
Lactic acid	240	179	61	149–151	1.074^{26}	1.4120^{26}
1-Hydroxyisobutyric acid	148	125	71	150 ± 0.3 (745)	1.0226^{26}	1.4034^{28}
Bisethylidene tartrate	—	—	—	122–128 (17)		
Dimethyl tartrate	184	172	81	137 ± 0.2 (16)	1.2306^{27}	1.4426^{27}
Diethyl tartrate	162	137	74	147 ± 0.5 (18)	1.1408^{25}	1.1438^{25}
Methyl malate	120	40	26	121–122 (15)	1.1975^{26}	1.4397^{26}
Ethyl malate	118	50	30	125–130 (17)	1.1215^{26}	1.4402^{26}
Dimethyl citrate	93	30	31	(m.p. 73°C)	—	—
Mandelic acid	82	62	64	142–144 (14)	1.1681^{25}	1.4145^{26}
Benzilic acid	86	64	68	198–200 (17) (m.p. 77°C)	—	—

[a]The reaction is carried out using 5.0 gm of a 55–65% soln. of BF_3 in methanol and 1.0 HgO dissolved in it by heating. The flask is cooled, the alcohol added, mixed well and dry acetylene added while stirring. The theoretical amount of acetylene is used.

[b]Mainly 1,2-isomer isolated by fractional distillation. The crude product contains 78% 1,2- and 22% 1,38-isomers.

[c][Reprinted from J. A. Nieuwland, R. R. Vogt, and W. L. Foohey, J. Amer. Chem. Soc. 52, 1018 (1930). Copyright 1930 by The American Chemical Society. Reprinted by permission of the copyright owner.]

$$ROH + HC\equiv CH \xrightarrow{\text{HgO—BF}_3} ROCH=CH_2 \xrightarrow{\text{ROH}} (RO)_2CH-CH_3 \qquad (13)$$

Some examples illustrating the utility of this reaction are shown in Table I and in Preparation 34-3.

The reaction of vinyl ethers with alcohols to give acetals has also been described in detail.

Monohydric alcohols other than methanol react with monoalkyl-acetylenes to afford polymers. However, the use of trichloroacetic acid causes the reaction to proceed smoothly to 2,2-dialkoxyalkanes.

The reaction of alcohols with conjugated vinylacetylene in the presence of CCl_3COOH catalyst gives β-alkoxy ketals. The use of only sodium methoxide as the catalyst causes the addition of only 1 mole of methanol to vinylacetylene at $100°C$ to give 4-methoxy-1-butyne. The use of $HgO-BF_3$ causes the addition of 3 moles of methanol to give 1,3,3-trimethoxybutane.

34-3. Preparation of 1,1-Dimethoxyethane [7, 8]

$$2CH_3OH + HC\equiv CH \longrightarrow (CH_3O)_2CH-CH_3 \qquad (14)$$

To a flask equipped with a mechanical stirrer, condenser, and gas addition tube and containing 10 gm of a 63% solution of boron tri-fluoride in methanol is added 1.0 gm of mercuric oxide and 200 gm (6.25 moles) of methanol. Then 70 gm (3.13 moles) of acetylene is bubbled into the reaction mixture with vigorous stirring at room temperature. After the reaction the catalyst is neutralized with aqueous potassium carbonate, the product is extracted into ether, dried, and distilled to afford 104 gm (37%), b.p. $64°-65°C$, n_D^{26} 1.3762.

REFERENCES

1. F. S. Wagner, Jr., *Encycl. Ind. Chem. Anal.* **4**, 62 (1967).
2. A. J. Showler and P. A. Darley, *Chem. Rev.* **67**, 427 (1967).
3. K. J. Persak. L. M. Blair, and Kirk-Othmer, *Encyclopedia of Chemical Technology* **1**, 112 (3rd Ed.) (1978).
4. H. Adkins and B. H. Nissen, *J. Amer. Chem. Soc.* **44**, 2749 (1922).
5. H. Adkins and B. H. Nissen, *Org. Syn. Coll.* **1**, 1 (1932).
6. A. I. Vogel, *J. Chem. Soc.* 624 (1948).
7. H. D. Hinton and J. A. Nieuwland, *J. Amer. Chem. Soc.* **52**, 2893 (1930).
8. J. A. Nieuwland, R. R. Vogt, and W. L. Foohey, *J. Amer. Chem. Soc.* **59**, 1018 (1930).

35

ANHYDRIDES

1. INTRODUCTION

The most convenient laboratory methods of preparing carboxylic acid anhydrides involve either the reaction of acyl chlorides with carboxylic acids or their sodium salts or the reaction of acetic anhydride with acids boiling higher than acetic acid. The thermal dehydration of carboxylic acids has its limitations since not all acids afford anhydrides. The use of ketene in the laboratory is inconvenient but it is valuable in giving mixed acetic–carboxylic anhydrides. Industrially, acetic anhydride is prepared by this method.

    The Diels–Alder reaction is valuable in adding maleic anhydride type structures to conjugated aliphatic and aromatic dienes.

From S. R. Sandler and W. Karo, *Organic Functional Group Preparations*, Vol. III, 2d ed. (San Diego, California, 1989), 86–128, by permission of Academic Press, Inc.

Many aliphatic and aromatic anhydrides, depending on their structures, undergo various condensation, substitution, addition, and oxidation or reduction reactions to give substituted anhydrides.

Polymeric anhydrides can be prepared by the above techniques and are beginning to find various commercial uses. These are the styrene–maleic anhydride copolymers and the alkyl vinyl ether–maleic anhydride copolymers. The use of maleic acid or itaconic acid is widely used in the polymer field to give copolymers with anhydride functionality.

The anhydride functionality is very important in organic synthesis and can be reacted to form amides, esters, hydrazides and semicarbazides, and so on.

2. CONDENSATION REACTIONS

A. ACYLATION OF CARBOXYLIC ACIDS BY ACYL HALIDES

The reaction of acyl halides with carboxylic acids affords good yields of simple and mixed anhydrides. The use of pyridine or triethylamine helps to remove the hydrogen chloride by-product. However, the preparation of benzoic anhydride from benzoic acid and benzoyl chloride has also been carried out in the absence of trialkylamines by heating under reduced pressure, in the presence of zinc chloride or by refluxing in chlorohydrocarbons such as methylene chloride.

Polymeric anhydrides have been reported to be prepared by the reaction of terephthaloyl chloride with various aromatic dibasic acids containing ether or amide groups (Eq. 1).

M.p. 205°–209°C

Where X = Br or Cl

A report in the literature describes the preparation of diacylium cations from tetrahaloterephthalic acids and their reactivity with free tetrahaloterephthalic acids to produce homopolymers and heteropolymers with an anhydride backbone. The perhalo polyanhydrides are surprisingly stable to hydrolysis and are stable at relatively high temperatures.

35-1. Preparation of Heptanoic Anhydride [1]

$$C_6H_{13}COCl + C_5H_5N \longrightarrow [C_6H_{13}CO^+C_5H_5N] \, Cl^- \xrightarrow{C_6H_{13}COOH}$$
$$(C_6H_{13}CO)_2O + C_5H_5\overset{+}{N}H\overset{-}{C}l \quad (3)$$

To a flask containing 15.8 gm (0.2 mole) of dry pyridine and 25 ml of dry benzene is rapidly added with stirring 14.8 gm (0.1 mole) of heptanoyl chloride. The reaction is only slightly exothermic and then 13.0 gm (0.1 mole) of heptanoic acid is added dropwise (5 min) causing the temperature to rise to 60–65°C. After stirring for 10 min, the solid (pyridinium hydrochloride) is quickly filtered, washed twice with 25 ml portions of dry benzene, the benzene washings concentrated under reduced pressure, and the residue distilled to afford 19–20 gm (78–83%), b.p. 155–162°C (12 mm Hg), or 170–173°C (15 mm).

Several other examples of this reaction are given in Table I.

B. DIACYLATION OF WATER

A related procedure involves the diacylation of water by the pyridinium salts of acyl halides to afford anhydrides (Eq. 4). The reaction was earlier described by Minunni to involve an acyl pyridinium complex.

TABLE I

Preparation of Anhydrides by the Reaction of Acyl Halides with Carboxylic Acids

RCOCl (moles) R =	R'COOH (moles) R' =	Reaction conditions				Yield (%)	B.p., °C (mm Hg) or m.p., °C	Ref.
		Amine (moles)	Solvent (ml)	Temp. (°C)	Time (hr)			
C_6H_{13} (0.1)	C_6H_{13} (0.1)	Pyridine (0.2)	C_6H_6 (25)	60–65	1/4	78–83	155–162 (12)	a
C_6H_5 (0.1)	(4-Cl-C_6H_4) Cl (0.1)	0.2	Ether (100)	25–30	1/4	69	66.5–70.0	b
(0.1)	(4-NO_2-C_6H_4) NO_2 (0.1)	0.2	C_6H_6 (100)	25–30	1/4	65	130.0–130.5	b

(continues)

TABLE I (*Continued*)

RCOCl (moles) R =	R'COOH (moles) R' =	Amine (moles)	Solvent (ml)	Reaction conditions		Yield (%)	B.p., °C (mm Hg) or m.p., °C	Ref.
				Temp. (°C)	Time (hr)			
RCOOCl								
C_2H_5 (0.1)	C_6H_5 (0.1)	$(C_2H_5)_3N$ 0.1	Toluene (150) Ether (50)	−5 to 0	1/2	—	—	c
(0.2)	$O_2NC_6H_4$ (0.2)	0.2	Ether (200)	−5 to 0	1/2	—	56–57	d
(0.1)	H (0.1)	0.1	Ether (200)	−15	1/2	—	—	e
$C_6H_5CH_2$ (0.02)	$p\text{-}O_2NC_6H_4$ (0.02)	0.02	Ether (200 gm)	−78 to −10	1	63	64.5–66.0	f
$(CH_3)_3C$ (0.065)	$p\text{-}O_2NC_6H_4$ (0.065)	0.065	Ether (300)	−15	1	27	92–93	g

[a] C. F. H. Allen, C. J. Kibler, D. M. McLachlin, and C. V. Wilson, *Org. Syn. Coll.* **3**, 28 (1955).

[b] J. M. Zeavin and A. M. Fisher, *J. Amer. Chem. Soc.* **54**, 3738 (1932).

[c] J. A. Price and D. S. Tarbell, *Org. Syn. Coll.* **3**, 285 (1963); N. A. Leister and D. S. Tarbell, *J. Org. Chem.* **23**, 1152 (1958).

[d] N. A. Leister and D. S. Tarbell, *J. Org. Chem.* **23**, 1149 (1958).

[e] T. Parasaran and D. S. Tarbell, *J. Org. Chem.* **29**, 3422 (1964).

[f] R. C. L. Chow and D. S. Tarbell, *J. Org. Chem.* **32**, 2188 (1967).

[g] C. J. Michejda and D. S. Tarbell, *J. Org. Chem.* **29**, 1168 (1964).

$$2\,RCOCl + 2\,C_5H_5N \longrightarrow 2\,[RCOCl{-}C_5H_5N] \xrightarrow{H_2O} (RCO)_2O + 2\,C_2H_5\overset{+}{N}H\bar{C}l \quad (4)$$

This reaction gives better yields when the complex is first prepared in the absence of water and then subsequently reacted with one-half mole of water per mole of acyl halide. Pyridine is preferred over the more basic triethylamine because the latter tends to effect dehydrohalogenation of aliphatic acyl halides to ketenes. Some typical examples of this method are summarized in Table II and also given in Preparation 35-2.

The benzoyl chloride–pyridine complex also reacts with hydrogen sulfide at about $-20°C$ to afford an 85% yield of dibenzoyl sulfide (Eq. 5).

$$2\,[RCOCl{-}C_5H_5N] + H_2S \longrightarrow (RCO)_2S + 2\,C_5H_5\overset{+}{N}H\bar{C}l \quad (5)$$

35-2. Preparation of Benzoic Anhydride [2]

$$2\,C_6H_5COCl + 2\,C_5H_5N \longrightarrow 2\,[C_6H_5COC_5H_5N]^+\,Cl^-$$
$$\downarrow H_2O \quad (6)$$
$$(C_6H_5CO)_2O + 2\,C_5H_5\overset{+}{N}H\bar{C}l$$

To a flask containing 14.0 gm (0.1 mole) of benzoyl chloride and equipped with a stirrer is added 40 ml of dry dioxane and it is cooled to 5°C. Then 10 ml (0.1 mole) of dry pyridine is rapidly added at 5–10°C followed by 1.0 ml (0.056 mole) of water. After stirring for 10–15 min at 0–5°C, the reaction mixture is then poured into a mixture of 75 ml of conc. hydrochloric acid, 75 gm of cracked ice, and 350 ml of water. The product is filtered, washed first for 1 min with a cold 5% solution of sodium bicarbonate and then with water. The product is dried to afford 11.0 gm (97%), m.p. 43°C (recrystallized from a mixture of ether–petroleum ether).

C. REACTION OF ACYL HALIDES WITH SALTS OF CARBOXYLIC ACIDS

The general reaction of acyl halides with carboxylic acid salts may be written as in Eq. (7).

TABLE II
Preparation of Anhydrides by the Diacyclation of Water

RCOCl (moles) R =	Pyridine (moles)	H_2O (moles)	Solvent (ml)	Reaction conditions		Yield (%)	B.p., °C (mm Hg) or m.p., °C	Ref.
				Temp. (°C)	Time (hr)			
(furyl structure) (0.05)	0.07	0.056	Pet. ether (60)	−20	1/12	64	71–73	a
C_6H_5—CH—C_2H_5 (1.0)	1.5	0.5	Dioxane (400)	5–10	1/3	85–99	42	a
C_4H_9CH (0.2)	0.25	0.11	C_6H_6 (150)	5–10	1/2	42.0–43.5	149–152 (8)	a
p-Cl—C_6H_4 (0.1)	0.6	5.6	—	80–100	1/12	96–98	192–193	b
o-Cl—C_6H_4 (0.1)	0.1	0.11	Dioxane (40)	5–10	1/2	50.0	77–79	a
p-I—C_6H_4 (0.05)	0.05	0.056	Dioxane (20)	5–10	1/2	46.0	227–229	a
CH_3 (0.50)	0.25	0.11	C_6H_6 (150)	5–10	1/2	20–24	136–139 (760)	a
C_2H_5 (0.40)	0.25	0.11	C_6H_6 (150)	5–10	1/2	29.0–31.0	163–166 (760)	a

[a] H. Adkins and Q. E. Thompson, J. Amer. Chem. Soc. 71, 2242 (1949).
[b] C. F. H. Allen, C. J. Kibler, D. M. McLachlin, and C. V. Wilson, Org. Syn. Coll. 3, 28 (1955).

$$RCOOM + R'COCl \longrightarrow RCOOCOR' + MCl \qquad (7)$$

where M = alkali metal or Tl; R may or may not be equal to R'; and R' = alkyl, aryl, or H (for one of the substituents).

35-3. Preparation of Benzoic Anhydride [3]

$$C_6H_5COONa + C_6H_5COCl \xrightarrow[\text{pyridine}]{H_2O} (C_6H_5CO)_2O + NaCl \qquad (8)$$

To a beaker containing 150 ml of an aqueous solution of 14.4 gm (0.1 mole) of sodium benzoate and two drops of pyridine is added slowly with stirring 14.0 gm (0.1 mole) of benzoyl chloride. The resulting solid is filtered, washed with water then petroleum ether, and dried to afford 22.1 gm (97.5%), m.p. 40–41°C.

D. ACETIC ANHYDRIDE METHOD

The reaction of mono-, di-, and polycarboxylic acids with refluxing acetic anhydride affords good yields of anhydrides,

35-4. Preparation of Benzoic Anhydride [4]

$$2 C_6H_5COOH + (CH_3CO)_2O \longrightarrow (C_2H_5CO)_2O + 2 CH_3COOH \qquad (9)$$

To a three-necked flask equipped with a mechanical stirrer, dropping funnel, fractionating column, and distillation head is added 150 gm (1.23 moles) of benzoic acid, 150 gm (1.47 moles) of acetic anhydride, and 0.1 gm of sirupy phosphoric acid. The mixture is stirred, heated, and distilled slowly, keeping the head temperature at 120°C or below. After 25 ml of distillate is removed another 25.0 gm (0.245 mole) of acetic anhydride is added and the distillation continued until the temperature of the reaction mixture in the flask reaches 270°C. The residue is distilled under reduced pressure to afford a fraction b.p. 210–220°C (19–20 mm Hg) which weighs 110–120 gm. Recrystallization of the crude benzoic anhydride from benzene–petroleum ether affords 100–103 gm (72%) of pure anhydride, m.p. 43°C.

REFERENCES

1. C. F. H. Allen, C. J. Kibler, D. M. McLachlin, and C. V. Wilson, *Org. Syn.* **26,** 1 (1946).
2. H. Adkins and Q. E. Thompson, *J. Amer. Chem. Soc.* **71,** 2243 (1949).
3. R. K. Smalley and H. Suschitzky, *J. Chem. Soc.* 755 (1964).
4. H. T. Clarke and E. J. Rahrs, *Org. Syn. Coll.* **1,** 91 (1932).

36

MONOALKYL
SULFATES

1. INTRODUCTION

Alkyl sulfates are usually prepared by the reaction of alcohols with SO_3 or its derivatives and by the reaction of unsaturated compounds with sulfuric acid.

This chapter is mainly concerned with methods of synthesis of sodium alkyl sulfates derived from alcohols. These compounds are useful as detergents and also have biological and biochemical importance. Sulfates are also synthesized in living systems, and their preparation is of interest in medicinal chemical research.

The most important methods of synthesizing sodium alkyl sulfates or monoalkyl sulfates are summarized in Scheme 1.

A less often used but important method for sulfating unsaturated alcohols is the use of sulfamic acid, as shown in Eq. (1).

$$ROH + HSO_3NH_2 \longrightarrow ROS\bar{O}_3\overset{+}{N}H_4 \tag{1}$$

From S. R. Sandler and W. Karo, *Organic Functional Group Preparations*, Vol. III, 2d ed. (San Diego, California, 1989), 129–161, by permission of Academic Press, Inc.

SCHEME I

METHODS OF SYNTHESIZING MONOALKYL SULFATES

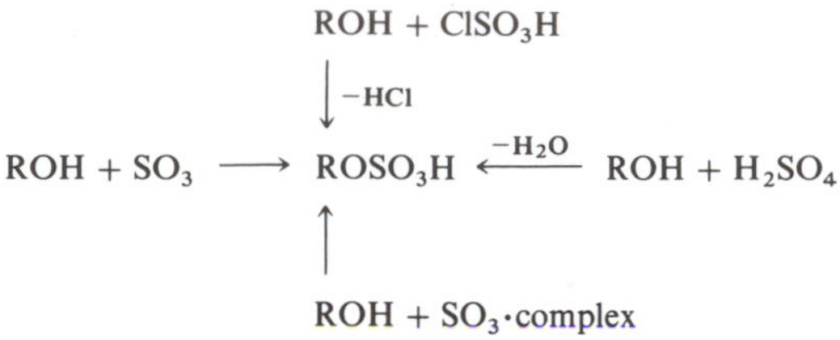

$$ROH + ClSO_3H$$

$$\downarrow -HCl$$

$$ROH + SO_3 \longrightarrow ROSO_3H \xleftarrow{-H_2O} ROH + H_2SO_4$$

$$\uparrow$$

$$ROH + SO_3 \cdot complex$$

Sulfur trioxide complexes have gained important industrial use in sulfating dyes, carbohydrates, sterols, and other sensitive heterocyclic compounds.

The nomenclature of the sulfate esters and their salts follows the *Chemical Abstracts* system. The following two compounds are examples:

$n\text{-}C_{12}H_{25}OSO_3H$

n-Dodecyl sulfate or lauryl sulfate

I

$n\text{-}C_{12}H_{25}OSO_3Na$

Sodium n-dodecyl sulfate or sodium lauryl sulfate

II

Sulfate esters and their sodium salts have relatively good stability except for those derived from tertiary alcohols or partially esterified alcohols, such as ethylene glycol monolaurate. Sodium alkyl sulfates hydrolyze under acid as well as basic conditions in aqueous solution at elevated temperatures. Basic hydrolysis of ethyl and propyl sulfates give no detectable olefin formation, but isopropyl and isobutyl sulfates afford minor amounts of olefins. Neopentyl and pinacolyl sulfates involve complete rearrangement during basic hydrolysis; isobutyl sulfate involves partial rearrangement and 1-ethylpropyl sulfate hydrolyzes without rearrangement. *sec*-Butyl sulfate hydrolyzes with inversion of configuration accompanied by only a small amount of racemization.

Sulfation of fatty alcohols and polyalkoxylates is the basis of commercial detergents and emulsifiers, which amounted to 111,000 metric tons in 1980.

Linear ethoxylates are used as the raw material for ether sulfates used in the production of biodegradable detergent formulations.

A brief review of sulfation by Knaggs, Nussbaum, and Schultz is worth consulting [1].

2. SULFATION OF ALCOHOLS

As noted in the introduction section, the five most important methods of sulfating alcohols involve the use of either sulfur trioxide, sulfur trioxide complexes, chlorosulfonic acid, sulfuric acid, or sulfamic acid. Chlorosulfonic acid is often used in the laboratory but has the disadvantage that in some cases the last traces of hydrogen chloride are difficult to remove. Sulfur trioxide tends to give pure products but may cause charring or colored products unless special precautions are taken. Sulfur trioxide complexes are relatively easy to prepare, and for some, the sulfation reactions can even be carried out in aqueous solution. This sulfation process is particularly important for unsaturated and acid-sensitive compounds.

In some cases, the amine complexes leave a difficult-to-remove amine odor. Sulfuric acid is used without any problems for most saturated primary alcohols, but oleum gives poor yields and poor color with some (tridecyl alcohol). Sulfamic acid is often used for acid-sensitive alcohols, but the reaction also requires elevated temperatures (125–145°C) to give the ammonium salt.

A. CHLOROSULFONIC ACID

Sulfation of alcohols with chlorosulfonic acid (Eq. 2) has been used in the laboratory and on an industrial scale to prepare sodium alkyl sulfates. The reaction is usually rapid in the presence or absence of solvents at 25–30°C and gives a product with little color. Unreactive solvents such as ether, dioxane, and halogenated hydrocarbons are commonly used in the sulfation reaction. Acetic acid has also been reported as a solvent for the sulfation of C_5–C_{19} long-chain secondary alcohols. Table I gives several examples of the type of alcohols sulfated with chlorosulfonic acid.

$$ROH + ClSO_3H \xrightarrow{-HCl} ROSO_3H \xrightarrow[-H_2O]{NaOH} ROSO_3Na \qquad (2)$$

Chlorosulfonic acid has the disadvantage that hydrogen chloride is evolved oftentimes incompletely, which later affords sodium chloride impurities in the neutralized product. This latter problem has hindered the early development of a continuous large-scale process.

TABLE I

SULFATION OF ALCOHOLS

Alcohol ROH, R =	Sulfating agent (approx. 1.0 mole/ mole alcohol)	Yield as $ROSO_3Na$ (%)	Purity (%)	Ref.
Lauryl	$ClSO_3H$	86–90	98	a
	SO_3	74	98	a
	Oleum (20%)	91	80–90	a
	H_2SO_4 (conc.)	49	60–70	a
	SO_3-pyridine	50	90	a
	H_2N—SO_3H	30	80–90	a
Hecadecyl	$ClSO_3H$	98	97	b
	SO_3	97	97	b
	Oleum (20%)	90	96	b
	H_2SO_4 (conc.)	70	91	b
Tridecyl	$ClSO_3H$	97	86	c
	SO_3	97–99	86	c
	Oleum (20%)	(Poor yield & color)		c
	H_2N—SO_3H	(Poor yield & color)		c
Ethoxylated tridecyl (4 moles ethylene oxide)	$ClSO_3H$	93	81	c
	SO_3	95	51	c
	Oleum (20%)	94	57	c
	H_2N—SO_3H	99	72	c

[a]Author's Laboratory (S.R.S.).
[b]Enjay Chem. Co., Tech. Infor. Sheet, DC-1 (date unknown).
[c]Enjay Chem. Co., Tech. Bull. C-21 (1960).

However, some continuous processes have been reported. Operating under vacuum or purging with nitrogen eliminates any residual hydrochloric acid. A batch process has been reported wherein one drum of chlorosulfonic acid is used per run.

36-1. Preparation of Sodium Lauryl Sulfate [2, 3]

$$n\text{-}C_{12}H_{25}OH + ClSO_3H \xrightarrow[]{-HCl} n\text{-}C_{12}H_{25}OSO_3H \xrightarrow[-H_2O]{NaOH} n\text{-}C_{12}H_{25}OSO_3Na \quad (3)$$

To a flask equipped with a stirrer, dropping funnel, condenser, and drying tube, and containing 250 gm (1.35 mole) of lauryl alcohol is added dropwise 156 gm (1.35 mole) of chlorosulfonic acid, keeping the temperature at about 25°C but not higher than 30°C. Nitrogen is passed through the reaction mixture to purge it of any remaining hydrogen chloride. The reaction mixture and 10% sodium hydroxide solution are simultaneously poured with good stirring into 1000 gm of cold water–crushed ice to give pH 7.7–8.0. The reaction mixture is worked up as follows: Add enough isopropyl alcohol (100–200 ml) to just form a solution. One gram of sodium carbonate is added as a buffer, and on standing, the final mixture separates into two layers. Sodium sulfate is added to just saturate the lower layer, and this layer is discarded. The product is obtained by concentration of the upper layer under reduced pressure to give 259 gm (90%) of product.

B. SULFURIC ACID

Sulfuric acid reacts with primary and secondary alcohols by a bimolecular displacement reaction with no alkyl-oxygen fission. The mechanism of the reaction is similar to the acid-catalyzed esterification of carboxylic acids. Sulfation of optically active 2-butanol gives a sulfate ester with complete retention of configuration. In addition, neopentyl alcohol sulfates without rearrangement.

Kinetic data indicate that primary alcohols, including neopentyl alcohol, were sulfated at comparable rates and were sulfated about 10-fold faster than the secondary alcohols. The kinetics followed the following rate expression:

$$\frac{d[ROSO_3H]}{dt} = k[ROH][H_2SO_4][H^+ \text{ activity}]$$

Several side reactions sometimes occur during the sulfation of alcohols, and they are shown in Eq. 4.

$$RCH_2{-}CH_2OH + H_2SO_4 \longrightarrow \begin{array}{l} (RCH_2CH_2O)_2SO_2 \\ RCH{=}CH_2 \\ (RCH_2CH_2)_2O \end{array} \tag{4}$$

The use of stoichiometric amounts of acid gives low yields, 40–60%, but the use of twice the stoichiometric amount shifts the equilibrium so that 80–90% sulfated product is obtained (Eq. 4), as in the case of 1-tetradecanol. The use of 20% fuming sulfuric acid also helps to shift the equilibrium toward sulfation by removal of water. Lauryl alcohol also gives improved yields by use of 20% fuming sulfuric acid (see also Table I). Other methods used to shift the equilibrium involve the azeotropic removal of water with carbon tetrachloride or boron sulfate.

$$ROH + H_2SO_4 \longrightarrow ROSO_3H + H_2O \tag{5}$$

Sodium lauryl sulfate is prepared industrially in one process that involves the continuous "flash" process utilizing 170% of theory of 99% H_2SO_4. The reaction time is 60 sec at 60–70°C, and the reaction mixture is quickly neutralized, affording 85–95% product and 10–15% of unreacted alcohol [4]

36-2. Preparation of Sodium Lauryl Sulfate Using Fuming Sulfuric Acid [2]

$$n\text{-}C_{12}H_{25}OH + H_2SO_4 \xrightarrow{-H_2O} n\text{-}C_{12}H_{25}OSO_3H \xrightarrow[-H_2O]{NaOH} n\text{-}C_{12}H_{25}OSO_3Na \tag{6}$$

To a 500-ml, three-necked flask equipped with a stirrer, dropping funnel, and drying tube is added 18.6 gm (0.1 mole) of lauryl alcohol dissolved in 100 ml of dioxane. The contents are cooled to 10–15°C and 12.2 gm (0.13 mole) of 20% fuming sulfuric acid is added. The mixture is stirred for $\frac{1}{2}$ hr at 10–15°C, neutralized with cold 10% sodium hydroxide, and concentrated under reduced pressure to afford 26.0 gm (91%) of product.

REFERENCES

1. E. A. Knaggs, M. L. Nussbaum and A. Sultz, in Kirk-Othmer's *Encyclopedia of Chemical Technology*, 3rd Edition, Vol. 22, 25–45 (1983).
2. Author's laboratory (S.R.S.).
3. A. Davidson and B. Milwidsky, "Synthetic Detergents," p. 115. Chemical Rubber Co. Press, Cleveland Ohio, 1968.
4. D. D. Whyte, *J. Amer. Oil Chem. Soc.* **32**, 313 (1955).

AMIDINES

1. INTRODUCTION
2. ADDITION REACTIONS
    37-1. Preparation of Benzamidine Hydrochloride

1. INTRODUCTION

The most important methods for preparing amidines are outlined in Scheme 1 on page 274.

2. ADDITION REACTIONS

37-1. Preparation of Benzamidine Hydrochloride [1]

$$C_6H_5CN + NH_4Cl \xrightarrow[150°C]{NH_3} C_6H_5\overset{\overset{\displaystyle NH_2Cl}{\|}}{C}-NH_2 \qquad (1)$$

To a stainless steel rocking autoclave equipped with a stirrer is added 103.0 gm (1.0 mole) of benzonitrile, 214.0 gm (4.0 mole) of ammonium chloride, and 306.0 gm (18.0 mole) of ammonia is introduced by means of a transfer bomb. The reaction mixture is heated at 150°C for 18 hr (pressure: 1300–6500 psig), cooled, and, with appropriate

From S. R. Sandler and W. Karo, *Organic Functional Group Preparations*, Vol. III, 2d ed. (San Diego, California, 1989), 239ff., by permission of Academic Press, Inc.

SCHEME 1
METHODS OF PREPARATION OF AMIDINES

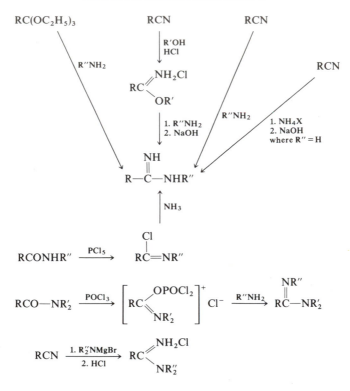

precautions for the safe control of the excess ammonia, is vented to atmospheric pressure. The reaction mixture is extracted with ether to remove approximately 5% unreacted benzonitrile, and then extracted with hot acetonitrile or ethanol to separate the amidine hydrochloride from the unreacted ammonium chloride. Concentration of the latter affords 120.5 gm (77%) of benzamidine hydrochloride, m.p. 161–163°C.

NOTE: The use of 200 ml of methanol in this preparation is also effective and reduces the reaction pressures to about 350–400 psig.

REFERENCE

1. F. C. Schaefer and A. P. Krapeho, *J. Org. Chem.* **27,** 1255 (1962).

IMIDES

1. INTRODUCTION

$\mathbf{S}$ome of the best synthetic methods for preparing cyclic imides are outlined in Scheme 1 on page 276.

2. CONDENSATION REACTIONS

38-1. Preparation of Phthalimide [1]

$$\left[\text{...} \right] \xrightarrow[300°C]{-2\,H_2O} \text{...} \quad (1)$$

From S. R. Sandler and W. Karo, *Organic Functional Group Preparations*, Vol. III, 2d ed. (San Diego, California, 1989), 281ff., by permission of Academic Press, Inc.

SCHEME 1
THE PREPARATION OF IMIDES

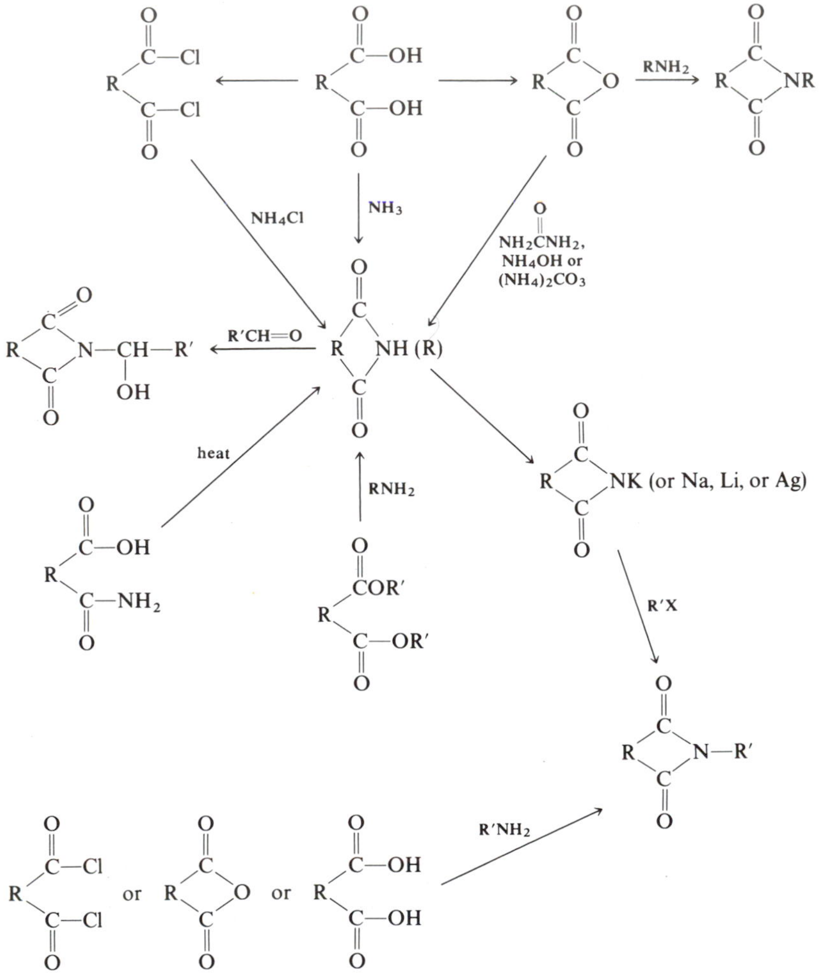

To a resin kettle fitted with an air condenser and mechanical stirrer and containing 100 gm (0.68 mole) of phthalic anhydride is added 80 gm (89 ml, 1.3 moles) of a 28% aqueous ammonia solution. The water is removed by slowly heating. Then the temperature is raised to 300°C. The sublimate in the condenser is pushed back into the reaction mixture. After 1 hr the crude product is poured out into a Pyrex dish, cooled, and weighed to give 94–96 gm (95–97%) of phthalimide (m.p. 238°).

NOTE: (a) Powdered ammonium carbonate or urea can be used in place of aqueous ammonia. (b) Phthalimide may be recrystallized from water. Solubility, 4 gm/liter water at 25°C.

38-2. Preparation of N-t-Butylphthalimide [2]

$$
\text{(phthalic anhydride)} + (CH_3)_3C-NH-\overset{\overset{\displaystyle O}{\|}}{C}-NH_2 \longrightarrow
$$

$$
\text{(N-t-butylphthalimide)} \quad N-C(CH_3)_3 + NH_3 + CO_2 \quad (2)
$$

To a 1-liter flask equipped with a condenser is added an intimate mixture of 35 gm (0.30 mole of t-butylurea and 100 gm (0.67 mole) of phthalic anhydride. The flask is immediately placed in an oil bath at 200°C and after about 10 min, when the initial vigorous effervescence subsides, the temperature is rapidly raised to 240°C and maintained there for 5 min. The product is cooled, 100 ml of ethanol is added, and the mixture is made alkaline with aqueous sodium carbonate. The mixture is diluted to 1 liter with water, filtered, and the solid pressed dry on filter paper and then heated with 500 ml of b.p. 60–68°C petroleum ether. The hot petroleum ether is filtered, dried, and concentrated to one third of its volume, cooled, and filtered to give 43.5 gm of product. The remaining filtrate is concentrated, cooled, and filtered to give 3.0 gm of product. Both product fractions melt at 59–60°C and their combination accounts for a 76% yield.

NOTE: N-t-Butylphthalimide can react with hydrazine hydrate and upon acidification of the reaction mixture t-butylamine is obtained in 89–94% yield [3, 4] (an example of the Ing–Manske reaction).

REFERENCES

1. W. A. Noyes and P. K. Porter, *Org. Syn. Coll.* **1**, 457 (1941).
2. L. I. Smith and O. H. Emerson, *J. Amer. Chem. Soc.* **67**, 1862 (1945).
3. H. R. Ing and R. H. F. Manske, *J. Chem. Soc.* **2348** (1926).
4. L. I. Smith and O. H. Emerson, *Org. Syn. Coll.* **3**, 141 (1955).

39
OXIMES

1. INTRODUCTION

The most common methods of preparing oximes are shown in Scheme 1. The two best-known methods are the oximation of carbonyl compounds and the nitrosation of active methylene compounds.

A. HAZARDS AND SAFE HANDLING PRACTICES

■ **CAUTION:** In 1974, L. J. Tyler reported two potentially devastating explosions during the preparation of ketoximosilanes of structure

$$RSi\left(O-N=C\begin{array}{c}C_2H_5\\CH_3\end{array}\right)_3$$

From S. R. Sandler and W. Karo, *Organic Functional Group Preparations*, Vol. III, 2d ed. (San Diego, California, 1989), 430ff., by permission of Academic Press, Inc.

SCHEME 1
PREPARATION OF OXIMES

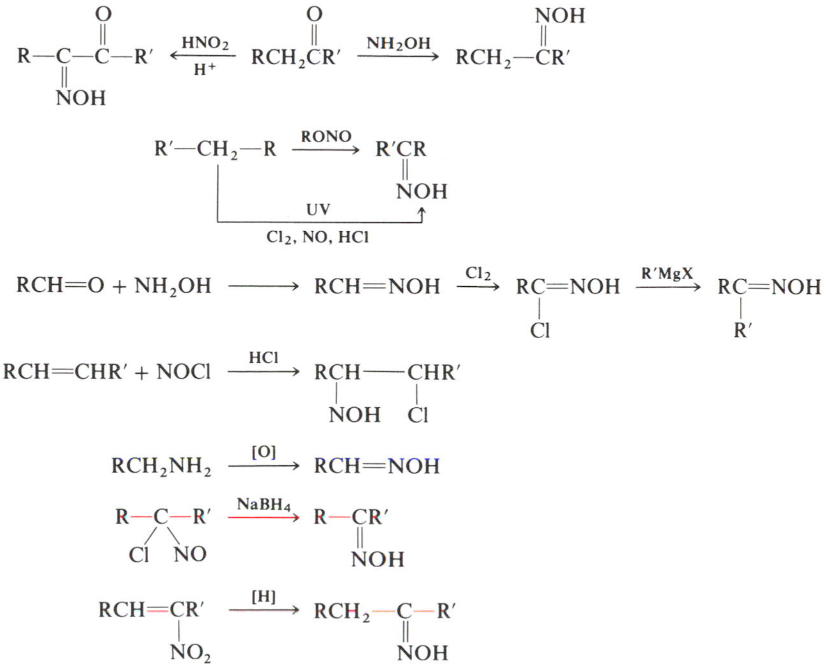

$$R-\underset{\underset{NOH}{\|}}{\overset{\overset{O}{\|}}{C}}-C-R' \xleftarrow[H^+]{HNO_2} RCH_2CR' \xrightarrow{NH_2OH} RCH_2-\overset{\overset{NOH}{\|}}{C}R'$$

$$R'-CH_2-R \xrightarrow{RONO} R'\underset{\underset{NOH}{\|}}{C}R$$

$$R'-CH_2-R \quad \underset{\underline{\qquad Cl_2, NO, HCl \qquad}}{\overset{UV}{\longrightarrow}}$$

$$RCH=O + NH_2OH \longrightarrow RCH=NOH \xrightarrow{Cl_2} \underset{Cl}{RC=NOH} \xrightarrow{R'MgX} \underset{R'}{RC=NOH}$$

$$RCH=CHR' + NOCl \xrightarrow{HCl} \underset{NOH \quad Cl}{RCH-\!\!-\!\!-CHR'}$$

$$RCH_2NH_2 \xrightarrow{[O]} RCH=NOH$$

$$\underset{Cl \quad NO}{R-C-R'} \xrightarrow{NaBH_4} \underset{NOH}{R-CR'}$$

$$\underset{NO_2}{RCH=CR'} \xrightarrow{[H]} RCH_2-\underset{\underset{NOH}{\|}}{C}-R'$$

The explosions were attributed to the inadvertent development of acidic conditions in the system, which may have led to highly exothermic Beckmann rearrangement conditions.

Further experiments showed that whereas methyl ethyl ketoxime can be distilled in the presence of impurities, the compound undergoes violent degradation when heated to $50-70°C$. Among the impurities that may cause explosions are acids, the hydrochloride salt of methyl ethyl ketoxime, and ferric chloride. A level of 2% of ferric chloride lowered the onset of violent degradation to $50°C$ [1].

We believe that, while this potential hazard needs further study, great care in the preparation of handling of oximes must be exercised. Precautions must be taken particularly because oximes are such common derivatives, prepared in the student laboratory for the identification of carbonyl compounds by the melting point of their oximes.

Furthermore, many of the syntheses involve the use of hydroxylamine or its salts. A report based on a Safety Newsletter of the National Safety Council states that hydroxylamine, its salts, and mother liquors contaning these materials have been known to explode on warming [2]. The causes and conditions for explosion are not well understood. Therefore, due precautions must be exercised in handling hydroxylamine, its salts, and solutions containing these compounds.

2. CONDENSATION REACTIONS

A classical method of preparing a standard solution of sodium hydroxylamine disulfonate is given here. Some disagreement exists in the literature as to the exact nature of this reagent. Sometimes it is referred to as sodium hydroxylamine monosulfonate [3, 4].

39-1. Preparation of Sodium Hydroxylamine Disulfonate [5]

$$NaNO_2 + NaHSO_3 + SO_2 \longrightarrow \left[HO-N \begin{matrix} SO_3 \\ SO_3 \end{matrix} \right]^{2-} + 2\,Na^+ \qquad (1)$$

In a 5-liter flask equipped with an efficient stirrer and surrounded with a large ice–salt bath is placed 4-liters of distilled water, 325 gm (3 moles) of anhydrous sodium carbonate, and 420 gm (5.8 moles) of 95% pure sodium nitrite. The mixture is stirred and cooled. When the temperature has reached $-5°C$, sulfur dioxide is passed into the solution at such a rate that the solution temperature never exceeds $0°C$. With efficient cooling, the reduction is completed in approximately 4 hr. The end of the reduction is indicated by a drop in the pH of the solution to the acid side. Just before the reaction turns acid, the solution becomes slightly brown in color. The solution is approximately 1.2 M in sodium hydroxylamine disulfonate and may be used directly for the preparation of oximes.

In using sodium hydroxylamine disulfonate, the reagent is treated with a carbonyl compound, warmed, and cooled. Only at this point is the reaction mixture neutralized with an alkaline solution [6]. This method is suitable for the preparation of a range of carbonyl compounds. Negative results were obtained in the case of benzil and quinone monoxime, either because the oximes are not sufficiently soluble in the reaction medium or because the oximes produced are not stable in the acidic reaction solution.

39-2. Preparation of Methyl Ethyl Ketoxime [6]

$$\underset{\text{CH}_3\overset{\text{O}}{\overset{\|}{\text{C}}}\text{CH}_2\text{CH}_3}{} + \left[\text{HON}\begin{matrix}\text{SO}_3\\\text{SO}_3\end{matrix}\right]^{2-} + 2\,\text{Na}^+ + \text{H}_2\text{O} \longrightarrow$$

$$\begin{matrix}\text{CH}_3\\\\\text{CH}_3\text{CH}_2\end{matrix}\text{C=NOH} + 2\,\text{NaHSO}_4 \quad (2)$$

With suitable precautions, to 1 liter of the sodium hydroxylamine disulfonate solution from Preparation 39-2 (approx. 1.2 moles) is added 72 gm (1 mole) of methyl ethyl ketone. The mixture is warmed to 70°C. Then the reaction flask is wrapped with insulation and allowed to cool slowly for 12 hr.

After neutralization with 48% sodium hydroxide solution, the oxime is extracted from the reaction mixture with benzene. The benzene solution is distilled fractionally. The product distills between 152° and 154°C; yield, 65 gm (75%).

39-3. Preparation of Phenylglyoxime [7]

$$\begin{matrix}\text{C}_6\text{H}_5\\\text{C=O}\\\text{C=NOH}\\\text{H}\end{matrix} \xrightarrow{\text{NH}_2\text{OH}} \begin{matrix}\text{C}_6\text{H}_5\\\text{C=NOH}\\\text{C=NOH}\\\text{H}\end{matrix} \quad (3)$$

To a solution of 50 gm (0.34 mole) of ω-oximinoacetophenone (ω-isonitrosoacetophenone) in 150 ml of ethanol is added to a solution of 48 gm (0.595 mole) of sodium acetate and 24 gm (0.345 mole) of hydroxylamine hydrochloride in 75 ml of water. The mixture is heated at reflux for 4 hr.

After cooling, a large portion of the solvent is evaporated off under reduced pressure. The mixed crude product, which precipitates, is filtered off, washed with water, and air-dried to afford 51 gm (92%), m.p. 150–158°C.

The separation of the three isomers by fractional crystallization procedures depends on the relative ratio of the components in the mixture, the concentration of the components in the solvent, the length of time the solvent–product mixture is heated, and so on. It is there-fore important that the progress of the crystallization be followed by thin-layer chromatography using freshly prepared tlc plates [Merck (Darmstadt) silica gel G is recommended as the coating]. The suggested tlc solvent is benzene–ethyl acetate (7:3), detection by iodine vapor; 0.45 R_f for *anti*-phenyl-*amphi*-glyoxime (I), 0.40 R_f for phenyl-*anti*-glyoxime (II), and 0.35 R_f for phenyl-*syn*-glyoxime (III).

Anti-phenyl-*amphi*-glyoxime (I) is separated from the mixed product by repeated recrystallization from acetone–chloroform (or alcohol–water), m.p. 178–180°C; UV, λ_{max} (95% ethanol) 230 mμ (ϵ 14,800).

Phenyl-*anti*-glyoxime (II) is isolated from the mother liquor of the first crystallization in the separation of (I). The residue from the evap-oration of this mother liquor is crystallized repeatedly from acetone chloroform. After sublimation, the properties of the material are as follows: sublimation temperature, 90°C (0.2 mm Hg); m.p 166–168°C; UV, λ_{max} (95% ethanol) 228 mμ (ϵ 14,380). Only about 1% of pure isomer is isolated from the starting product.

Phenyl-*syn*-glyoxime (III) is isolated by crystallizing the starting product from a dilute solution of ethyl acetate, m.p. 168–170°C; UV, λ_{max} (95% ethanol) 252 mμ (ϵ 12,200).

39-4. Preparation of *syn*-2-Pyridine Aldoxime from 2-Picoline in Liquid Ammonia [8]

$$(4)$$

In a hood, in a suitably cooled flask equipped with an explosion-proof stirrer, a dry ice-cooled reflux condenser, is placed 400 ml of anhydrous liquid ammonia and 0.5 gm of ferric nitrate nonahydrate. To the flask is added, with vigorous stirring 46 gm (2.0 gm-atom) of sodium metal. The metal is permitted to dissolve completely and, after approximately 20 min, the blue color changes to gray. At this point, 279 gm (3 moles) of 2-picoline is added over a 30-min period. After stirring for 1 hr, a solution of 103 gm (1.0 mole) of butyl nitrite in 110 ml of diethyl ether is added over a 45-min period (intense red mixture forms). The reaction is quite rapid and considerable amounts of solids form. After 1 hr, a solution of 198 gm (1.5 moles) of ammonium sulfate in 300 ml of water is added rapidly. A thick precipitate of sodium sulfate replaces the solid previously formed and the color disappears. While 500 ml of ether is being added, the ammonia is allowed to evaporate. When the temperature has reached room temperature, the liquids are separated and filtered free of solids. The solids are washed repeatedly with ether. The ether layers are combined, the aqueous layer is separated, and repeatedly extracted with ether. The ether extracts are combined and distilled, first at atmospheric pressure, then at reduced pressures up to 80°C (2 mm) to remove butanol and unreacted 2-picoline. Finally the product is distilled at 110°C (0.9 mm Hg) through a wide-diameter distillation head to afford 91.5 gm (75% based on requirement of 2 gm-atom of Na per mole of product), m.p. 105–111°C. On recrystallization from benzene, the melting point is raised to 113–113.5°C.

REFERENCES

1. L. J. Tyler, *Chem. & Eng. News.* page 3 (Sept 2, 1974).
2. Anon., Chemical Processing, p. 30 (December 2, 1963).
3. W. L. Semon and V. R. Damerell, *Org. Syn. Coll.* **2,** 204 (1943).
4. W. L. Semon and V. R. Damerell, *J. Amer. Chem. Soc.* **47,** 2033 (1925).
5. R. Adams and O. Kamm, *J. Amer. Chem. Soc.* **40,** 1281 (1918).
6. W. L. Semon and V. R. Damerell, *J. Amer. Chem. Soc.* **46,** 1290 (1924).
7. J. V. Burakevich, A. M. Lore, and G. P. Volpp, *J. Org. Chem.* **36,** 1 (1971).
8. S. E. Forman. *J. Org. Chem.* **29,** 3323 (1964).

40

HYDROXAMIC ACIDS

1. INTRODUCTION

The variety of substituted hydroxamic acids is surprisingly large. If we consider only the structures which may be written upon alkylation and/or acylation of hydroxylamine but ignore the tautomeric derivatives, the following compounds result.

$$
\underset{\text{I}}{\text{R}-\overset{\overset{\text{O}}{\|}}{\text{C}}-\text{NHOH}}
\qquad
\underset{\text{II}}{\underset{\overset{|}{\text{R}'}}{\text{R}-\overset{\overset{\text{O}}{\|}}{\text{C}}-\text{N}-\text{OH}}}
\qquad
\underset{\text{III}}{\underset{\overset{|}{\text{OH}}}{\text{R}-\overset{\overset{\text{O}}{\|}}{\text{C}}-\text{N}-\overset{\overset{\text{O}}{\|}}{\text{C}}-\text{R}'}}
$$

From S. R. Sandler and W. Karo, *Organic Functional Group Preparations*, Vol. III, 2d ed (San Diego, California, 1989), 482ff., by permission of Academic Press, Inc.

$$R\overset{\overset{\displaystyle O}{\|}}{-C}-NOHR'$$
$$|$$
$$H$$

IV

$$NH_2O\overset{\overset{\displaystyle O}{\|}}{C}-R$$

V

$$R\overset{\overset{\displaystyle O}{\|}}{-C}-N-OR'$$
$$|$$
$$R''$$

VI

$$R'NHO\overset{\overset{\displaystyle O}{\|}}{C}-R''$$

VII

$$\begin{array}{c} R-\overset{\overset{\displaystyle O}{\|}}{C} \\ \diagdown \\ N-O\overset{\overset{\displaystyle O}{\|}}{C}-R' \\ \diagup \\ R'-\overset{\underset{\displaystyle O}{\|}}{C} \end{array}$$

VIII

$$\begin{array}{c} R' \\ \diagdown \\ N-O\overset{\overset{\displaystyle O}{\|}}{C}-R \\ \diagup \\ R'' \end{array}$$

IX

$$\begin{array}{c} R-\overset{\overset{\displaystyle O}{\|}}{C} \\ \diagdown \\ N-OR'' \\ \diagup \\ R'-\overset{\underset{\displaystyle O}{\|}}{C} \end{array}$$

X

$$R\overset{\overset{\displaystyle O}{\|}}{C}-N-O\overset{\overset{\displaystyle O}{\|}}{C}-R'$$
$$|$$
$$H$$

XI

$$R-\overset{\overset{\displaystyle O}{\|}}{C}-N-O\overset{\overset{\displaystyle O}{\|}}{C}-R'$$
$$|$$
$$R''$$

XII

Compounds such as IV, VI, and X are esters of hydroxamic acids and in the case of structure IV are named alkyl hydroxamates.

A. HAZARDS AND SAFE HANDLING PRACTICES

■ **CAUTION:** Reports exist that severe explosions have taken place with reactions in which hydroxylamine or its common salts have been used. These observations were made upon scale-up of small laboratory preparations. Explosions with pure hydroxylamine hydrochloride and with hydroxylamine sulfate were noted on heating these reagents [1]. The factors involved in these explosions are not known, consequently adequate precautions should be taken. We suggest that hydroxylamine, its salts, or solutions containing any of these materials should never be exposed to elevated temperatures; the presence of peroxides or hydroperoxides (as in ether solutions) must be avoided; possibly all reactions should be carried out in inert atmospheres. All other precautions against explosive hazards should also be taken. In view of this hazard, all preparations involving hydroxylamine, its salts, or its derivatives, are given here for reference only and should not be construed as recommendations of safe reactions.

2. CONDENSATION REACTIONS

40-1. Preparation of Benzohydroxamic Acid in Pyridine [2]

$$\underset{}{\text{C}_6\text{H}_5}\overset{\displaystyle O}{\overset{\|}{\text{C}}}\!-\!\text{OC}_2\text{H}_5 + \text{NH}_2\text{OH} \xrightarrow{\text{KOH-pyridine}} \underset{}{\text{C}_6\text{H}_5}\overset{\displaystyle O}{\overset{\|}{\text{C}}}\!-\!\text{NHOH} + \text{C}_2\text{H}_5\text{OH} \quad (1)$$

To a rapidly stirred suspension of 19.6 gm (0.35 mole) of powdered anhydrous potassium hydroxide in 120 ml of pyridine, maintained at 0–5°C, is a added a solution of 13.9 gm (0.2 mole) of hydroxylamine hydrochloride in 100 ml of pyridine.

While maintaining the reaction temperature at 0–5°C, 15 gm (0.1 mole) of ethyl benzoate is added. Vigorous stirring is continued at room temperature for 6 hr. Then the solids are filtered off. The solids are washed with cold water to remove inorganic coproducts. The remainder, recrystallized from aqueous ethanol, represents a 94% yield of potassium benzohydroxamate.

This salt is triturated with cold 0.01 N hydrochloric acid. From this mixture, by the usual procedures, 12.5 gm (91% overall) of benzohydroxamic acid is isolated, m.p. 131°C (from aqueous alcohol).

By a similar procedure, a quantitative yield of laurohydroxamic acid (not described further) is said to have been prepared from methyl laurate. p-Nitrobenzohydroxamic acid was prepared in only a 17% yield by this method.

The general procedure of reacting aliphatic esters with hydroxylamine as described has been criticized because it does not lend itself well to the preparation of large quantities (100-gm scale). By careful attention to details such as the order of addition, adding the ester to a strongly alkaline hydroxylamine solution at such a rate that the ester dissolves completely in the reaction medium (with additional ethanol if necessary), the reaction is said to be completed within a few minutes at least to the extent of 90% [3].

Fletcher and Lipowski [4, 5] reacted polyacrylonitrile with hydroxylamines to form a polyamidoxime, which, on hydrolysis, gave a polymeric hydroxamic acid.

40-2. Preparation of Poly (Hydroxamic Acids) [5, 6]

$$\left(\text{CH}_2-\text{CH}\atop\quad\ \ \text{CN}\right)_n + \text{NH}_2\text{OH}\cdot\text{H}_2\text{SO}_4 \xrightarrow{\text{OH}^-} \left(\text{CH}_2-\text{CH}\atop\begin{array}{c}\text{C}=\text{N}-\text{OH}\\ \text{NH}_2\end{array}\right)_n \tag{2}$$

$$\left(\text{CH}_2-\text{CH}\atop\begin{array}{c}\text{C}=\text{N}-\text{OH}\\ \text{NH}_2\end{array}\right)_n \xrightarrow{\text{H}_2\text{O}} \left(\text{CH}_2-\text{CH}\atop\begin{array}{c}\text{C}-\text{NHOH}\\ \text{O}\end{array}\right)_n \tag{3}$$

A mixture of 80 g of poly(acrylonitrile), 300 g of hydroxylamine sulfate, and 2500 ml of water, to which had been added 140 g of sodium hydroxide, is heated with stirring at 90°C for 12 hr. The mixture is cooled to room temperature, and the polymeric amidoxime is filtered off, then washed with water until the wash water is neutral.

To 20 g of the polyamidoxime is added 100 g of cold concentrated (37%) aqueous hydrochloric acid. The mixture is stirred for 5 min at 10°C and then mixed with 200 g of ice.

The gel that forms is separated, washed with deionized water and dried to a brittle poly(hydroxamic acid) containing 14.8% total nitrogen.

In another procedure Vernon and Eccles first hydrolyzed cross-linked poly(acrylonitrile) with 50% sulfuric acid. The resulting polymer, after thorough washing, was treated with hydroxylamine in a sodium acetate solution and then worked up [6, 7].

REFERENCES

1. Based on information from Safety Newsletter of National Safety Council, *Chem. Proc.* 30 (Dec. 2, 1963).
2. G. B. Bachman and J. E. Goldmacher, *J. Org. Chem.* **29,** 2576 (1964).
3. W. N. Fishbein, J. Daley, and C. L. Streeter, *Anal. Biochem.* **28,** 13 (1969).
4. C. A. Fetscher and S. A. Lipowski, U.S. Pat. 3,345,344 (October 3, 1967); *Chem. Abstr.* **67,** P 109273a (1967).
5. C. A. Fetscher and S. A. Lipowski, U.S. Pat. 3,367,959 (February 6, 1968); *Chem. Abstr.* **69,** P 2,528b (1968).
6. F. Vernon and H. Eccles, *Anal. Chim. Acta* **83,** 187 (1976).
7. F. Vernon and H. Eccles, *Anal. Chim. Acta* **82,** 369 (1976).

41

POLYMER
PREPARATIONS

1. POLYESTERS

41-1. Preparation of Poly(ethylene maleate) [1]

$$
\begin{array}{c}
\underset{\substack{\text{CO}\\\text{CO}}}{\overset{\text{CO}}{\Big\rangle}}\text{O} + \text{HOCH}_2\text{CH}_2\text{OH} \xrightarrow{\text{heat}} \left[\underset{\text{COOCH}_2\text{CH}_2\text{O}-}{\overset{\text{CO}-}{\Big\rangle}}\right]_n + 2\,\text{H}_2\text{O}
\end{array}
\qquad (1)
$$

To a polymer tube is added 32.5 gm (0.33 mole) of maleic anhydride and 18.6 gm (0.30 mole) of ethylene glycol. The tube is heated to $195°–200°C$ for 4 hr. Then the heating is continued at $210°–215°C$ at reduced pressure (0.2 mm) for 3 hr. The polymer tube is cooled and ethylene chloride added. The polymer, 40 gm (94%), separates as an oil which later changes to a white powder when cooled for 2 hr at $5°–10°C$, m.p. $88°–95°C$. After drying the polymer has m.p. $>250°C$ and is insoluble in most common solvents including ethylene dichloride.

This polymer has also been obtained by reacting silver maleate with ethylene dibromide [2].

In a related manner Carothers prepared poly(ethylene phthalate), poly(trimethylene phthalate), poly(hexamethylene phthalate), poly-(decamethylene phthalate), and poly(ethylene fumarate) [1].

41-2. Preparation of Poly(ethylene terephthalate) [3]

$$
\text{CH}_3\text{O}\overset{\text{O}}{\overset{\|}{\text{C}}}-\underset{}{\big\langle\!\!\big\rangle}-\overset{\text{O}}{\overset{\|}{\text{C}}}-\text{OCH}_3 + 2\,\text{HOCH}_2\text{CH}_2\text{OH} \xrightarrow{-2\,\text{CH}_3\text{OH}}
$$

$$
\text{HOCH}_2\text{CH}_2\text{O}\overset{\text{O}}{\overset{\|}{\text{C}}}-\text{C}_6\text{H}_4-\overset{\text{O}}{\overset{\|}{\text{C}}}-\text{OCH}_2\text{CH}_2\text{OH}
$$

$$
\text{heat} \;\Big|\; -\text{HOCH}_2\text{CH}_2\text{OH} \;\downarrow
$$

$$
\left[-\text{OCH}_2\text{CH}_2\text{O}\overset{\text{O}}{\overset{\|}{\text{C}}}-\text{C}_6\text{H}_4-\overset{\text{O}}{\overset{\|}{\text{C}}}-\right]_n
\qquad (2)
$$

■ **CAUTION:** This reaction should be carried out in a hood behind a protective shield.

Polyesters from S. R. Sandler and W. Karo, *Polymer Syntheses*, Vol. I, 2d ed. (San Diego, California, 1992), 68ff., by permission of Academic Press, Inc.

To a weighed thick-walled glass tube with a constricted upper portion for vacuum tube connection and equipped with a metal protective
sleeve is added 15.5 gm (0.08 mole) of dimethyl terephthalate (see Note
a), 11.8 gm (0.19 mole) of ethylene glycol (Note b), 0.025 gm of calcium acetate dihydrate, and 0.006 gm of antimony trioxide. The tube
is warmed gently in an oil bath to melt the mixture and then a capillary
tube connected to a nitrogen source is placed in the melt. While heating
to 197°C a slow stream of nitrogen is passed through the melt to help
eliminate the methanol. The tube is heated for 2–3 hr at 197°C, or until all the methanol has been removed (Note c). The side arm is also
heated to prevent clogging by the condensation of some dimethyl terephthalate. The polymer tube is next heated to 222°C for 20 min and
then at 283°C for 10 min. The side arm is connected to a vacuum pump
and the pressure is reduced to 0.3 mm Hg or less while heating at 283°C
for $3\frac{1}{2}$ hr. The tube is removed from the oil bath, cooled (Note d) behind a safety shield, and then weighed. The yield is quantitative if no
loss of dimethyl terephthalate has occurred. The polymer melts at
about 270°C. (The crystalline melting point is 260°C.) The inherent
viscosity of an 0.5% solution in *sym*-tetrachloroethanol/phenol
(40/60) is approx. 0.6–0.7 at 30°C.

NOTE: (a) Dimethyl terephthalate is the best grade or is recrystallized
from ethanol, m.p. 141°–142°C, (b) The ethylene glycol is anhydrous
reagent grade or prepared by adding 1 gm of sodium/100 ml, refluxing
for 1 hr in a nitrogen atmosphere, and distilling, b.p. 196°–197°C.
(c) Failure to remove all the methanol leads to low molecular weight
polymers. (d) On cooling the polymer contracts from the walls and
this may cause the tube to shatter.

2. POLYAMIDES

The Schotten–Baumann [4] reaction can be applied to the preparation of polyamides using bifunctional reagents. Since the reaction
is very rapid at room temperature, it can be carried out at low temperatures (a) in solution or (b) by an interfacial polycondensation
technique.

Polyamides from S. R. Sandler and W. Karo, *Polymer Syntheses*, Vol. I, 2d ed. (San Diego,
California, 1992), 111ff., by permission of Academic Press, Inc.

SCHEME 1

THE MAJOR PREPARATIVE METHODS FOR THE SYNTHESIS OF POLYAMIDES

$$H_2N(CH_2)_xCOOH \qquad (CH_2)_x \overset{C=O}{\underset{NH}{\diagup}}$$

$$-H_2O \searrow \qquad \swarrow \text{catalyst}$$

$$[-NH(CH_2)_xCO-]_n$$

$$\uparrow -2\,H_2O$$

$$\left[\begin{array}{c} O \quad\quad O \\ \parallel \quad\quad \parallel \\ -OC(CH_2)_xCO- \end{array} \right]^{2-} \quad H_3\overset{+}{N}(CH_2)_x\overset{+}{N}H_3$$

$$R(COZ)_2 + R'(NH_2)_2 \xrightarrow{-HZ} \left[\begin{array}{c} O \quad\quad O \\ \parallel \quad\quad \parallel \\ -C-R-CNHR'NH- \end{array} \right]_n$$

$$(Z = Cl, OR'', \text{ or } NH_2)$$

$$R(COOH)_2 + R'(NHCOR'')_2 \longrightarrow RCO\left[\begin{array}{c} O\;O \\ \parallel\;\parallel \\ -HNR'NHCRC- \end{array} \right]_n OH + R''COOH$$

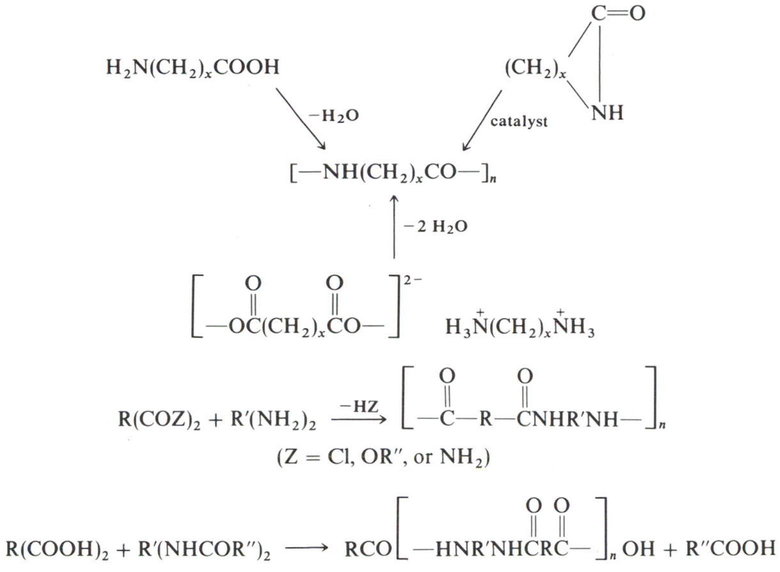

SCHEME 2

MISCELLANEOUS METHODS FOR THE SYNTHESIS OF POLYAMIDES FROM NITRILES

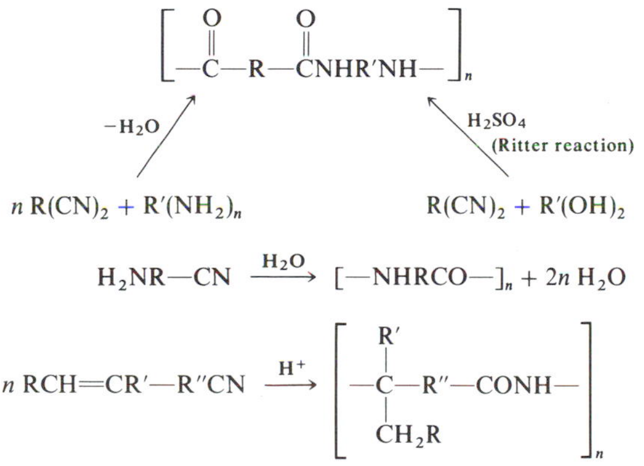

$$\left[\begin{array}{c} O \quad\quad O \\ \parallel \quad\quad \parallel \\ -C-R-CNHR'NH- \end{array} \right]_n$$

$$-H_2O \nearrow \qquad \nwarrow \begin{array}{c} H_2SO_4 \\ \text{(Ritter reaction)} \end{array}$$

$$n\,R(CN)_2 + R'(NH_2)_n \qquad\qquad R(CN)_2 + R'(OH)_2$$

$$H_2NR-CN \xrightarrow{H_2O} [-NHRCO-]_n + 2n\,H_2O$$

$$n\,RCH=CR'-R''CN \xrightarrow{H^+} \left[\begin{array}{c} R' \\ | \\ -C-R''-CONH- \\ | \\ CH_2R \end{array} \right]_n$$

In the solution polycondensation method the reaction is carried out in a single inert liquid in the presence of an acid acceptor. The polymer may precipitate out of the solution or it may be soluble.

In the interfacial polycondensation method the reaction is carried out at the interface of two immiscible solvents. The amine is dissolved in water and the acid chloride is dissolved in the hydrocarbon layer. In 1938 Carothers [5] reported on the potential use of this reaction to prepare polyamides. However, it was not until the work of Magat and Strachan [6] and later Morgan and co-workers [7–10] that this reaction was exploited for the preparation of various polyamides.

41-3. Preparation of Poly(hexamethyleneadipamide) (Nylon 6–6) by the Interfacial Polymerization Technique [7]

$$(CH_2)_6(NH_2)_2 + (CH_2)_4(COCl)_2 \longrightarrow \left[-HN-(CH_2)_6-NH-\overset{O}{\underset{\|}{C}}-(CH_2)_4-\overset{O}{\underset{\|}{C}}- \right]_n \quad (3)$$

In an ice-cooled blender jar containing a solution of 3.95 gm (0.034 mole) of hexamethylenediamine dissolved in 200 ml of water containing 3.93 gm (0.070 mole) of potassium hydroxide is added with agitation a solution of 6.22 gm (0.034 mole) of adipoyl chloride dissolved in 200 ml of xylene. The addition takes about 5 min and the speed of the blender agitation is slow at first and then speeded up toward the end of the addition period. The product is filtered, washed with water, and dried to afford 5.6 gm (73%), $\eta_{inh} = 1.16$.

41-4. Preparation of Poly(hexamethylenesebacamide) (Nylon 6–10) by the Interfacial Polymerization Technique [6]

$$(CH_2)_6(NH_2)_2 + (CH_2)_8(COCl)_2 \longrightarrow \left[-HN-(CH_2)_6-NH\overset{O}{\underset{\|}{C}}-(CH_2)_8-\overset{O}{\underset{\|}{C}}- \right]_n \quad (4)$$

To a tall-form beaker is added a solution of 3.0 ml (0.014 mole) of sebacoyl chloride dissolved in 100 ml of distilled tetrachloroethylene. Over this acid chloride solution is carefully poured a solution of 4.4 gm (0.038 mole) of hexamethylenediamine (see Note) dissolved in 50 ml of water. The polyamide film which begins to form at the interface of these two solutions is grasped with tweezers or a glass rod and slowly pulled out of the beaker in a continuous fashion. The process stops when one of the reactants becomes depleted. The resulting "rope"-like polymer is washed with 50% aqueous ethanol or acetone, dried, and weighed to afford 3.16–3.56 gm (80–90%) yields of polyamide, $\eta_{inh} = 0.4$ to 1.8 (m-cresol, 0.5% conc. at 25°C), m.p. 215°C (soluble in formic acid).

NOTE: In this experiment excess diamine is used to act as an acid acceptor.

3. POLYURETHANES

SCHEME 3

PREPARATION OF POLYURETHANES

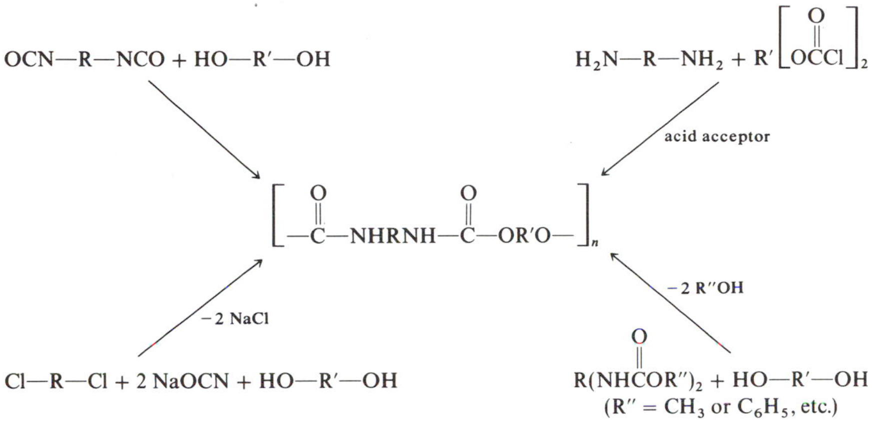

$$OCN-R-NCO + HO-R'-OH \qquad\qquad H_2N-R-NH_2 + R'\left[\underset{O}{\overset{O}{\underset{\|}{O}CCl}}\right]_2$$

acid acceptor

$$\left[-\overset{O}{\underset{\|}{C}}-NHRNH-\overset{O}{\underset{\|}{C}}-OR'O-\right]_n$$

-2 NaCl -2 R''OH

$$Cl-R-Cl + 2 NaOCN + HO-R'-OH \qquad R(NHCOR'')_2 + HO-R'-OH$$
$$(R'' = CH_3 \text{ or } C_6H_5, \text{ etc.})$$

41-5. Preparation of Poly[ethylene methylene bis(4-phenylcarbamate)] [11, 12]

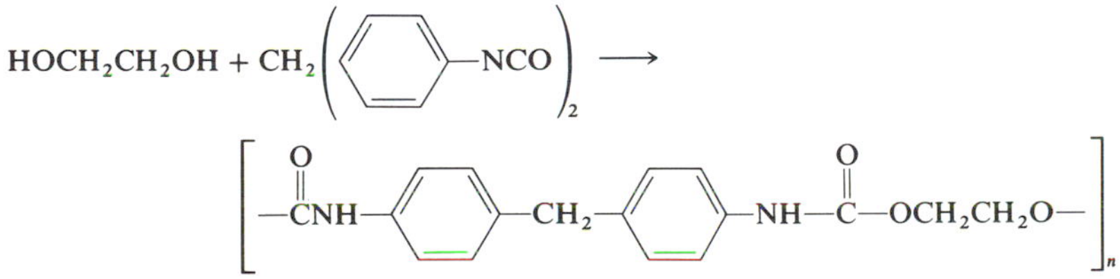

$$HOCH_2CH_2OH + CH_2\left(\underset{}{\bigcirc}-NCO\right)_2 \longrightarrow$$

$$\left[-\overset{O}{\underset{\|}{C}}NH-\bigcirc-CH_2-\bigcirc-NH-\overset{O}{\underset{\|}{C}}-OCH_2CH_2O-\right]_n \quad (5)$$

To a flask equipped with a mechanical stirrer, condenser, and dropping funnel and containing 40 ml of 4-methylpentanone-2 and 25.02 gm (0.10 mole) of methylene bis(4-phenyl isocyanate) (MDI) is added all at once 6.2 gm (0.10 mole) of ethylene glycol in 40 ml of dimethyl sulfoxide. The reaction mixture is heated at 115°C for $1\frac{1}{2}$ hr, cooled, poured into water, and filtered. The white polymer is chopped

Polyurethanes from S. R. Sandler and W. Karo, *Polymer Syntheses*, Vol. I, 2d ed. (San Diego, California, 1992), 232ff., by permission of Academic Press, Inc.

up in a blender, washed with water, and dried under reduced pressure at 90°C to afford 29.6–31.2 gm (95–100%), $\eta_i = 1.01$ (0.05% solution in DMF at 30°C), polymer melt temperature, 240°C.

4. ACRYLIC POLYMERS

In the casting of polymer sheets, bulk polymerization of initiator-containing monomers may be used, although "prepolymers" (i.e., solutions of polymers in their monomers, usually prepared by bulk polymerizing a monomer until the viscosity of the mixture has reached a desired level) [13] are more commonly used commercially to reduce problems arising from shrinkage, not to mention leakage of monomer through the flexible gasketing usually used.

41-6. Generalized Procedure for the Preparation of Poly(methyl methacrylate) Sheets

A casting mold is constructed by clamping a length of polyethylene tubing between three sides of two sheets of carefully cleaned plate glass with spring paper clips of suitable size. It is most convenient to make the mold of plate glasses cut to the same width but unequal lengths (e.g., one piece 10 × 10 cm and one 10 × 15 cm) and forming the gasket along three sides. The excess glass section will facilitate the filling of the mold.

The mold is supported in an inclined position with the larger side forming the lower part of the mold. A solution of methyl methacrylate containing 0.5% of benzoyl peroxide is carefully poured into the mold to fill approximately two-thirds of the available space.

The mold is then supported in an upright position, preferably in a shallow dish. The top of the mold may be closed by forcing a length of tubing along the open edge. The assembly is then placed in an *explosion-proof*, high-velocity air oven and heated at 70°C for 72 hr. The curing time varies with the overall size of the casting. In the case of thick cross-section castings, the curing time is considerably longer.

After the polymerization has been completed, the casting is cooled gradually. The sheet is removed by removing the glass plates. Because of the inhibiting effect of air, the top portion of the sheet may be

Acrylic Polymers from S. R. Sandler and W. Karo, *Polymer Syntheses*, Vol. I, 2d ed. (San Diego, California, 1992), 333ff., by permission of Academic Press, Inc.

somewhat soft. The plastic sheet may be finished by conventional plastic shaping techniques.

Similar molding techniques may be used with casting syrups, although great care must be exercised to allow trapped bubbles to rise prior to the final curing stages. With sizable glass molds, special arrangements have to be made to prevent bulging of the glass plates under the hydrostatic pressure of the monomer.

41-7. Solution Copolymerization of Glycidyl Methacrylate and Styrene [14]

To 228 gm of xylene maintained at 136°C is added with stirring over a 3 hr period a solution of 453 gm of styrene, 80 gm of glycidyl methacrylate, and 11 gm of di-*tert*-butyl peroxide. After the addition has been completed, the solution is heated and stirred for an additional 3 hr at 136°C (to 100% conversion). After cooling, the polymer solution may be diluted to 54% solids by the addition of 228 gm of methyl isobutyl ketone. The relative viscosity of the copolymer (1% solution in 1,2-dichloroethylene at 25°C) is 1.175.

41-8. Redox Emulsion Polymerization of Ethyl Acrylate [15]

In a 1 liter resin kettle fitted with a stainless steel stirrer, a thermometer which extends well into the lowest portion of the reactor, a reflux condenser, a dropping funnel, and a nitrogen inlet tube attached to the apparatus is prepared a solution of 376 ml of deionized water and 24 gm of Triton X-200. While the nitrogen flow is on, with stirring, 200 gm of uninhibited ethyl acrylate, 4 ml of a solution freshly prepared by dissolving 0.3 gm of ferrous sulfate heptahydrate in 200 ml of deionized water, and 1 gm of ammonium persulfate are added. After stirring for 30 min, the mixture is cooled to 20°C and 1 gm of sodium metabisulfite and 5 drops of 70% *tert*-butyl hydroperoxide are added. The polymerization starts rapidly and the temperature in the flask rises to nearly 90°C. After approximately 15 min, the polymerization is complete. The latex is cooled and passed through a nylon chiffon strainer.

If inhibited monomers are used, an induction period is observed in this process although the rate of polymerization remains about the same. Variations of procedure such as the gradual addition of monomer to a seed polymer are also possible. The gradual addition of reducing agent solution along with gradually added monomer may also be used as a technique of redox latex polymerization.

41-9. Preparation of Isotactic Poly(methyl methacrylate) [16]

In a 500 ml four-necked flask with suitable adapters to carry a mechanical stirrer, addition funnel, thermometer, inlets and outlets for dry, oxygen-free nitrogen, and an addition port covered with a rubber serum-bottle cap, is placed 300 ml of anhydrous toluene. Inhibitor-free, dry methyl methacrylate (15 gm, 0.15 mole) is placed in the addition funnel and nitrogen is bubbled through both the monomer and the solvent for 3 hr. After cooling the flask contents to 3°C with an ice bath, 3.6 ml of a 3.3 M solution of phenyl-magnesium bromide in diethyl ether is injected by means of a hypodermic syringe and, with moderate stirring, the monomer is added. Stirring is continued for 4 hr. The viscous solution is then poured slowly into 3 liters of vigorously stirred petroleum ether. The solid is collected on a filter and dried under reduced pressure. Inorganic impurities are removed by digestion with aqueous methanol containing hydrochloric acid. Yield, 11.2 gm (74.6%) of isotactic poly(methyl methacrylate), viscosity average molecular weight 480×10^3. After annealing at 115°C, crystalline characteristics of the polymer can be demonstrated by X-ray diffraction.

By similar procedures, isopropyl acrylate, ethyl acrylate, and methyl acrylate have been polymerized [17]. Detailed directions for the preparation of isotactic poly(isopropyl acrylate) and of syndiotactic poly(isopropyl acrylate) are given in *Macromolecular Syntheses* [18, 19], by Sorenson and Campbell [20], and by Kiran and Gillham [21].

5. ACRYLONITRILE, ACRYLAMIDE, AND N-VINYLPYRROLIDONE POLYMERS

41-10. Suspension Copolymerization of Styrene and Acrylonitrile [22]

To a stirred dispersion of 60 gm of deaerated water, 1 gm of powdered hydroxyapatite (see Note), and 0.008 gm of sodium oleate are added under a nitrogen atmosphere, with stirring, a solution of 0.07 gm of benzoyl peroxide in 36 gm of styrene and 4 gm of acrylonitrile. The dispersion is heated at 90°C for 16 hr. The polymer is isolated by fil-

Acrylonitrile and *Acrylamide* from S. R. Sandler and W. Karo *Polymer Syntheses*, Vol. I, 2d ed. (San Diego, California, 1992), 388ff., by permission of Academic Press, Inc.; *N-Vinylpyrrolidone* from S. R. Sandler and W. Karo, *Polymer Syntheses*, Vol. II, (New York, 1977), 253ff., by permission of Academic Press, Inc.

tration, washed with warm water, and dried under reduced pressure at 45°–45°C.

NOTE: To prepare hydroxyapatite, $3Ca_3(PO_4)_2 \cdot Ca(OH)_2$, to a slurry of lime in water is added orthophosphoric acid until the mixture is only slightly basic. Obviously by varying concentrations, temperatures, addition rates, and the order of addition, the nature of the calcium phosphate complexes produced may be controlled.

41-11. Suspension Copolymerization of Acrylamide and Acrylic Acid [23]

$$n\ CH_2{=}CH{-}\overset{\overset{\displaystyle O}{\|}}{C}{-}NH_2 + m\ CH_2{=}CH{-}\overset{\overset{\displaystyle O}{\|}}{C}{-}OH \longrightarrow$$

$$\left(\begin{array}{c}{-}CH_2CH{-} \\ | \\ C{=}O \\ | \\ NH_2\end{array}\right)_n \left(\begin{array}{c}{-}CH_2{-}CH{-} \\ | \\ C{=}O \\ | \\ OH\end{array}\right)_m \quad (6)$$

In an apparatus fitted for azeotropic removal of water and with a stirrer, to a refluxing mixture of 225 gm of benzene, 1 gm of carbon tetrachloride and 14 gm of ethoxylated tallow fatty acid amides, is added over a period of approximately 2 hr a solution of 91.8 gm of acrylamide, 39.2 gm of acrylic acid, 16 gm of 25% aqueous solution of sodium vinylsulfonate, 10 gm of urea, 30 gm of water, 37 gm of 25% aqueous ammonia, and 0.02 gm of potassium persulfate. Heating is continued until all of the water has been removed. The bead polymer is filtered off and dried under reduced pressure at 50°C. Yield: 134 gm of a water-soluble copolymer.

■ **CAUTION:** Since acrylamide is supplied as a white crystalline material (m.p. 84°–85°C) its appearance is similar to many well-crystallized, pure, and innocuous organic products.

Only relatively recently has it been discovered that acrylamide, and possibly also the substituted acrylamides, are toxic in a unique manner. Neuromuscular disorders of varying severity have been reported. The monomer may be absorbed through the intact skin, by inhalation, or by ingestion. There may be a degree of cumulative toxicity in animals. It is suggested that reference [24] be consulted for recommendation of safety precautions that should be taken when handling acrylamide and related compounds.

41-12. Polymerization of Acrylamide in Aqueous Solution [25]

In a 1 liter flask, to a solution of 27 gm of acrylamide and 54 gm of sodium sulfate in 210 ml of deoxygenated water under nitrogen is added with stirring 0.012 gm of 3,3′,3″-nitrilotris(propionamide) and 0.012 gm of ammonium persulfate. The reaction mixture is warmed at 60°C for 30 min. The solution is then cooled. In the product, 94% of the monomer has been converted to polymer.

41-13. Continuous Solution Polymerization of N-Vinylpyrrolidone [26]

The reaction is carried out in a 3050-ml glass flask fitted with an efficient stirrer, two inlets for reactants, an outlet for the product solution, a thermometer, and a nitrogen inlet. Means of controlling the flow rates of reactants also are provided.

The following reactants are prepared: reactant 1, consisting of a solution of 11 liters of N-vinylpyrrolidone, 11 liters of isopropanol, and 154 ml of 28% aqueous ammonia solution; and reactant 2, consisting of approximately 9 liters of 0.12% aqueous hydrogen peroxide.

Into the stirred reactor, maintained during the preparation at 70°C, while a slow stream of nitrogen bubbles through the equipment, Reactant 1 is fed in at a rate of 190 ml/hr, while simultaneously, reactant 2 is fed in at a rate of 80 ml/hr. The overflow is collected as the product solution. After 42 hr the product solution has 87% of the monomer converted to polymer with a Fikentscher K value between 45 and 50 and no sign of gel formation. When the preparation is continued for 110 hr similar results are observed.

If water is substituted for isopropanol in reactant 1, gelation interferes with this continuous process.

6. VINYL ACETATE POLYMERS

41-14. Suspension Polymerization of Vinyl Acetate (Control of Molecular Weight by Variation in Initiator Level) [27, 28]

(a) Preparation of Low-Molecular Weight Poly(Vinyl Acetate). In a 3-liter reaction kettle fitted with a mechanical stirrer, reflux condenser, thermometer, and an addition funnel, 800 ml of distilled water

Vinyl Acetate Polymers from S. R. Sandler and W. Karo, *Polymer Syntheses*, Vol. III, (New York, 1980), 215ff., by permission of Academic Press, Inc.

and 0.8 gm of the sodium salt of an equimolar copolymer of styrene and maleic anhydride (German trade name Styromal) are heated with agitation to 80°C. Meanwhile a solution of 600 gm of vinyl acetate, 5.4 gm of dibenzoyl peroxide, and 3 gm of ethyl acetate is prepared.

To the warm aqueous suspending agent is added 100 gm of the vinyl acetate solution. The stirred mixture is brought up to 80°C by external heating. Once the polymerization has started, heating and cooling is applied as required while the remainder of the monomer solution is added over a 5-hr period. After the addition has been completed, heating is continued for an additional 3-hr period. The residual monomer is removed by steam distillation with agitation. The aqueous dispersion is cooled with agitation to 4°C. The polymer beads are filtered off or centrifuged and washed repeatedly with water at 5°C to remove the suspending agent. The polymer is then dried under reduced pressure at 30°C. The dry product is glass clear. The product is reported to have MW 110,000.

(b) Preparation of an Intermediate-Molecular Weight Poly(Vinyl Acetate). For this preparation, the procedure used is the same as that given in Preparation (*a*) except that the reactants used are 800 ml of distilled water and 0.8 gm of Styromal (sodium salt) to which is added by the described gradual addition technique a solution of 600 gm of vinyl acetate and 1.2 gm of dibenzoyl peroxide.

The final product has a molecular weight on the order of 1,000,000.

(c) Preparation of a High-Molecular Weight Poly(Vinyl Acetate). For this preparation, the procedure used is the same as that given in Preparation (*a*) except that the reactants used are 800 ml of distilled water and 1.2 gm of Styromal (sodium salt) to which is added by the described gradual addition technique a solution of 600 gm of vinyl acetate and 0.18 gm of di-*o*-toluyl peroxide.

The final product has a molecular weight on the order of 1,500,000.

A patented procedure for the suspension polymerization of vinyl acetate which is claimed to produce a nonsticky bead polymer [29] uses an aqueous phase consisting of 53.7 gm of distilled water, 0.25 gm gum tragacanth, and 0.10 gm sodium dioctylsulfosuccinate (Aerosol OT). The monomer charged consists of 690 gm of vinyl acetate and 0.69 gm of dibenzoyl peroxide.

41-15. Continuous Addition Emulsion Polymerization of Vinyl Acetate, Potassium Persulfate Initiated [30]

In a 1-liter, four-necked resin kettle fitted with a mechanical stirrer, reflux condenser, a pressure-equilizing addition funnel, and a thermometer, with provisions for maintaining a nitrogen atmosphere in the system, is heated (using a water bath which has been arranged for convenient and safe raising and lowering) 210.75 gm of deionized water, 0.75 gm of Tergitol-7 (sodium heptadecyl sulfate, an anionic surfactant from Union Carbide), 12.5 gm of Cellosize WP-09 (a protective colloid from Union Carbide) and 1 gm of potassium persulfate and is blended under nitrogen with agitation. Then 25 gm of vinyl acetate is added and the mixture is heated to 70°C. When the polymerization has started, with stirring, over a 2- to 3-hr period, 250 gm of vinyl acetate is added. The temperature is maintained at 70°–80°C. After the addition has been completed, heating at 70°–80°C is continued for an additional 30 min. The residual monomer content of this latex is less than 1%.

More complex gradual addition procedures have been reported. For example, in Vona *et al.* [31], a procedure for the gradual addition of monomer and initiator is given. In Gulberkian [32], three ingredients are added simultaneously at a steady rate: a monomer–surfactant solution, the persulfate initiator solution, and a sodium hydroxide solution.

REFERENCES

1. W. H. Carothers, *J. Amer. Chem. Soc.* **51,** 2560 (1929); R. H. Kienle and F. E. Petke, *ibid.* **62,** 1083 (1940).
2. L. H. Flett and W. H. Gardner, "Maleic Anhydride Derivatives," p. 176. Wiley, New York, 1952.
3. J. R. Whinfield and J. T. Dickson, British Patent 578,079 (1946); J. R. Whinfield, *Nature (London)* **158,** 930 (1946).
4. C. Schotten, *Ber. Deut. Chem. Ges.* **15,** 1947 (1882); C. Schotten and J. Baum, *ibid.* **17,** 2548 (1884); C. Schotten, *ibid.* p. 2545; **21,** 2238 (1888); **23,** 3430 (1890); E. Baumann, *ibid.* **19,** 3218 (1886).
5. W. H. Carothers, U.S. Patent 2,071,250 (1937).
6. E. Magat and D. R. Strachan, U.S. Patent 2,708,617 (1955).
7. E. L. Wittbecker and P. W. Morgan *J. Polym. Sci.* **40,** 289 (1959).
8. P. W. Morgan and S. l. Kwolek, *J. Polym. Sci.* **40,** 299 (1959); P. W. Morgan, *SPE (Suc. Plast. Eng.) J.* **15,** 485 (1959).
9. P. W. Morgan and S. L. Kwolek, *J. Chem. Educ.* **36,** 182 (1959).

10. R. G. Beaman, P. W. Morgan, C. R. Koller, and E. L. Wittbecker, *J. Polym. Sci.* **40,** 329 (1959).
11. D. J. Lyman, *J. Polym. Sci.* **45,** 49 (1960).
12. H. C. Beachell and J. C. Peterson, *J. Polym. Sci., Part A-1* **7,** 2021 (1969).
13. "The Manufacture of Acrylic Polymers," Tech. Bull. SP-233. Rohm and Haas Company, Philadelphia, Pennsylvania, 1962.
14. J. A. Simms, *J. Appl. Polym. Sci.* **5,** 58 (1961).
15. "Emulsion Polymerization of Acrylic Monomers," Bull. CM-104 J/cg. Rohm and Haas Company, Philadelphia, Pennyslvania.
16. W. E. Goode, F. H. Owens, R. P. Fellmann, W. H. Snyder, and J. E. Moore, *J. Polym. Sci.* **46,** 317 (1960).
17. W. E. Goode, F. H. Owens, and W. L. Myers, *J. Polym. Sci.* **47,** 75 (1960).
18. C. F. Ryan and J. J. Gormley, *Macromol. Syn.* **1,** 30 (1963).
19. W. E. Goode, R. P. Fellmann, and F. H. Owens, *Macromol, Syn.* **1,** 25 (1963).
20. W. R. Sorenson and T. W. Campbell, "Preparative Methods of Polymer Chemistry." 2nd ed., pp. 284–286. Wiley (Interscience), New York, 1968.
21. E. Kiran and J. K. Gillham, *Polym. Prepr., Amer. Chem. Soc., Div. Polym. Chem.* **13** (2), 1212 (1972).
22. J. M. Grim. U.S. Patent 2,594,913 (1952); *Chem. Abstr.* **46,** 7822 (1952).
23. H. Spoor, German Patent 2,064,101 (1972); *Chem. Abstr.* **77,** 127438u (1972).
24. "Chemistry of Acrylamide." Process Chemicals Department, American Cyanamid Co., Wayne, New Jersey, 1969.
25. American Cyanamid Co., Netherlands Patent Appl. 6,505,750 (1965).
26. J. F. Volks and T. G. Traylor, U.S. Patent 2,982,762 (1961).
27. H. Bartl, *in* "Houben-Weyl Methoden der organischen Chemie" (E. Müller, ed.), 4th ed., Vol. 14, Part 1, pp. 905ff. Thieme, Stuttgart, 1961.
28. M. Bravar, J. S. Rolich, N. Ban, and V. Gnjatovic, *J. Polym. Sci., Polym. Symp.* **47,** 329 (1974).
29. H. M. Collins, U.S. Patent 2,388,601 (1945).
30. "Typical Emulsion Polymerization Recipes," Prod. Bull. F-40039B. Carbide and Carbon Chem. Co., New York, 1956.
31. J. A. Vona, J. R. Constanza, H. A. Cantor, and W. J. Roberts, *in* "Manufacture of Plastics" (W. M. Smith, ed.) Vol. 1, p. 216. Van Nostrand-Reinhold, Princeton, New Jersey, 1964.
32. E. V. Gulberkian, *J. Polym. Sci., Part A-1* **6,** 2265 (1968).

APPENDIX

Documentation of Product and Process Research and Development

T he careful record of the scientist's activities is crucial to experimental science. From the standpoint of the individual, it is necessary to preserve the reasons for doing the work, the details of the actual work, the observations, the results (whether as anticipated or not, even if negative), and a summary, which may serve as a guide to future work. All of this, and more, is recorded in the notebook of the experimenter. Recent news reports on possible scientific misconduct indicate the importance of maintaining accurate records and establishing the date of the entries [1].

The employer of the chemist also has an interest in the individual's notebook. To a large extent the employer's technical know-how is to be found there. Patenting a product or process may depend on the dates found in a notebook. Priority claims in a patent interference case may require an examination of notebooks. Therefore, dates, signatures, and witnesses to the work must be carefully recorded. These are some of the reasons for maintaining and protecting notebooks.

A detailed discussion of these topics is found in reference [2]. One interesting aspect discussed is the hardware to be used in connection with proper note keeping: bound notebooks, preferably with pages made of acid-free paper; permanent black ink or permanent black ballpoint pens; adhesives for attaching charts, etc; suggested methods of notebook storage; and other physical aspects of note keeping.

Notebook entries, detailed in the documents below, should be legible, concise, and relatively neat. The matter of neatness is rarely discussed realistically. The notebook definitely must not have the appearance of a fair copy, (i.e., pages copied at some later date from miscellaneous notes which may no longer be available). The notebook should be kept as the work is being carried out, preferably with the setup used for the experiment being documented. A small amount of spillage on a page is no major catastrophe—quite the contrary, it lends verisimilitude to the notes.

The notes should be written in a clear and concise manner. Another qualified person should be able to read and understand the entries so that he/she may serve

---

The following documents have been reformatted for reproduction in this book. Their content has not been altered.

as a witness to the notes. Enough details should be given so that the work can be understood even when it is evaluated several years later. We suggest that as soon as a new entry is started, its title be entered immediately in the notebook's table of contents. Experience has shown us that if this step is delayed, the table of content entry may never be made. As to writing style, the conventional passive voice should be avoided (note how convoluted this sentence reads). References [3] and [4] are specific about proper writing styles, covering grammar, symbols, abbreviations, permissions, and so on.

The American Chemical Society has prepared a record keeping fact sheet that gives guidelines for keeping a notebook record (see Document 1).

For comparison, we discussed the above ideas with several firms active in the chemical field. Both Rohm and Haas Company and E. I. Du Pont de Nemours and Company very kindly furnished material about their procedures. We thank Dr. Fred H. Owens of the Rohm and Haas Company, Dr. George N. Sausen of Central Research and Development Department of the Du Pont Company, and Dr. K. K. Likhyani of the Fiber Division also of Du Pont for making their material available to us. Dr. Owens, in particular, gave us many ideas and suggestions.

The practice at Rohm and Haas is to supply new employees with the *Instructions for Keeping Laboratory Records* (see Document 2). Another reference is given that gives considerable details on maintaining records for patent protection [5]. This publication is unusual in that it does not recommend gluing graphs, spectra, and other bits of paper directly into the notebook but recommends their storage in an Accopress® with cross-referencing to the notebook. Ultimately, the material in the Accopress binder is permanently bound.

The Central Research and Development Department of the Du Pont Company offers guidelines emphasizing the use of archival quality materials for the preservation of records (see Document 3). Evidently, their notebooks are collected regularly for microfilming. Detailed comparisons between Documents 2 and 3 are left to the reader for the development of his own record keeping methodology.

The American Chemical Society Committee on Patents and Related Matters has produced a guide for electronic record keeping for patent purposes (see Document 4). The use of computers in keeping scientific records is expanding rapidly. In the 7th Annual Scientific Computing & Automation Conference and Exposition (1991), K. K. Likhyani presented a paper entitled "The Authentication of Electronic Record for Legal Defensibility, Part II." As a result, an industry consortium is being formed to study the development of solutions to design strategies for handling computer records, the authentication of dates, authorship, and so on.

Obviously there are many problems associated with trying to keep scientific records in a computer. For reasonable permanence of the stored records, we presume that diskettes (and in the future ROM discs) will be used. Diskettes will need to bear information for future reference such as the type of computer used; the choice of format, 360K, 720K, 1.2M, or others; the software program and disc operating system used; the method used to copy to the disk; to mention only a few. With the rapid changes in the computer industry, it is difficult to predict what other factors will need to be considered. All of this has an impact on record keeping and communicating information between computers. Adding this to the problems arising from patent law, it is reasonable to assume that some time will elapse before the problems are resolved.

REFERENCES

1. A. J. S. Rayl, *The Scientist.5.* no. 22, 18. November 11, 1991. and David P. Hamilton, *Science* (1991), **251,** 1168.
2. H. M. Kanare, "Writing the Laboratory Notebook." (1985), American Chemical Society, Washington, D.C.
3. Louis F. Fieser and Mary Fieser, "Style Guide for Chemists." (1960), Reinhold, New York.
4. Janet S. Dodd, ed. "The ACS Style Guide." (1986), American Chemical Society, Washington, D.C.
5. Omri M. Behr, Build strong patents on strong notebooks. *Chemtech.* (1982), **12,** 490.

DOCUMENT 1

RECORD KEEPING FACT SHEET

The American Chemical Society's Committee on Patents and Related Matters has prepared this fact sheet as a guideline for maintaining complete research records. Such records are crucial to the advancement of invention and to the protection of intellectual property rights.

1. **DO** keep the record factual.
   DO record novel concepts and ideas relating to the work project.
   DON'T editorialize.
2. **DO** use a record book with a permanent binding.
   DON'T use a loose-leaf, spiral-bound or otherwise temporarily bound book that provides for page deletions and insertions.
3. **DO** enter data and information including formulas and/or drawings directly into the record book promptly as generated.
   DO sign and date each page of the record book at the time the page is completed.
   DON'T rely on memory or use informal loose sheets for entries with the intention of later putting these into the bound record book.
   DON'T leave any completed page unsigned and undated.
   DON'T postpone signing and dating all completed pages.
4. **DO** use a permanent ink, preferably black, which will produce well when photocopied in black and white.
   DON'T use a pencil or non-permanent inks.
   DON'T use colored inks.
5. **DO** write legibly.
   DON'T make entries in handwriting that later on can be subject to interpretation, translation or wrong meaning.
6. **DO** identify errors and mistakes and explain them.
   DON'T ignore errors and mistakes.
   DON'T obliterate, delete or otherwise render errors unreadable.

The Record Keeping Fact Sheet, copyright 1988 is reprinted with the permission of the American Chemical Society (ACS), prepared by the ACS Committee on Patents and Related Matters, and published by the American Chemical Society, Department of Government Relations and Science Policy, 1155 16th Street, N.W., Washington, D.C. 20036.

7. **DO** completely fill each page.
   DO sign and date each page immediately after the last entry.
   DO draw vertical lines through unused portions of a page where an experiment takes less than a full page.
   DON'T leave part of a page blank.
8. **DO** attach support records to the record book where practical; where volume and size prohibit this action, store such records, after properly referencing and cross-indexing, in an orderly form in a readily retrievable manner.
   DON'T file supporting records in a haphazard, helter-skelter manner without any record of their relationship or connection to the research reported in the record book.
9. **DO** use standard accepted terms; avoid abbreviations, code names, trademarks, trade names or numbers if possible; if abbreviations or code names, trademarks, trade names or numbers are used, make certain these are defined at least once in every record book.
   DON'T use any abbreviation, code name or number without giving its meaning or definition, or identifying the compound of the trademark and/or trade name and source.
10. **DO** keep the record book clean; avoid spills and stains.
    DON'T subject the pages of the research notebook to chemical or physical destruction from spills.
11. **DO** see that the record is promptly witnessed by a knowledgeable person who understands what is being reported and, preferably, who assisted in or witnessed the work, but who is not a contributor to the research being conducted.
    DON'T postpone having notebooks witnessed.
    DON'T have notebooks witnessed by someone who is not technically skilled in the art being reported and who does not understand the contents of the record.
    DON'T use as a witness someone who has contributed professionally, conceptionally or technically to the work being reported.
12. **DO** maintain the confidentiality of the record until properly released.
    DON'T treat the record book as a publication which is freely available to the public.
13. **DO** maintain control of an assigned record book at all times, keeping it in a fireproof safe, file or vault when not in use.
    DON'T let the book lay open around the laboratory when not in use.
    DON'T remove the record book from the company's or institution's premises.
14. **DO** index and close out the record book as soon as it is filled or a project is completed and check it back in for filing and storage to the person who issued it.
    DO reference the location where the book is being stored to assure ready retrieval.
    DON'T keep a closed out and completed record book in the possession of the author.
    DON'T file or store a book without referencing its location.
15. **DO** remember the record book is a legal document and should be treated as such and made available to your legal and patent counsel if needed.
    DON'T keep a record book beyond the company's or institution's established record retention policy for such record.

DOCUMENT 2

INSTRUCTIONS FOR KEEPING LABORATORY RECORDS

INTRODUCTION

It can seldom be foreseen, at the time a notation is made, that a specific bit of information will become important in proceedings before the U.S. Patent and Trademark Office or a court. A fact which seemed unimportant at the time it was recorded has often weighted the balance of a judgment involving large sums of money or our future operations in that research area. That the record was properly and promptly written, dated, signed by the recording worker and witnessed may become of great significance. The prudent course of action, then, is to handle every record of an idea or fact with utmost care.

To prevail in a contest with one or more other patent applicants or patent holders to obtain a patent—in a procedure in the U.S. Patent and Trademark Office called "interference"—you must establish that, under the legal rules, you are the first inventor. To do so, you may be required to prove the following:

1. On a certain date you (alone or with others) made a discovery or thought of an original idea;
2. On a certain date you disclosed the discovery or idea to one or more other persons who were competent to understand it and in sufficient detail for those persons to comprehend your discovery or idea;
3. On a certain date you completed your invention by running one or more appropriate experiments.

In some cases, you may also have to prove that you proceeded from step 1 to step 3 in a reasonably continuous manner.

The following methods have been established at Rohm and Haas Company to secure uniformity in practice and to assure maximum protection for our discoveries. Our Patent Department and Research supervision desire that all research workers and others who may conceive and develop ideas in which the Company may have an interest follow these methods explicitly.

Instructions for Keeping Laboratory Records is reprinted with permission from Rohm and Haas Research Division, copyright 1986.

ISSUANCE AND NUMBERING OF NOTEBOOKS

All workers in research or sales-service activities must record their experimental notes in an official, bound Research Notebook; others having a proper need may also be assigned an official Notebook. Notebooks that have cross-hatched ruling are also available, as are binders for loose sheets.

Every Research Notebook bears a unique issue number which is a member of a single consecutive series. A record is kept of the person to whom each book is assigned, his or her department, and the date of issue. (Affiliated and subsidiary companies and divisions other than Research may elect to maintain independent registration and control procedures. In such instances, they may obtain unnumbered Research Notebooks from the Spring House library.)

You obtain a Research Notebook from the local Research Library. A formal entry will be made in a log at the Library. You will be asked to fill out the first three lines of the title page of the Notebook in the presence of the issuing Library clerk, i.e. the Notebook registration number, your name and department, and issue date. You may, in addition, identify your Notebook by means of a numbering system of your own.

Research Notebooks and their contents are the property of Rohm and Haas Company. The department of origin or its successor retains permanent control over a Notebook regardless of subsequent transfers of projects or personnel.

Accopress[1] Binders—Loose material such as tables, photographs, instrument charts, computer output and other data sheets may be assembled and bound in Accopress binders. These binders are obtained from the Research Libraries. Like bound Research Notebooks, they bear consecutive issue numbers. They are logged out and a record is kept in the same manner as stitched Notebooks.

Accopress binders should be limited to one inch in thickness to permit permanent stitching after completion. Accopress binders containing information developed in conjunction with a product, intermediate or process are accountable and must receive the same care as bound Research Notebooks.

RECORDING OF NOTES

Any single Notebook or Accopress binder should be used by only one individual; where exceptions are necessary, see "Multiple Users", below. All Notebook entries should be made in permanent black ink. Sheets accumulated in an Accopress binder must be written in ink and must be numbered consecutively on entering into the binder.

Dates and the Signature—All entries must be signed and dated. Record the date at the beginning of each page. If an experiment is continued over more than one day, enter the date of each day's work. Sign each page with your full signature. If you start another experiment or enter another idea on a lower portion of a page which already contains information relating to another experiment or idea, re-enter the date at that point.

---

[1] Accopress is a trademark of Acco International Inc., Chicago, IL 60519

Margins—In the standard Research Notebook, allocation of space is clearly indicated. Avoid making notations in the margins of any form of Notebook, especially in the binding edge; entries in such areas can be lost in the microfilming process.

Pagination—The pages of stitched Notebooks are prenumbered. You should utilize both sides of the Notebook pages, in consecutive order. However, if you prefer, it is permissible for you to use only one side of the Notebook page for notes, with the other side (usually the left, or even-numbered page as the book lies open) being reserved for arithmetic calculations, temporary notations, contemplative formulas and diagrams, planning, etc. If you adopt this style, it must be your consistent practice throughout the Notebook. In the case where one side is reserved for calculations and miscellaneous notes, the witness statement should read something like "Witness for subject matter of odd-numbered pages 5–9, inclusive, disclosed to me and understood."

Loose sheets compiled in a binder must be numbered in chronological sequence upon entry into the binder. Once bound—taped or stitched—no page may be removed.

Transcription from Notes—In recording notes, avoid using odd books or loose sheets of paper, unless you must record data under conditions which would endanger the Notebook. In such instances, copy the data accurately into the Notebook on the very day on which they were taken, and then destroy the loose notes. (The problem of preserving charts, photographs, etc. is considered below.)

Negative Results—Not all experiments turn out as planned. However, few experiments are designed to fail. It is important to avoid pejorative, derogatory or belittling language in describing results and the conclusions inferred from them. A result may be surprising or unexpected or not the one foreseen. To conclude that an experiment "fails" because the results are other than those anticipated could impair the ability to obtain a valid patent. Indeed, inappropriate language in describing negative results could be used as evidence of a failure to recognize your own invention.

Describe negative results as honestly and fully as affirmative results. Unexpected results can lead to new discoveries. But don't declare a product or test a failure merely because, for instance, it does not measure up to commercial standards; further improvement in temperature control, catalyst, or the like, may yield a product or test result closer to target criteria.

Multiple Users—To consolidate data and maintain continuity when more than one person is involved in identical experiments in shift work, co-workers are permitted to write in the same Notebook, but only if each enters date, time of day, and initials at the margin when he or she starts and finishes. As with single-user Notebooks, you must enter full signatures and dates at the beginning and at completion of the experiment.

Standard Procedures—When standard methods or apparatus are used and permanent descriptions thereof exist, references identifying such methods or apparatus may be used to avoid lengthy repetition. A preferred way of making such a record is to describe in detail the test apparatus the first time it is used in the Notebook and thereafter in that Notebook state: "Test performed as described on page(s) _____ of this Notebook." Clearly indicate and explain any departure from the described procedure.

Codes—Code designations, such as BMA or X-970, may be employed to describe chemicals or products used. Precisely the same considerations apply as in the case of an apparatus or test method. At the first appearance the chemical or product should be identified by its proper chemical name, and thereafter reference can be made to the Notebook page where the definition appears. For example, the first time you refer to, say, X-970 you would write: "1,3-butylene glycol dimethacrylate (X-970)". Thereafter you might use the expression "X-970 (page __)."

Analytical Results—Analytical results or other data that are copied from other Notebooks—your own or another worker's—must be accompanied by a reference to the original record. If you obtain analyses or application tests from another laboratory, enter the submission of the sample, with date. Enter the results when received, again with date. Retain the report itself and reference it.

Calculations—Show all calculations in the Notebook. At the least, the data and the method of calculation should be clearly identified.

Corrections—Under no circumstances may entries in a Notebook be erased or obliterated. If a correction is necessary, draw a single line through the word or number to be deleted, then initial and date the correction. If alterations must be made after witnessing, these must be re-witnessed.

Avoid additions and interlineations. Instead, make a forward reference to the next blank page; enter the addition or insertion there, with appropriate cross-reference back to the first location; sign and date the addition.

Inserts—Special charts, such as infrared and GLC graphs, photographs (see below), or standard data sheets are often developed in the course of work. It is not always practical nor desirable to copy them into Notebooks. It is important, however, to transfer essential information on which you base your conclusions, so that your inferences are supported by your Notebook entries.

If only a few loose sheets are to be dealt with, they may be stapled to a blank page of the Notebook. Take care not to obscure any entries beneath. Mark the insert with the Notebook number and page, in case the insert becomes separated from the Notebook.

If there are many loose sheets, assemble them, and bind them separately in an Accopress binder issued by the Research Libraries for this purpose (see Issuance and Numbering of Noteboks, above). The bound sheets, now a single volume, are treated the same as a Research Notebook. (See the earlier section on Accopress binders. See also "Security and Preservation of Records," to follow.)

Graphs—Graphs which are produced in several colors should be avoided as Notebook entries, because color distinctions vanish in the mircofilmed record. Where colors exist, the color-coded curves or areas or legends should also bear explicit labels.

Photographs—Photographs are very effective if properly made and documented. All photographs by Research Photography are assigned reference numbers which correspond to the dates listed on the order form. Photographs procured through any other means must show the circumstances, the identity of the subject and the photographer, as well as the date. It is essential, however, that the Notebook carry independently an adequate description of the experiment and the results obtained.

Blanks—Fill in Notebook pages chronologically. Do not leave blank space on any page, nor skip any pages to be filled in later—i.e. after intervening pages have been used. Analytical or test data pertaining to a given experiment, which are not available at the time the experiment is being performed, are best recorded on a later page in proper chronological order, when received, with a suitable reference back to the Notebook pages on which the experiment was described. A cross-reference to this analytical or test result may be inserted at the earlier location in the book. (See "Alternative Method of Entering Analytical or Test Results," below.)

Entries in the Notebook relating to work proposed but not yet carried out should clearly show that the experimental follow-up is to come later.

When you wish to terminate an entry that does not fill a page and go on to the next page, sign, date and end the page by drawing a line across the page immediately below the signature, another line across the bottom of the page, and a third line on a diagonal to connect the two horizontal lines in the form of a "Z".

Leave at least half a page blank at the end of the Notebook for the close-out stamp. See "Security and Preservation of Records" which appears later.

Computer Output—In many experiments, acquisition and recording of observations and frequently reduction of the data are automated by means of electronic transducers, recorders and minicomputers. There may even be no human observer. In such cases, the computer-generated sheet should be signed by whoever set up the experiment and witnessed by someone who can verify the date of execution of the run. Number the sheets individually with reference to your primary Notebook and collect the sheets in Accopress binders in the manner described above, under "Inserts". Enter all results and conclusions into the your primary Notebook with appropriate reference to the computer-generated output sheet.[2]

Alternate Method of Entering Analytical or Test Results—If it is desired, a tabular form may be inserted at the time the experiment is written for the later insertion of the analytical or test results. In this case:

1. It must be noted that the sample has been submitted for analysis or test, to whom, the date, and that the space is being held for later completion.

---

[2] It is important to recognize the distinction between data and other information that may be in a computer data base, and a permanent record of that information. It is not uncommon to acquire primary data electronically and accumulate information in data bases. These data bases are sometimes described as electronic notebooks. Such a method has the merit of enabling rapid retrieval and facile work-up of data. But there is no way to guarantee the integrity of a data base, neither as regards content, time, person nor place in the absence of some kind of hard copy. Only a timely print-out of a computer file or a data base—signed, dated, witnessed and secured—can serve as a record. In the future, courts may accept other criteria for valid records. For the present, Research Notebooks shall be in paper or other permanent, physical form.

2. The signature and witness statement must bear such words as "except for the table on p. _____ that is being held open for analytical results."
3. At the completion of the table, it must be signed, dated, and witnessed—preferably by the same witness as before, in which case initialing and dating is sufficient.

NOTE THAT THIS REQUIRES SEPARATE AND INDEPENDENT WIT-NESSING, SIGNING AND DATING FOR EACH SET OF DATA ON THE PAGE.

Long-term Experiments—Where a long-term experiment requires daily recording of either numerical data or subjective observations, then:

1. At the beginning of an experiment, sufficient space would be allowed to record all of the data and observations. The description of the experiment should, at the beginning, state: "Pages _____ to _____ are being reserved for recording data pertaining to this experiment."
2. Data and observations would be recorded daily, with the date, and initialed by the one making the entry and by a witness.
3. At the end of the experiment, the page at the end would be stamped, witnessed with full signature, and dated by the same witness or witnesses who had initialed the daily entries. Any remaining unused space reserved for the experiment would be crossed out in the manner prescribed under "Blanks," above.

WITNESSING OF NOTEBOOKS

Witnessing of Notebooks is necessary in order to establish legal proof that the witnessed notes and data were duly recorded on or before the indicated dates of the signatures of the worker and the witnesses.

Witnessing serves two purposes:

1. to **authenticate** the physical integrity of the Notebook page and confirm the entry of the information on it; and
2. to **corroborate** the contents of the page.

A witness attests to more than just that the formalities of filling pages have been met, i.e. that the pages are signed and without blanks and that the work was done on a certain date, by a certain person. He or she should also have some comprehension of the work being performed and its significance. Corroboration implies a degree of understanding of the content of the Notebook entries being witnessed. It is for this reason that the witness stamp for use with loose sheets and the witness statement printed in the standard Notebook read "Subject matter of pages _____ to _____ disclosed to me **and understood**" (emphasis added).

Every scientist, engineer, and technician is obligated to make proper entries in his or her Research Notebook on the same day in which the work is performed. Witnessing should occur that same day. For it is the signing and witnessing which establishes the **provable** date of the work. In the event of a contest between patent applicants, any patent is awarded to the person who can prove he or she made the invention first. Tardy witnessing can jeopardize your claim to your invention.

If there are any corrections or interlineations evident on the pages witnessed, both you and the witness must initial and date each correction individually. If you have made any entries or corrections after your notes have been first witnessed, you must sign and date each such new entry and have each witnessed and dated individually at the time any such supplementary entry is made—preferably by the original witness.

Because a witness may be called upon to give testimony at a legal hearing, it is highly desirable that he or she be competent to understand and have a reasonably clear knowledge of the material witnessed. The ideal witness is a technical worker, either exempt or non-exempt, who, by contact with the researcher, is acquainted with the work being done and its day-to-day progress, who actually sees the worker at the bench or other place of work at frequent intervals, and who observes and knows enough of what is being done that he or she can, at a hearing, testify not only to the written notes, but also to his or her understanding of the work and knowledge of its performance. Since generally inventors cannot witness their own work, a technician, laboratory or technical assistant who may be a coinventor should not be a witness. But technical assistants who may be familiar with, but are not directly engaged in, the work or project are excellent candidates as witnesses.

Signature—Regular Research Notebooks bear a witness block on every page. The witness normally signs the witness statement in full. However, if a consecutive block of pages is being witnessed, the witness needs sign in full only the last page of the block, provided he or she initials and dates the intervening pages. For non-standard Notebooks not having imprinted witness statements and for loose sheets, a form such as the following is recommended for witnessing and is available as a rubber stamp from the local Research Library.

Signed _____ Date _____

Subject matter of pages _____ to _____ disclosed to me and understood.

Witness _____ Date _____

Wording of this stamp must be modified under certain conditions. See "Pagination," "Alternate Method of Entering Analytical or Test Results" and "Long-Term Experiments."

Inspection—Supervisors must inspect Notebooks from time to time. The Research Section Manager or Senior Research Associate is directly responsible for seeing that Research Notebooks are kept correctly in his or her group, especially as regards signatures, dates, and timely witnessing. It is also the responsibility of the Research Section Manager or Senior Research Associate, or someone higher in the line of supervision, to approve completed Notebooks for final registration by the Research Library.

CONCEPTION OF AN INVENTION

It is important for patent purposes that ideas and conceptions of inventions be properly documented. For this reason, ideas must be promptly recorded in the Notebook, at least in outline form. A desirable approach is to:

1. point out what is now lacking in the relevant technology,
2. show how this deficiency is overcome by your invention,
3. describe your invention in detail and explain how it works,
4. point out any other advantages or disadvantages, and
5. state what you believe to be new.

You should promptly show and explain the written record, signed and dated, to others and have it witnessed.

DILIGENCE, REDUCTION TO PRACTICE, AND CORROBORATION

Diligence, reduction to practice, and corroboration are legal terms. Their significance to you as an inventor will be evident from the paragraphs which follow.

An idea or conception, alone, does not constitute an invention. Under U.S. patent law, an inventor must maintain momentum in proceeding from conception to an experiment which illustrates that the discovery works, i.e. "completes" the invention. The technical expression is to "exercise diligence." The experimental confirmation of the idea is called "reduction to practice" of the inventive concept. The significance of these terms is that you may not merely record an idea and then allow it to remain fallow; you must follow through by actually performing, or having someone else perform, the necessary experiment to demonstrate that the idea works.[3]

Moreover, it is necessary that the experiment establishing reduction to practice be independently confirmed—"corroborated." Your record, by itself, is not sufficient. As an inventor you cannot corroborate your own work; joint inventors cannot corroborate one another. In an industrial research environment, a scientist usually instructs a technician to conduct the experiment which demonstrates an invention. This technician's record is an excellent form of corroboration (provided he or she is not a co-inventor).

It should be clear that all work must be corroborated, either from the record of a subordinate who carried out the work, or from that of a fellow worker who repeated the work at your request.

Corroboration can also be accomplished through the testimony of witnesses and through circumstantial evidence. It is the signing, dating, and proper witnessing, however, which usually form the basis for corroboration.

---

[3]There is a legal alternative known as "constructive reduction to practice" which consists of the act of filing a patent application.

SECURITY AND PRESERVATION OF RECORDS

Storage—Research Notebooks shall be handled as company-confidential material. Overnight and weekends, and at all other times when they are not in use, Notebooks and pertinent charts or standard data sheets in your custody must be kept in locked files meeting corporate specifications for security containers.

Close-out Procedure—When your Notebook is full or if you transfer to another department or your department is reorganized, submit the Notebook without delay to the Research Section Manager, Senior Research Associate, or anyone above in line of supervision for inspection and approval. If the Notebook meets with approval for final registration and storage, the supervisor should stamp, sign, and date on the last page used or on the last page of a full Notebook. The stamp should read:

```
┌─────────────────────────────────────────────────┐
│                                                 │
│  Notebook (or Binder) by                        │
│  ─────────────────────────────────────────────  │
│                                                 │
│  Closed and Approved for Library Registration   │
│  ──────────────────────────  ─────────────────  │
│                                                 │
│  Name                        Date               │
│                                                 │
└─────────────────────────────────────────────────┘
```

Present the approved Notebook promptly to the Research Library for completion of registration, at which time the date of its last entry will be recorded. The Notebook will then be microfilmed and filed in the record vault. If you need it for further reference, you may borrow it from the Research Library.

The same practice applies to Accopress-bound volumes of charts, photographs, computer-output and standard data sheets.

Other forms of Notebooks which may not bear a serial number should also be turned in for proper storage.

Listing of Contents—It is desirable to maintain a list of contents in each Notebook. Regular Notebooks have pages reserved for this purpose in the front.

Transfers and Separations—When you are transferred to another department or leave the Company, your Notebooks and/or volumes of charts bound in Accopress binders are to be signed, witnessed, and delivered to the local Research Library for final close-out registration, recording and storage.

Copying—Copying of any portion of a Research Notebook is **strictly forbidden**. Copies of certain pages, that may be required for legal or other sufficient reason, may be made on the written request of the manager of the research department of origin or of its successor or by the Patent Department, but such copying shall be executed **only** by a Research Library.[4] The Library will make an appropriate notation on the title page so that a complete record of copying is part of the Notebook.

---

[4]To accommodate a remote location, the manager of the Research Library having cognizance over Notebooks may choose to authorize, in writing, a local person (e.g. a member of the Patent Department) to log and copy matter from Research Notebooks.

Loans—Notebooks which have completed final registration in the Research Library and have been microfilmed may be borrowed. If they are needed for reference in the laboratory, Notebooks may be placed on permanent loan in care of the Research Department Manager or Senior Research Associate.

Filed books and bound volumes of charts may be borrowed from the Library vault when required by the author or by others. Requests for personal loans must be endorsed by your Research Department Manager or Senior Research Associate. If you are not a member of the department of origin or its successor, then the manager of that originating department must grant a release.

If you wish to borrow your own Notebooks after a recall, your supervisor's endorsement is unnecessary, provided you remain in the same department.

Recalls—The Research Division limits the active life of a Notebook to one year. The Research Library periodically submits lists of Notebooks or binders open more than a year to the Research Department Managers and Senior Research Associates, with a request to have the Notebooks returned for close-out. So-called "idea" books, occasionally used by managers and supervisors, are also recalled for filming, but they need not be closed out.

Lost Notebooks—Notebooks which cannot be found or accounted for must formally be declared missing. To do so the manager of the department of origin must address a written notice to the local Research Library with a copy to the Patent Department. This notice must identify the Notebook number and the author. It should state that the Notebook cannot be found after a diligent search, and indicate whether or not it contained information or records, the loss of which is of significance to our technology or pending patent matters.

> Approved by:
> Research Information Services & Research Records
> Research Patent Liaison
> Patent Department

DOCUMENT 3

CR&D LABORATORY NOTEBOOKS

COMPLETE

A notebook is "complete" after:

1. All pages are filled or the author decides that no further entries will be made and,
2. It has been properly signed and witnessed and,
3. It has been indexed and stamped as a "complete" book in the Technical Records Center.

It will be microfilmed as a complete book after the above requirements are satisfied.

Make no further entries in a notebook that is stamped:

```
INDEXED
CR&D TECHNICAL RECORDS CENTER
```

ACTIVE

A notebook is active when it is not yet filled to the last page and/or the author is currently making entries.

All other books should be indexed and microfilmed as complete at the next microfilming session. You will have the opportunity to indicate this change on the form that accompanies your notebooks to TRC.

The Employee Relations Department questions the active status of books which are filled to the bottom of the last page or have not had any new entries for several years. Since active books are microfilmed every year until the author declares them complete, both time and film are wasted when these inactive books are remicrofilmed each year.

ADDITIONAL GUIDES

Inserts

(a) Use inserts only when they are vital to the record, interpretation of this data should be recorded in legible writing in the notebooks.

(b) Inserts not fastened to pages in the notebooks will not be microfilmed.

(c) Inserts (one to a page) should be pasted onto a blank page and at the bottom of the insert the words, "No Writing Underneath" should be written or stamped in ink.

(d) Inserts larger than a notebook page should be pasted at the outside edges of the pages and folded in such a way that, when unfolded the insert material, will fall within the boundaries of the open notebook. It is also important that the **page** number of the notebook not be obscured by an insert.

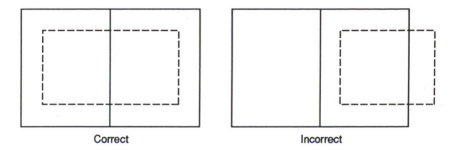

Correct   Incorrect

Margins—Leave one-half inch margins on all outside edges of the notebook pages.

Signature and Witnessing—Signatures should be written at the time the data are entered and should be witnessed as soon as possible. Books that are assigned to technicians (after permission is filed with TRC) can be signed by that technician with their scientist's signature below.

First notebook page

Title ———————————————— Date ————— Page No. ———
Purpose ————————————————————————— Book No. ———
Experimenter (Technician's signature)   Date
Work directed by (Scientist's signature)
Witnessed by   Date

Witnesses must be someone who is able to comprehend the research but is not involved with the project. Scientists should not witness any pages that their technicians had signed.

RULES FOR RECORDING EXPERIMENTS

1. Original records are to be kept in officially issued notebooks in nonfading water and solvent resistant black ink. Ball-point pens supplied through Company storerooms meet these specifications. **Do not use felt-tip markers.**
2. Each notebook page when filled in, including pages with inserts, must be signed and dated by the experimenter in the space provided. The scientist must sign for all work done under his direction, including the work of the technician. The name of the technician helping in an experiment should be noted, and the technician should sign any entry he makes.
3. Each notebook page when filled in must be read and signed by a witness not directly involved in the work performed, but who understands the purpose of the experiment and results obtained. The witness should sign on the same day as the experimenter; or if this is impossible, as soon thereafter as feasible.
4. Entries should be chronological as far as possible. Where the record of the experiment extends over several pages which are not consecutive, (e.g., as when analyses are returned after another experiment has been started) a cross reference should be added to the original page. When entries on a single experiment do not completely fill a page, the remainder of the blank page should be ruled out.
5. The bound notebook must be preserved intact. In no case should any page or part of a page be removed.
6. No erasures are permitted in the record. Any corrections or changes must be made by cancellation, leaving the original entry legible.
7. Any inserts should be signed and witnessed and preferably attached to a blank page. They should be taped, or glued in place, preferably with Duco® cement and the words "No Writing Underneath Insert" should be written on the page in ink. Notebook and page numbers should be recorded on each insert. If it is necessary to attach inserts such as analytical slips and test reports to a page on which data have already been recorded, they should be attached so that no writing on the page is obscured.
8. Inserted material such as graphs, charts, typed data, films, samples, etc., should have writing on one side only, and preferably, should not be larger than one notebook page. **Any insert larger than a page should be folded so that it lies within the boundaries of the open notebook and should not cover the writing on the page when folded back.**
9. Entries should not be made within the margins provided on the page. The margins are provided for clamps used in the annual microfilming for all uncompleted notebooks.

Document 3, *CR&D Laboratory Notebooks*, is used with permission from the Du Pont Company.

ELECTRONIC RECORD-KEEPING FOR PATENT PURPOSES
CAUTIONS AND PITFALLS

The American Chemical Society's Committee on Patents and Related Matters (CP&RM) recognizes that computers are increasingly being used as a means of gathering original data directly from instrumentation, storing data, and maintaining research records. This pamphlet was prepared to call the reader's attention to pitfalls and precautions in maintaining research records in computers and data storage devices; some of these may be obvious, but others may not have occurred to the reader.

Important legal considerations in maintaining original research records are:

- Records must be accurate, permanent, and contemporaneous with research.
- Records must be protected from compromise, including addition or revision at a later date.
- Records, including exact dates of discovery, are critical to establish priority of claims in patent litigation.

There is no precedent in patent case law where computer-stored data have been used to establish priority of a claim. Therefore, computer record-keeping cannot be assumed to serve as a substitute for maintaining an original, permanently bound, handwritten research notebook. (In other publications, the Committee has prepared materials on how to keep an original handwritten research notebook to meet the above requirements.)

Nonetheless, computers are convenient as word processors to prepare summary reports of original research and to store original experimental data in compact form. Computer records also can be easily and rapidly searched for key words and phrases. Compared to traditional notebooks, however, these labor- and time-saving features are accompanied by risks of reduced permanence, lack of contemporaneity, inaccuracies, loss, or compromise. There also is a concern that the electronically stored data is not easily identified as being written by the author, while handwriting identifies a notebook as being authored by the person. This issue has not been tested in priority claims.

Electronic Record-Keeping for Patent Purposes: Cautions and Pitfalls, copyright 1990, is reprinted with the permission of the American Chemical Society (ACS), prepared by the ACS Committee on Patents and Related Matters, and published by the American Chemical Society, Department of Government Relations and Science Policy, 1155 16th Street, N.W., Washington, D.C. 20036.

PERMANENCE, ACCURACY, CONTEMPORANEITY, AND PROTECTION AGAINST LOSS

Storage of computer records uses magnetic or optical devices. Computers and magnetic storage media must be protected from voltage surges and declines, lightning strikes, and magnets operating in the same area or carried by people. Remember that magnetic fields are generated by most operating electric and electronic equipment, including some telephones, and can be harmful to magnetically stored records.

Since accidents, malfunctions, and sabotage do occur, all data to be stored electronically must be backed up, labeled with clear identification of the data, dated, witnessed, and stored in a place safe from magnetic fields or intervention. When original data are computer-generated or computer-collected, both the disk and a hard copy print-out should be labeled, signed, dated, and witnessed on each page (or on each disk). The disk should be write protected. Hard copy should be referenced in and attached to the handwritten notebook.

In addition, records can be destroyed or rendered inaccessible by computer viruses that might be imported with affected programs brought into computer systems. Moreover, disgruntled former employees have been known to sabotage data systems with such viruses, even by telephone access or by dated "timebombs" buried in programs. Therefore, it is important to periodically backup the stored data and also to take precautions to avoid importation of viruses.

In maintaining a written research notebook, accuracy of observations is stressed, and when errors are recorded, corrections must be made and noted without erasing or obliterating the original error. This same principle should apply to electronic research records. In using computers as word processors for original reports, accuracy and integrity of observations are at risk for several reasons. The computer station is frequently at a location other than that of the experiment, so recording may not be made directly into the computer as an observation is made, but after some time interval following the observation, perhaps at the end of the day. Moreover, one of the strengths of word processing is the ease of editing material, especially with magnetically stored data. Editing of original research descriptions runs the risk that original meanings may unintentionally be altered or biased, and some seemingly unimportant observations may be edited out of the record. Thus, errors, unintentional bias, or inadvertent omissions can occur in such records.

With magnetic media, alterations are made easily by either the original worker or another individual. Schemes can be developed and implemented to reduce the risk, but they require maintenance and the involvement of one or more noninvolved individuals.

Programs also can be developed so that a file containing original data cannot be altered directly; alterations of the originally saved file would produce an additional file which would not replace the original, but would overlay the original information input. Additional generations of overlays can contain many subsequent changes (each without replacing its predecessor), but to avoid unnecessary waste of storage media, data management responsibilities may devolve upon persons other than those inputting the data.

Alternative steps can be taken, however, to ameliorate these risks. Systems are available, such as certain laser optical compact disks, that are written upon physically and permanently by lasers (WORM—Write Only, Read Many times). Although the technology also has been developed to allow erasure and rewriting on laser discs, WORM technology might provide valued record permanence and history, especially if earlier versions could not be read except with special equipment.

PROTECTION AGAINST COMPROMISE

Data stored in computers may be compromised in several ways, so preventive steps must be taken, especially when the computer station is part of a computer network system.

Research data should not be stored in computer networks without security systems for protection against unauthorized access and computer hacks. Key locks and screens are available; some computers have removable magnetic bulk storage that can be locked in vaults and cabinets when not in use.

Log-on procedures should involve usercodes and passwords. Care must be exercised to protect one's terminal, usercode, and password. Usercodes and passwords should be a proper name or a nonsensical or misspelled word; codes that contain blanks or other non-alphabetic characters can help prevent the use of computer dictionaries in schemes to "crack" passwords. Passwords should be changed frequently and never shared with another person. In addition, usercodes and passwords of employees who have left the department or organization should be immediately removed from the system. System service usercodes and passwords should be removed immediately from the system upon its installation. Periodic spot checks should be made to ascertain who has accessed and used the system.

There have been cases where intelligence operatives have used sophisticated equipment to intercept and decipher fluctuations in magnetic fields produced by computer equipment. If the work is highly sensitive, steps might be taken to shield the computer equipment and the cables connecting the network. In the alternative, fiber optic cabling can be used.

PRIORITY OF CLAIMS

Again, no precedent in case law exists for use of computer records in establishing priority of claims, so the necessity of maintaining an original handwritten, dated, and witnessed research notebook remains important. Steps, however, can be taken that might assist in using electronically stored data to supplement the notebook in a priority claim. Maintaining backup records that are labeled, dated, witnessed, sealed, and properly stored might help as might periodic hard-copy printouts similarly dated, signed, witnessed (on each page), and properly stored. In addition, a handwritten statement on the witnessed hard copy, such as "On this date I have made an invention," also might be helpful.

The use of a storage medium that is Write Only, Read Many times (WORM) is available to help convince someone that electronically stored data is original and permanent. Compact disks (CD), because of their large storage capacity, also might be suitable for entering pictorial and diagrammatic representation of records. Password witnessing can be used on documents stored on these disks.

CONCLUSION

The Committee takes a conservative position on the matter of research record-keeping, by emphasizing the necessity of maintaining a traditional handwritten research notebook. But CP&RM also recognizes the convenience and efficiency of computer technology in association with a standard notebook.

This publication has attempted to identify some of the pitfalls of electronic record-keeping and has offered suggestions that might help avoid problems. Electronic record-keeping is a relatively new subject, so there are undoubtedly additional issues that should have been identified here, and there likely will be cases in patent law that will influence the subject. Hence, the Committee solicits comment, suggestions, and advice that might be incorporated in future revised editions of this pamphlet.

Electronic record-keeping is a developing area of patent law, and the reader is instructed to obtain current legal counsel from a patent attorney.

ACKNOWLEDGMENTS

This brochure was developed and prepared by the Education Subcommittee of the ACS Committee on Patents and Related Matters (CP&RM) under its chairman, Bob A. Howell (Central Michigan University). A special thanks goes to William F. Little (University of North Carolina) who drafted the publication. Others who deserve recognition for their contributions to this publication include the subcommittee members: C. Kenneth Bjork (Dow Chemical Company), Lucile E. Decker (Chemical Abstracts Service), Gustave K. Kohn, Howard M. Peters (Phillips, Moore, Lempio, & Finley), K. B. Raut (Savannah State College), George N. Sausen (E. I. du Pont de Nemours), and Rochelle K. Seide (Weil, Gotshal, & Manges). Thanks also go to J. Dolf Bass (Kodak Research Laboratories), Allen Bloom (The Liposome Company), Maurice U. Cahn (Leydig, Voit, & Mayer), Roger G. Ditzel (University of California), Hubert E. Dubb (Fliesler, Dubb, Meyer, & Lovejoy), Herbert B. Roberts (Haverstock, Garrett, & Roberts), Rod S. Berman (Spensley, Horn, Jubas & Lubitz), and J. O. Thomas (U.S. Patent & Trademark Office) for their review of the publication.

DISCLAIMER

The materials contained in this manual have been compiled by recognized authorities from sources believed to be reliable and to represent the best opinions on the subject. This manual is intended to serve only as a starting point for good practices and does not purport to specify minimal legal standards or to represent the policy of the American Chemical Society. No warranty, guarantee, or representation is made by the American Chemical Society as to the accuracy or sufficiency of the information contained herein, and the Society assumes no responsibility in connection therewith. Users of this manual should consult with their own legal counsel when implementing notebook-keeping procedures.

NAME REACTION INDEX

SUBJECT INDEX

Survey of Org. Synthesis
— Buehler + Pearson